Y0-AGI-387

Form 306

TEXTBOOK LABEL

Book No. **6**   Cost **2 80**

BOARD OF EDUCATION
Property of
DISTRICT NO. 301

(Name of school)

INSTRUCTIONS: The pupil to whom this book is loaned will
sign in space provided below. He or she will be required to pay
to this school the cost price of this book if it is lost or damaged
during the period for which it is loaned. Allowance will be made
for wear caused by careful use.

Signature of Pupil	Date Loaned	Date Returned	Condition

© NATIONAL SCHOOL METHODS, INC., 205 W. WACKER DR., CHICAGO 6, ILL.

McGRAW-HILL RURAL ACTIVITIES SERIES

W. A. ROSS, *Consulting Editor*

SHOPWORK ON THE FARM

McGraw-Hill

RURAL ACTIVITIES SERIES

W. A. Ross, *Consulting Editor*

Gustafson · Using and Managing Soils

Jones · Shopwork on the Farm

Jull · Successful Poultry Management

Jull · Raising Turkeys, Ducks, Geese, Game Birds

Peters and Deyoe · Raising Livestock

Turner and Johnson · Machines for the Farm, Ranch, and Plantation

[Other books in process]

SHOPWORK ON THE FARM

BY

MACK M. JONES, M. S.

Professor of Agricultural Engineering, University of Missouri;
Member, American Society of Agricultural Engineers

MCGRAW-HILL BOOK COMPANY, INC.

New York   London

1 9 4 5

SHOPWORK ON THE FARM

COPYRIGHT, 1945, BY THE
McGRAW-HILL BOOK COMPANY, INC.

PRINTED IN THE UNITED STATES OF AMERICA

*All rights reserved. This book, or
parts thereof, may not be reproduced
in any form without permission of
the publishers.*

X

PREFACE

THIS BOOK has been prepared to serve as a text and reference book in farm shopwork. It deals simply and directly with tools, materials, operations, and processes, or activities, rather than with jobs or projects. It can be used, therefore, with any jobs or projects that suit individual interests, abilities, or needs. Such jobs may be related to the home, the farm, the school, or to the student's major project.

Emphasis is placed upon correct methods together with the underlying reasons, or the *why;* for only by understanding the why can a student master a subject or use acquired knowledge in the solution of his own problems.

Illustrations have been used generously. These show the proper methods of using tools and performing the basic shop operations, and often practically duplicate in picture form the explanations given in the text. There are also many summarized lists of practical points on various topics, such as sharpening saws, hack sawing, drilling holes in metal, and welding.

For satisfactory achievement in shopwork, a student must do some reading and studying as well as working with his hands. As he proceeds with his work and doubts or difficulties arise, he will get effective help by consulting his textbook. Self-help thus rendered develops resourcefulness and self-confidence. Furthermore, it is not possible for an instructor to give individual instruction to all students of a large class at just the time needed. A textbook that can be used to supplement the work of the instructor is an invaluable teaching aid. It is to provide such an aid that this book has been prepared.

This book is intended primarily for the use of vocational agriculture students and other farm youth, who will be the farmers of tomorrow, and for college students taking their first course in shopwork. The book should also be valuable to farmers who desire to improve their shop facilities and to keep their machinery and equipment in the best operating condition with a minimum of expense for repairs and maintenance.

MACK M. JONES.

ACKNOWLEDGMENTS

THE AUTHOR gratefully acknowledges valuable assistance from the following in the preparation of illustrations: H. P. Richter, author of "Practical Electrical Wiring"; Thomas E. French, author of "Engineering Drawing"; Thomas E. French and Carl L. Svenson, authors of "Mechanical Drawing for High Schools"; Fred R. Jones, author of "Farm Gas Engines and Tractors"; H. P. Smith, author of "Farm Machinery and Equipment"; W. A. Ross and Don Critchfield, authors of "Painting Farm Buildings and Equipment"; University of Minnesota; Successful Farming; Republic Steel Corporation; Stanley Tools; Henry Disston & Sons, Inc.; E. C. Atkins and Company; Simonds Saw and Steel Company; The Carborundum Company; Kester Solder Company; Lead Industries Association; Linde Air Products Company; The Deming Company; General Electric Company; Deere & Company; The Prime Manufacturing Co.; Harry Ferguson, Inc. Special acknowledgment is due the Portland Cement Association for the material used in Chapter 8 on Concrete Work.

CONTENTS

EDITOR'S FOREWORD

The American farmer is an unspecialized mechanic. In growing crops, raising animals, and marketing the products of his toil, many mechanical jobs are encountered. Such jobs must be done in order to accomplish his primary purpose. Along with these responsibilities is that of maintaining an efficient, satisfying farm home and surroundings. All this involves planning and some degree of mechanical skill; otherwise, the necessary repair, replacement, reconditioning, and construction work, of many types and kinds, cannot be taken care of properly.

Since this mechanical work must be performed with the aid of tools, a well-equipped farm workshop has long been a necessity on every farm. The steady increase in mechanization over the past 25 years has been appreciably accelerated by wartime activities, and the new shop equipment recently made available in thousands of rural communities bears testimony to the increased interest of farm people in farm shopwork. The modern farm must be kept in good running order or food production decreases and operation costs become excessive. Time, labor, and money are saved, and inconvenience avoided with a good farm workshop efficiently used.

Outstanding features of this book are the practical direct approach, the logical activity basis of organization, the scope of the subject matter and the understandable language used, the story told by the many well-selected illustrations, the unique ways included for performing various skills, and the ease with which the desired information can be located in the chapters by the reader.

The author, Prof. Mack M. Jones, is well known in the fields of agricultural engineering and farm shopwork. His writings include books, bulletins, and technical articles. He has been engaged for years in training teachers of mechanical subjects on both the college and the secondary levels. "Shopwork on the Farm," coming from the pen of an author with such a background, should be welcomed by student, farmer, teacher, and householder with equal enthusiasm.

W. A. Ross.

1. Providing and Equipping a Farm Workshop

I<small>N ORDER</small> to keep the farm home, buildings, machinery, and equipment in good repair, a workshop of some kind is essential. It need not be elaborate and expensive, but it should be orderly and systematic. Ample room is necessary for working on machines, and a system should be followed for storing and protecting the shop tools and keeping them where they can be readily found when needed. A knowledge of how to use tools is not enough. Unless the farmer as a farm mechanic is systematic and orderly and has an appreciation of the value of tools, they will soon become broken, dulled, or lost and, therefore, of little or no value when needed.

With the increased mechanization of farms, it has become necessary for the successful modern farmer to be proficient in the use, repair, and maintenance of mechanical equipment of various kinds. Although some farmers are expert mechanics, the majority of farmers need be only general mechanics—not specialists or experts. It is usually much better for the average farmer to depend upon well-equipped commercial shops for his specialized needs, such as the complete overhaul of tractors and the repair of complicated electrical equipment.

Although the farmer needs to be an unspecialized mechanic, rather than a specialized mechanic, he should nevertheless be a good one. He should be thorough and systematic. Slovenly or slipshod methods have no more place on the farm than in other businesses or occupations. Machinery that works well, gates that open and shut easily, and buildings and fences that are orderly and in good repair not only save time and money for the farmer, but contribute to morale and pride of ownership.

MAJOR ACTIVITIES

1. Selecting the Site

2. Planning the Building

3. Selecting Tools, Equipment, and Supplies

4. Arranging the Interior and Storing Tools

5. Storing Supplies and Materials

1. Selecting the Site

The location for the farm shop will depend on the location and arrangement of buildings, fences, and lots on the farmstead and the desires of the individual farmer. The shop should be handy to machines and implements as they are taken to or brought from the fields. It is desirable also to have the shop near paths or routes that are regularly used by the farmer as he goes about his daily chores, so that small pieces of equipment can be readily taken there for repairs. On many farms, an ideal location for the shop is in or near the machine and implement shed. Other factors that should be kept in mind when selecting the site for the shop are general appearance of the farmstead, location of windbreaks, and the possibility of future additions.

In case a large complete shop cannot be afforded, the shop can often be started in some seldom used building, such as an old granary or perhaps a part of a barn. A place for a bench and the orderly arrangement of some tools and supplies, although not wholly adequate, will be much better than no shop at all. As finances or conditions warrant, the shop can then be enlarged and possibly moved to more adequate quarters. In fact, starting with a few good tools and a place to keep and use them, and then adding to the collection from time to time, is a very practical way of acquiring satisfactory shop facilities on a farm. It is not often that a farmer can afford to build and outfit a shop completely at the outset.

2. Planning the Building

Although farm shop buildings are not standardized as to type, there are two general classes: separate shop buildings and shops combined with other buildings such as machine sheds or garages. Each of these plans has its advantages. The separate building makes a very satisfactory arrangement, if it is large enough to permit bringing in farm machines. In many cases, a shop made on one end of a machine shed will prove just as satisfactory, and the cost will usually be lower.

A combination shop and garage is also satisfactory if it is large enough. The shop may be located in one end of the garage, or along one side. The car, truck, or tractor that is normally housed in the garage can usually be removed for a few hours, or even a day or two, when extra space is needed for repair or construction work. Such arrangements are shown in Figs. 1 and 2.

Building the shop onto one end of a machine shed is generally better than combining it with a garage. Arrangements of this type are shown in Figs. 3 and 4. The machine shed is usually nearer the center of shop activities on the farm, and the extra space in the shop

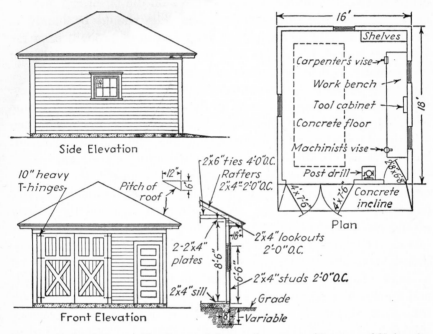

Fig. 1.—A plan for a combination workshop and garage. (Courtesy of University of Minnesota.)

can often be well used for temporary storage of machinery, seed, supplies, or equipment.

Determining the Size of the Shop. Although the floor-space requirements of a shop are flexible, a shop about 14 by 20 ft. is considered adequate for most farms. A shop 10 by 14 ft. is about the smallest that is generally satisfactory, and one 16 by 26 ft. ordinarily is considered ideal as to size.

Planning the Foundation and Floor. The foundation should be of concrete, extending about 18 in. into the ground. A footing will not be required if the soil is firm and solid and if the foundation is made 10 to 12 in. thick at the base. That part of the foundation

aboveground should be at least 6 in. thick.   If a 4- by 12-in. footing
is used, the foundation may be 6 in. thick all the way up.   The
foundation should extend at least 6 in., preferably higher, above the
ground line.

Concrete makes an ideal floor for the shop and should be used
where possible.   Where some saving in cost is imperative, a tamped-
earth floor can be used.

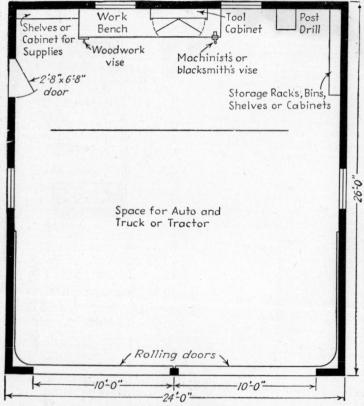

Fig. 2.—A floor plan for a double garage with a shop in the end.

Choosing Building Materials.   The farm shop can be of frame
construction or of brick, tile, or concrete blocks.   The choice of
materials will be determined by comparative costs, the preference of
the owner, the materials used in other buildings on the farmstead,
and the materials available.   Brick, tile, and concrete are more fire-
resistive than wood construction, although the fire hazard in any shop
can be reduced by careful planning, by prompt disposal of rubbish,
and by keeping the shop clean and orderly.

Fig. 3.—A combination workshop and machine shed. (Courtesy of Successful Farming.)

Providing Windows and Doors. Provide plenty of light and ventilation in the shop by putting windows on at least two sides. Windows on three sides is even better. Place one window over or near the workbench. If there is a vise on each end of the bench, a window over each end is desirable. For an average size shop, four windows, each with six 10- by 12-in. lights, is usually adequate.

An essential requirement for a good farm shop is an easy grade entrance and large doors for bringing in machinery. The door opening should be at least 9 ft. wide and 8 ft. high, preferably wider and higher. The doors may be of the large double type, which swing out,

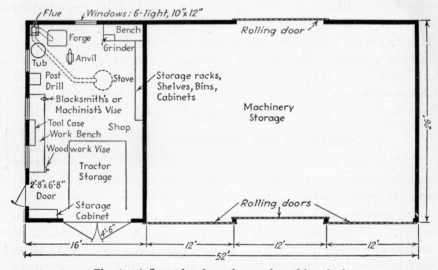

Fig. 4.—A floor plan for a shop and machine shed.

or of the rolling type. In either case, the doors should be tight-fitting when closed so as to keep out drafts and birds. It is essential that the doors be easily opened and closed. In most cases, it will probably be more satisfactory to provide rolling doors than hinged doors. Hinged doors, if they are well made so as not to warp or sag, give good service in areas where deep snow and freezing do not hamper opening during the winter.

In addition to the large doors for admitting implements and machines, the farm shop should have an ordinary service door about 2 ft. 8 in. wide, so that the large doors will not have to be opened every time one enters the shop.

Providing Electricity for Lights and Power. If electricity is available on the farm, the shop should, of course, be wired for it.

Good lights will encourage and facilitate work on dark, rainy days when outside work cannot be done. Provide good shaded lights over the workbench and other work centers. Also, install several convenient outlets for electrically operated tools and equipment, such as drills and grinders. At the time of wiring, there is frequently a tendency to underestimate the needs for convenient outlets. It is therefore best to install plenty of them, or to make provision for easy extension of the wiring later if this should prove desirable. (See pages 445 to 447 for information on electric wiring.)

Heating the Shop. Heat in a shop is a great convenience on cold days. Adequate heat can often be provided by means of an old stove, or a well-made heater fashioned from discarded oil drums. If both a forge and a stove are to be installed in the shop, it is better to have a separate flue for each, although both can be connected to the same flue. If both are connected to one flue, install dampers in the separate pipes before they enter the flue or a common pipe. It will then be possible to regulate the drafts better.

3. Selecting Tools, Equipment, and Supplies

A practical way to equip a shop is first to secure a few of the more common and most needed tools, and then to add others from time to time as special needs develop or as finances permit. Woodworking tools, such as a saw, hammer, square, brace and bits, are most often needed and probably should be bought first if a complete set of equipment is not to be bought at the beginning. Some cold-metal working tools, such as a vise, hack saw, and files, may well be included in the first installment also. More complete equipment, possibly blacksmithing tools or power-driven equipment, could be added later.

Consideration should be given to quality when tools are being purchased. Most tools are available in two or three grades or qualities, with the quality indicated, to some extent at least, by the selling prices. It is usually false economy to buy the cheapest grade of tools for the farm work shop, and in most cases it is difficult to justify the purchase of the highest quality. Normally, an intermediate quality tool is quite satisfactory.

If many tools are to be purchased at one time, it is a good plan to obtain prices from two or more sources of supply. Prices on tools sometimes vary considerably, even on tools of comparable quality. Where possible, prices should be obtained on specified makes or brands and model numbers of tools. Otherwise, one quotation may

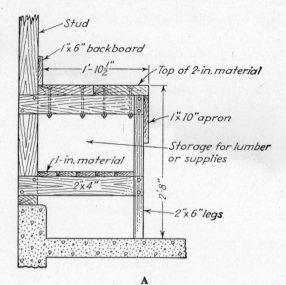

A

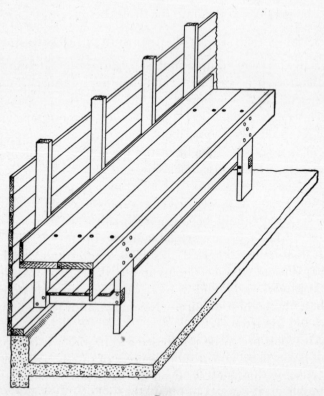

B

Fig. 5.—A substantial built-in workbench can be made easily in the shop.

be on one grade of tool, and another on quite a different grade or quality.

Providing Homemade Shop Equipment. Much of the shop equipment like workbenches, sawhorses, nail and tool boxes, miter boxes, and storage racks, shelves, and drawers can be made easily in the shop.

Make workbenches rugged and sturdy, and build them into or attach them permanently to the wall if possible. Sometimes the wall

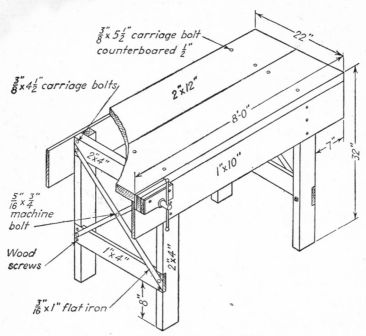

Fig. 6.—A plan for a movable workbench.

studs can be used to support the back of the bench. Figure 5 shows the construction of such a bench. It is best to use 2-in. lumber for the top, legs, and main supports of the bench. Use bolts or screws rather than nails for fastening the parts of a bench together.

Make the benches 24 to 30 in. wide, particularly if they are to be fastened to a wall. Benches are frequently made too wide, making it difficult to reach tools or supplies in cabinets or on shelves above the bench. Also, with a wide bench there is more tendency to pile tools and materials on it instead of returning them to their proper places. Make the benches 30 to 34 in. high. The length of a bench may be

6 to 12 ft., depending upon the space available and the desires of the owner. Benches are often made longer than necessary. It is much better to have a small bench with cabinets, bins, and shelves arranged conveniently about it, than to have a large bench that is cluttered with tools and materials because there is no regular place for them.

Fig. 7.—A good way to mount a blacksmith's vise. Cut a section from an old range boiler or large pipe and set it in the floor. Then on top make a small table of steel plate or of wood covered with sheet iron.

Check Lists of Tools and Equipment for a Farm Workshop

WOODWORKING TOOLS

1 saw, hand crosscut, 8 point, 26 in.
1 saw, rip, 6 point, 28 in.
1 steel square, 16 by 24 in.
1 nail hammer, 16 oz.
1 level, carpenter's, 26 in.
1 brace, ratchet, 8 or 10 in.
6 bits, auger, Nos. 4, 6, 8, 10, 12, 16
1 bit, countersink
1 bit, expansive, $\frac{7}{8}$ to 3 in.
1 bit, screw driver
2 chisels, socket firmer, $\frac{1}{2}$ and 1 in.
1 drawknife, 8 in.

1 plane, jack, 14 in.
1 oilstone, 2 by 8 in.
1 rule, folding, 4 ft.
1 mallet, wooden
1 miter box (homemade)
3 screw drivers, 3, 6, and 8 in.
1 tool grinder
1 grinding wheel dresser
1 glass cutter
1 putty knife
1 vise, woodworker's, 4 by 7 in.
1 saw set, pistol grip
1 saw vise (homemade)
1 file, auger bit
6 files, saw, slim taper, 5 and 6 in.
1 file, half round, wood, 10 in.
1 rasp, wood, 10 in.
1 file card

METALWORKING TOOLS

1 hack saw, 8 to 12 in.
1 blacksmith's post drill, $\frac{1}{2}$ in.
7 twist drills, bit stock, $\frac{1}{8}$ to $\frac{3}{8}$ in.
6 twist drills, blacksmith's, $\frac{1}{4}$ to $\frac{1}{2}$ in.
11 twist drills, straight round shank, $\frac{1}{16}$ to $\frac{1}{4}$ in.
1 hand drill, $\frac{1}{4}$ in.
1 hammer, machinist's ball-peen, 1 lb.
6 cold chisels and punches, assorted sizes (possibly homemade)
1 screw plate (tap and die set), National Coarse $\frac{1}{4}$ to $\frac{1}{2}$ in.
1 pair tinner's snips, 3 in.
1 blow torch, 1 qt.
1 soldering copper, $1\frac{1}{2}$ lb.
1 pair pincers, 12 in.
1 pair pliers, 6 in.
3 wrenches, crescent adjustable, 6, 8 and 10 in.
4 wrenches, double-end, open end, $\frac{3}{8}$-$\frac{7}{16}$, $\frac{1}{2}$-$\frac{9}{16}$, $\frac{19}{32}$-$\frac{11}{32}$, $\frac{5}{8}$-$\frac{3}{4}$
4 wrenches, box end, 12 point, 45 deg. offset heads, $\frac{3}{8}$-$\frac{7}{16}$, $\frac{1}{2}$-$\frac{9}{16}$, $\frac{5}{8}$-$\frac{11}{16}$, $\frac{3}{4}$-$\frac{25}{32}$
1 set socket wrenches, 12 point, $\frac{1}{4}$ to $\frac{3}{4}$ in.
2 pipe wrenches, 14 and 18 in.
1 vise, machinist's, 4 in.
2 reamers, repairman's taper, $\frac{1}{2}$ and 1 in.

1 file, flat bastard, 12 in.
1 file, mill bastard, 10 in.
2 files, round bastard, 6 and 10 in.

FORGING TOOLS

1 forge, geared blower
1 anvil, 100 or 125 lb.
1 hammer, blacksmith's cross-peen, 2 lb.
1 hardie to fit anvil
1 pair tongs, bolt, $\frac{3}{8}$ by 20 in.
1 pair tongs, straight lip, 20 in.
1 vise, blacksmith's leg, $4\frac{1}{2}$ in.

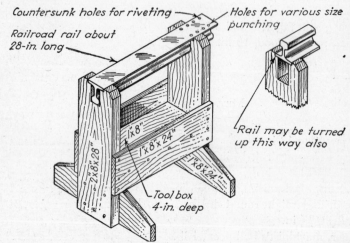

Countersunk holes for riveting
Holes for various size punching
Railroad rail about 28-in. long
Rail may be turned up this way also
2"x8"x28"
1"x8"
1"x8"x24"
1"x8"x24"
Tool box 4-in. deep

Fig. 8.—A piece of discarded railroad iron mounted on a homemade stand makes a practical lightweight anvil.

CONCRETE TOOLS

1 edger
1 float, wooden (homemade)
1 trowel
1 sand screen (homemade)
1 tamper (homemade)

PLUMBING TOOLS

1 pipe cutter, $\frac{1}{2}$ to 2 in.
1 piper reamer, $\frac{1}{4}$ to 2 in.
1 set pipe threading dies and stock, $\frac{1}{2}$ to $1\frac{1}{4}$ in.
1 pipe vise, $\frac{1}{8}$ to $2\frac{1}{2}$ in.

Leatherworking Tools

1 awl with assorted blades
1 doz. needles, harness sewing (assorted)
1 leather punch
1 riveting machine
1 stitching clamp (homemade)

Providing Shop Supplies and Materials. It will be impossible to foresee the various jobs that will be done in the shop and to provide

Fig. 9.—Many needed pieces of shop equipment like sawhorses, nail boxes, and tool boxes are easily made in the shop.

in advance all the supplies and materials that will be required. Experience indicates, however, that it is best to have on hand some of the more commonly needed supplies and to add to them from time to time as needs arise. The following lists may help in making initial selections of supplies and materials:

Check Lists of Shop Supplies

Woodworking Supplies

Glue, casein or liquid
Lumber

Nails
Sandpaper
Screws, wood, flathead bright
Glaziers' points
Putty

METALWORKING AND FORGING SUPPLIES

Smithing coal
Welding compound
Bar iron, $\frac{1}{4}$ to $\frac{1}{2}$ in. round, $\frac{1}{4}$ to $\frac{3}{8}$ in. flat, assorted widths
Bolts, carriage and machine, assorted sizes and lengths
Nuts, assorted
Hack-saw blades
Rivets, iron, $\frac{3}{16}$ and $\frac{1}{4}$ in., $\frac{1}{2}$ to 1 in. long
Solder, 50–50, bar or wire
Soldering flux
Sal ammoniac, cake
Washers, steel, assorted
Threading oil
Cap screws, assorted
Cotter keys, assorted
Lock washers, assorted

LEATHERWORKING SUPPLIES

Harness oil
Rivets, copper, assorted
Rivets, tubular, assorted
Thread, linen, No. 10

4. Arranging the Interior and Storing Tools

There are probably as many ways of arranging benches and other farm-shop equipment like a forge, anvil, or post drill as there are ways of arranging furniture in a room. Certain principles, however, should be kept in mind in choosing locations for the larger pieces of shop equipment. In general, place the larger, heavier, pieces along a side, or at one end, or both along one side and at one end of the shop, in order to leave a large unobstructed space to accommodate implements being repaired or larger pieces of equipment being constructed.

It is usually a good plan to assemble woodworking tools and equipment about one bench and metalworking and blacksmithing about another; or if only one bench is to be used, woodworking equipment at one end and metalworking equipment at the other end.

Locations of windows and doors will have an important bearing on the locations of benches and the larger pieces of equipment, particularly when old buildings are being adapted to use as a shop.

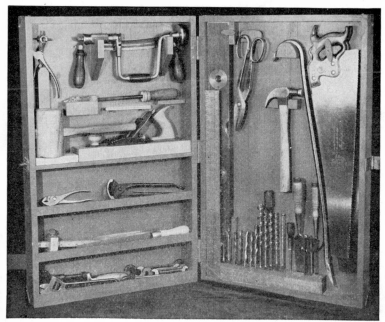

A

B

Fig. 10.—Two excellent types of wall-tool cabinets. (Courtesy of University of Minnesota.)

Good light at the work centers is important. Also, it is important to locate the work centers so as to avoid unnecessary steps and backtracking in getting tools from cabinets to the bench, replacing them, etc.

Orderliness and system contribute much to the value of a shop. Keep tools where they can be easily and quickly found when needed and easily and quickly replaced after a job is finished. Locate tool cabinets or racks near the benches or other places where they will be most frequently used.

Providing Tool Cabinets and Tool Boards. Closed cabinets and open tool boards or racks are both used with success in farm shops

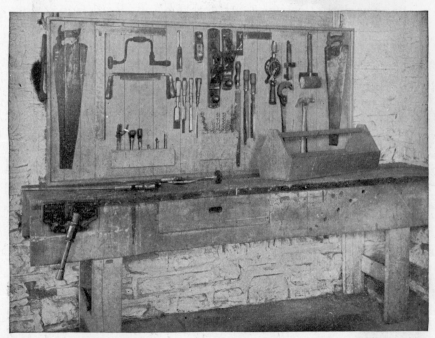

Fig. 11.—A workbench with an open tool board above it. The painted silhouettes behind the tools greatly facilitate returning them to their places.

(see Figs. 10 and 11). Cabinets that can be closed are usually to be preferred, particularly for the smaller and more expensive tools, as they give better protection against dirt and dust and can be locked if desired. Plain wooden cabinets are easily constructed in the shop and fastened to the walls. Hooks, shelves, and racks can be arranged inside the cabinets, even on the inside of the cabinet doors. Thus, many tools can be stored in a small space.

Larger tools, like spades, shovels, hoes, axes, and large wrenches and hammers, can be hung on racks or hooks on the walls or stored in closets.

Marking the Places for Tools. It is a good plan to mark the outlines for tools stored in cabinets and for many tools hung on walls or tool boards. This can be done quickly and easily with a heavy black lead pencil or with crayons. It is better, however, to take a little more time and paint neat silhouettes of the tools (see Fig. 11). Such marking or silhouetting greatly facilitates replacement of tools to their proper places and helps keep tools from becoming misplaced or lost.

In arranging places for tools, give due consideration to cutting tools, particularly keen-edged woodworking tools. Such tools should not be hung so as to endanger dulling or damaging the sharp edges.

Fig. 12.—A good way to keep tools like shovels, hoes, and forks from getting underfoot.

Do not crowd them together, nor hang them in such a place or way that they might fall or be dropped in getting them from their places, or in returning them.

5. Storing Supplies and Materials

Systematic and orderly storage of supplies and materials, as well as of tools, contributes a great deal to the value and usefulness of a shop. Keep supplies and materials where they can be easily and readily found, and do not allow them to clutter up the benches, work spaces, or floor.

Storing Miscellaneous Small Supplies. Small bins or drawer cabinets are ideal for storing nuts, bolts, screws, nails, and similar supplies. Bins for bolts are not hard to make in the shop. If small drawers are not available, screws may be kept in small glass jars, or left in their original paper box packages and arranged in order on shelves. Well-marked containers placed on shelves also make a good system for storing small quantities of many miscellaneous supplies.

Storing Lumber. If wall space is available, small amounts of lumber may be stored on racks built along the walls of the shop. This is usually easy to do in shops of frame construction. In many shops, overhead storage is also practical for certain supplies, as well

Fig. 13.—A vertical iron rack makes an excellent place to store rods and bars.

as lumber. In case of overhead storage, be sure to provide a convenient and safe ladder or stairs for use in putting materials away and getting them down.

Usually only small amounts of lumber will need to be stored in the shop for long periods. Where considerable lumber is to be stored, some place in a machine shed or barn is usually better than the shop

Lumber stacked on the floor requires considerable space and often handicaps work in the shop.

Storing Iron.   Rods and bars of iron for forging or cold-metal work can usually best be stored by standing them on end.   A rack with small compartments for keep-ing the various sizes separate is easily constructed (see Figs. 13 and 14).   Short lengths as well as long ones are easily stored in such a rack.   Furthermore, the iron re-quires very little floor space when thus stored, and the rack is easily kept in order.

Storing Cement. Always store cement in a dry place, as it has a tendency to absorb moisture and become lumpy. It may be advisable to keep a sack or two of cement on hand for an occasional repair job, but it is usually best to buy larger quantities of cement only shortly before it is to be used. Keep stored cement up on boards away from contact with either the floor or the ground; otherwise it will harden and become worthless.

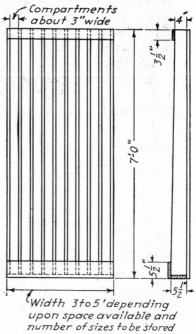

Fig. 14.—Plans for a vertical iron rack like that shown in Fig. 13.

Important Points in Providing, Equipping and Using a Farm Workshop

1. The workshop need not be elaborate and expensive, but it should be orderly and systematic.
2. Locate the shop near paths regularly used by the farmer as he goes about his daily chores and where it will be easily accessible to field machines and implements.
3. Make the shop large enough; 14 by 20 ft. is a good size, and 16 by 26 ft. is even better.
4. The shop may be a separate building, or it may be combined with the machine shed or garage.
5. A concrete floor is ideal for the shop.
6. Four windows, each with six 10- by 12-in. lights, are usually adequate. Place windows on at least two sides of the shop, preferably three, to give good cross ventilation as well as light.

7. If electricity is available, wire the shop for lights and power. Provide plenty of convenient outlets.

8. Make large doors and an easy grade entrance for bringing in machinery. Provide a small door also, so the large one will not have to be opened every time one enters the shop.

9. Heat in the shop is a great convenience on cold days. An old stove will usually do.

10. A practical way to equip a shop is first to secure a few of the more common and most needed tools, and then to add others from time to time.

11. Locate benches and stationary shop equipment along the side or ends of the shop, leaving a large unobstructed space in the center for repair and construction work.

12. Have a place for all tools. Use closed cabinets or open tool boards as preferred.

13. Painted silhouettes will greatly help to keep tools in their places.

14. Small bins or drawer cabinets are ideal for storing nuts, bolts, screws, and similar small supplies.

15. Well-marked containers arranged in order on shelves make a good system for storage of miscellaneous small supplies.

16. Keep the shop clean and orderly. Do not allow rubbish or junk to accumulate.

Jobs and Projects

1. Make a general survey or inspection of the farmstead on your home farm or some other farm in your community, and decide upon a good location for a farm workshop.

   Make a simple map or sketch of the layout of buildings, and indicate your choice of locations for the shop. Give reasons for your choice.

2. Make a list of advantages of the separate shop building as compared with a combination shop and garage, or a shop and machine shed. What advantages do the combination buildings have?

3. Make an inspection of the school shop and such farm workshops and commercial repair shops as are available, and note the features you would like to incorporate in your own shop. List them.

4. Consult catalogues, and make a short list of the more important tools you would recommend for starting a farm workshop. Tabulate, and include catalogue numbers, brief descriptions, sizes, prices, etc.

5. Make a classified list of homemade shop equipment and tools you would recommend for a farm workshop.

6. Check those pieces of needed equipment listed under No. 5 that you believe you can and should make.

7. List usable shop tools and equipment that are already available on your farm and that could be assembled and put into a shop.

2. Sketching and Drawing

A working knowledge of sketching and drawing is a most valuable asset to anyone who deals with mechanical equipment. It enables him not only to read and interpret plans, drawings, and blueprints made by others, but also to clarify and transmit his own ideas regarding changes, repairs, or new construction work. The modern farmer deals more and more with mechanical equipment, and he particularly needs some knowledge and skill in making and reading sketches and drawings. Farm papers, periodicals, books, and bulletins contain a wealth of new and up-to-date information, much of which is in the form of diagrams and drawings. Improvements and repairs on machinery and buildings, which a farmer can make or have made, often depend upon his skill in reading or making sketches and drawings.

MAJOR ACTIVITIES

1. Making Freehand Sketches

2. Making Pictorial Sketches and Drawings

3. Reading Working Drawings and Blueprints

4. Making Working Drawings

5. Lettering Sketches and Drawings

6. Making Floor Plans for Buildings

7. Making Out Bills of Materials

8. Writing Specifications to Accompany Drawings

1. Making Freehand Sketches

For many jobs around the farm, the home, or the farm shop, simple rough sketches will do practically as well as finished drawings. Every farmer should learn how to make sketches of top, side, and end views of an object and also pictorial sketches, regardless of whether he does his own construction work or has it done by others.

Sketching Lines.   Use a soft, well-sharpened pencil, and hold it lightly and not too close to the point.   Use a series of short, rapid strokes, making a series of short, light lines joined together with a little overlapping.   Make vertical lines from the top downward, and mainly with a finger movement (see Fig. 15).   Make horizontal lines

Fig. 15.—In sketching a vertical line, use short, rapid strokes, making a series of short, light lines joined together with a little overlapping. Sketch vertical lines from the top downward with a finger motion.

Fig. 16.—Sketch horizontal lines with a side-sweeping wrist motion.

from left to right, and mainly with a side-sweeping wrist motion (see Fig. 16).

The principles of sketching and drawing may be better understood by imagining that the object to be drawn is placed in a glass box (see Fig. 17).   The top view is what would be seen by looking straight down through the top of the glass box directly from above.   The front view is what would be seen when the object is viewed directly from the front.　Likewise, a right or left end (or side) view is what would be seen from the right or left end.

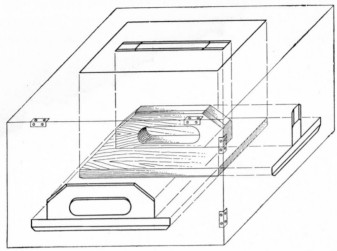

Fig. 17.—A wood float in a glass box with top, side, and end views projected onto the top, side, and end of the box.

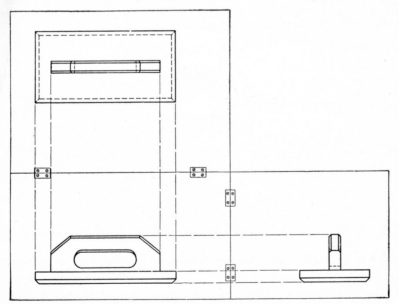

Fig. 18.—The glass box opened to show positions of top, side, and end views.

The proper position of the various views on a sketch or drawing may be determined by imagining that the top and the ends of this glass box are hinged and swung out as shown in Fig. 18.　The top view occupies the upper left part of the drawing, the front or side view is directly below, and the right end view is directly to the right of the side view.

Choosing the Views. Several different views of an object may be shown, but usually only two or three are necessary. Select those views which will best and most easily show the shape and size of the

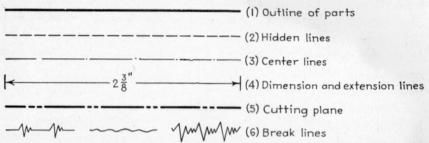

Fig. 19.—Alphabet of principal lines used in drawing and sketching.

various parts of the object. Usually a top view, a side view, and one end view are sufficient, and, in many cases, only a side view and an end view are needed.

As the sketching proceeds, use care to make the various lines of suitable length, so as to keep the different parts of the drawing in

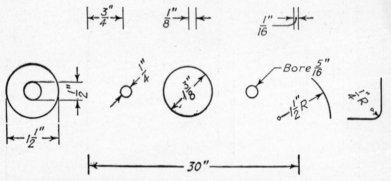

Fig. 20.—Examples of correct methods of dimensioning.

proper proportion. Make the outlines of the main views first, then add the minor lines, and finally the dimensions and notes.

Using the Right Kind of Lines; Alphabet of Lines. Use normal full-weight lines to show the main outlines of an object which are visible in any particular view. Use dashed lines, called *dotted* or *broken lines*, to show those parts which are not visible. The various kinds of lines commonly used in sketching and drawing are shown in

Fig. 19. Since each kind of line has a definite meaning or significance, it should always be used for its particular purpose.

Dimensioning Sketches and Drawings. Always show all measurements or dimensions of an object that are essential for its construction. Make light lines, called *extension lines* (see Fig. 19), extending out from the edges or other parts to be dimensioned. Then sketch or draw light dimension lines, with arrows at the ends, between these extension lines. Make neat sharp arrows, and make the points come exactly to the extension lines or other parts to

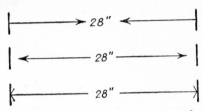

Fig. 21.—Examples of incorrect methods of dimensioning.

be dimensioned. Leave space in the middle of the dimension lines for insertion of the figures. Examples of correct methods of dimensioning are shown in Fig. 20. and some examples of incorrect methods are shown in Fig. 21.

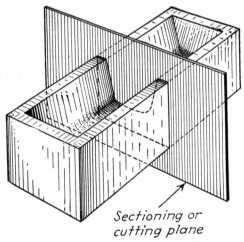

Sectioning or cutting plane

Fig. 22.—An imaginary sectioning or cutting plane making a cross section of a concrete trough.

Making Sectional Views. Sectional views, often called simply *sections*, are used to show the shape of complicated parts, particularly the interior of parts, and objects composed of several pieces. Dotted lines can be used to show the shape or construction of such pieces, but they sometimes unduly complicate a drawing and make it difficult to read.

To make a sectional view, imagine the object is cut in two and the front part removed (see Fig. 22). Then make a view of the exposed end, and show the solid portions cut by the sectioning plane by crosshatching, that is, with lightweight diagonal lines (see Fig. 23).

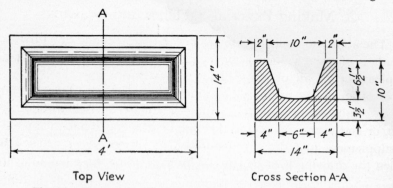

Top View   Cross Section A-A

Fig. 23.—Top view and cross section of trough shown in Fig. 22.

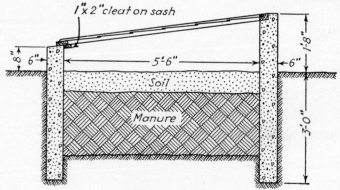

Fig. 24.—A cross-section view of a hot bed. Note the methods of showing sections of the different materials.

Fig. 25.—Revolved sections to show the shape of the hammer handle at different places.

The exact position of the section is indicated on one of the main views by means of a line made up of a series of long dashes separated by two short dashes (see Fig. 23).

Instead of crosshatching, other symbols are commonly used for materials like concrete, earth, and wood (see Fig. 24).

A *revolved section* is a small sectional view shown on one of the main views. Such a sectional view shows the shape of a cross section at that particular place (see Fig. 25).

2. Making Pictorial Sketches and Drawings

The ability to make a pictorial view either freehand or with instruments is of great value in conveying a general idea of the shape or appearance of an object. Pictorial drawings show three sides or faces of an object in one view, much as a picture would. In farm shopwork, a pictorial sketch is often adequate in itself without side, top, or end views. A pictorial sketch or drawing is often valuable as a supplement to a drawing composed of various views, particularly when the drawing is to be read or interpreted by those not altogether

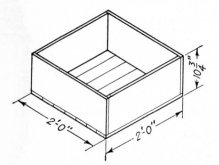 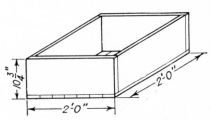

Fig. 26.—An isometric drawing.   Fig. 27.—An oblique drawing.

familiar with the principles of drawing. Complete, detailed dimensions are difficult or impossible to show on pictorial drawings. Such drawings, therefore, are not generally used alone to show a complicated object, particularly if it must be made to exact dimensions.

Pictorial sketches and drawings may be grouped into three main classes: isometric, oblique, and perspective.

Making Isometric Sketches and Drawings. These are made about three lines or axes, one vertical, and two at 30 deg. with the horizontal (see Fig. 26). Such a drawing can show a top, a side, and an end view of an object. The dimensions of only the main parts of an object can be indicated on an isometric drawing.

Making Oblique Sketches and Drawings. These are made about three axes, much in the same manner as isometric sketches and drawings, except that one of the axes is always horizontal, one vertical, and the third at any convenient angle (see Fig. 27). One view is thus the same as an ordinary front view. If one side of an object contains curved or irregular lines, this side should be selected for the front view.

Making Perspective Sketches and Drawings. These show objects just as they appear to the eye. They have a more pleasing appearance than isometric or oblique drawings, but are more difficult to make. A simplified type of perspective drawings, however, may be made similar to isometric drawings. Such a drawing is made like an isometric drawing, except the lines at an angle to the horizontal

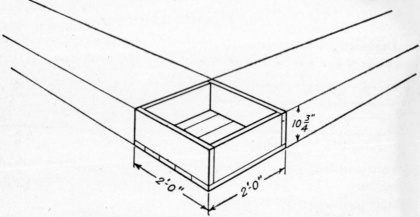

Fig. 28.—A simplified type of perspective drawing.

are made to converge toward vanishing points, instead of being made parallel (see Fig. 28).

3. Reading Working Drawings and Blueprints

Skill in reading working drawings and blueprints comes with practice. Drawing is a language of its own. In order to read drawings and to understand them, one must know something of the principles of drawing as outlined in this chapter. As in most shop work, it is advisable to become well grounded on the principles by thoughtful practice. Then the application and use of the principles become easy.

In reading a drawing, first look at its title or nameplate, to see what it represents or what the purpose of it is. Then look at the various views and identify them. By so doing, you will learn the general shape of the object. By comparing the various views and studying them in relation to each other, one can picture in his mind's eye the shape of the object. For example, a hole or a projection on the front of an object appears exactly the same on the front view. A glance at the top or side view quickly reveals whether it is a hole or a projection.

After learning the general shape of the object, then proceed to find out details and dimensions. First note the over-all size or dimensions and then the dimensions of the smaller parts or subdivisions of the main object. If detail or separate drawings of certain parts are provided, study these to determine the shape and size of the various parts.

4. Making Working Drawings

Drawings are made in much the same manner as outlined in the preceding pages for making sketches. The main difference is that

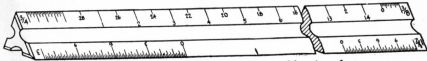

Fig. 29.—A mechanical engineer's or architect's scale.

lines are made straight with the use of a T square and triangles, instead of freehand. The lines are also measured and made exactly proportional in length to the parts of the object they represent. It is usually not practical to make the drawing the same size as the object.

Drawing to Scale. Making a drawing proportional in size to the object is known as drawing to scale. The scale of a drawing is commonly indicated by such terms as $\frac{1}{4}$ in. = 1 in., $\frac{1}{2}$ in. = 1 ft., or $\frac{1}{8}$ in. = 1 ft.

A mechanical engineer's or an architect's scale (see Fig. 29) is commonly used for measuring lines in making or reading a

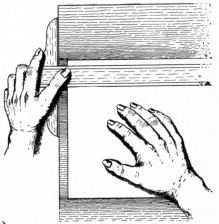

Fig. 30.—Aligning the paper with the edges of the drawing board.

drawing. If such a scale is not available, however, an ordinary ruler may be used. A ruler is particularly easy to use when some common division, such as $\frac{1}{8}$, $\frac{1}{4}$, or $\frac{1}{2}$ in. equals 1 in. or 1 ft.

Laying Out and Developing the Views. To start a drawing, first fasten the paper to the drawing board with thumbtacks or tape, using the T square to align the paper with the edges of the board (see Fig. 30). Next, choose a suitable scale and locate and mark off the spaces for the various views. Lay out the main lines of the drawing,

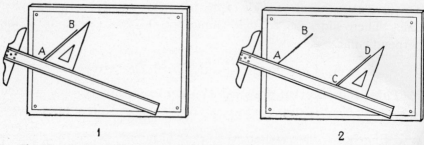

Fig. 31.—To draw a line parallel to a given line: adjust the T square and triangle to align with the given line AB; then slide the triangle along the T square to the desired position and draw the required line CD.

and then add the minor ones. Develop all views along together, projecting from one view to the other with the T square and triangles. Add dimension lines and notes last.

Always use a *well-sharpened* pencil so that the lines will be light and sharp and located accurately. Rule vertical lines from the

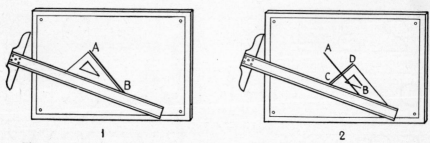

Fig. 32.—To draw a line perpendicular to a given line: adjust the T square and the triangle to align with the line AB, with the hypotenuse of the triangle against the T square. Then slide the triangle along into position and draw the required line CD.

bottom up and horizontal lines from left to right. Lines at various angles can be made with the T square and triangles. Hold the head of the T square firmly against the edge of the drawing board, and hold the triangles firmly against the T square.

To make a line parallel to a given line, move the T square about so that when the base of the triangle is against the T square, one side of the triangle coincides with the line. Then slide the triangle along into position for the new line, holding the T square firmly in place (see Fig. 31).

To make a line perpendicular to a given line, place the triangle with the *hypotenuse* against the T square (see Fig. 32) and shift the T square about until one side of the triangle coincides with the given line. Then slide the triangle along the T square and draw the perpendicular.

5. Lettering Sketches and Drawings

Titles and notes on drawings are lettered (printed)—not written in script. Neat lettering adds materially to the general appearance

Fig. 33.—Vertical style of lettering.

of a drawing or sketch, as well as making it more legible and easily and quickly read.

Neat rapid lettering comes first from careful attention to approved ways of making each letter, and then from patient practice. Care should also be given to the spacing of the words and to the spacing of letters within words. Use light guide lines for lettering drawings. This practice is used by even skilled and rapid draftsmen.

Two general styles of lettering are commonly used—the vertical and the inclined. The vertical style is probably easier to make. Probably the fastest way of acquiring a reasonable proficiency in lettering is to use capital letters altogether. It is the mark of poor or careless workmanship to use capital letters and small (lower case) letters indiscriminantly, or to use various styles of lettering on the same drawing.

Fig. 34.—Inclined style of lettering.

Figures 33 and 34 show how to make two different styles of letters and figures.

6. Making Floor Plans for Buildings

Before drawing a floor plan for a building such as a barn, some preliminary estimating and figuring will need to be done. If a barn is to be remodeled, the main dimensions will already be determined, and the problem will be to make the best arrangement of such units as bins, stalls, and feed alleys. If a new building is being planned, it may be better to design the floor plan from units, allowing for a certain number of stalls of a given size, a certain number of bins, etc., and thus arrive at the required over-all, or outside, dimensions. Barns, as well as many other farm buildings, are commonly made in certain more or less standard widths. Reference to standard plans and bulletins will usually suggest suitable widths, as well as other principal dimensions. It is usually easier and more practical to make modifications in a common or somewhat standard plan than to make an altogether new design.

Once the over-all or main dimensions are determined, lay off the outside lines of the floor plan, using a suitable scale. Next, lay off the lines showing the insides of the walls. Then locate the doors, windows, inside partition walls, stall partitions, mangers, feed alleys, etc., using standard or other suitable dimensions for these various units. Finally, add dimensions and notes and such smaller details as may be desired. Figure 35 shows a floor plan for a general-purpose barn.

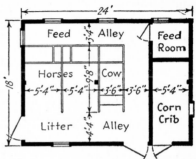

Fig. 35.—A floor plan for a small general-purpose barn.

7. Making Out Bills of Materials

Plans for buildings and for larger appliances are usually accompanied by lists of materials required for their construction. When such lists or bills of materials do not accompany the plans, then they should be made before construction is started.

A bill of materials is usually in tabular form and indicates just what part of the building or project the various pieces are to be used for, as well as the number of pieces, and the size, length, kind, etc. In case of a larger building or project, the bill of materials should give a summarized list of the total requirements of all pieces of the same kind and size. Descriptions of materials should be exact and complete, so that they can be easily bought or ordered from the usual sources of supply.

Although the making of bills of materials for any except the simpler and smaller buildings is beyond the scope of a course in beginning shopwork, every student should include lists of materials with the job plans for the jobs he expects to do in the shop.

The following is such a bill of materials for the workbench shown in Fig. 6.

Bill of Materials for Workbench

LUMBER

2 pcs.	$2'' \times 12''$—8'	No. 1 yellow pine	Top
4 pcs.	$2'' \times 4''$—30''	No. 1 yellow pine	Legs
2 pcs.	$2'' \times 4''$—22''	No. 1 yellow pine	Top cross pieces
2 pcs.	$1'' \times 4''$—22''	No. 1 yellow pine	Lower cross pieces
2 pcs.	$1'' \times 10''$—8'	No. 1 yellow pine	Aprons

HARDWARE AND IRON

8	$\frac{3}{8}'' \times 5\frac{1}{2}''$	Carriage bolts
10	$\frac{3}{8}'' \times 4\frac{1}{2}''$	Carriage bolts
2	$\frac{5}{16}'' \times \frac{3}{4}''$	Machine bolts
20	$1\frac{1}{2}''$ No. 10	Flathead, cadmium-plated, wood screws
1 pc.	$\frac{3}{16}'' \times 1''$—11'	Flat iron

Figuring Board Measure. Since lumber is commonly sold by the *board foot*, or foot, *board measure*, a bill of material usually indicates the number of board feet of various kinds of lumber required. A board foot is the amount of lumber in a piece 1 in. thick, 1 ft. wide, and 1 ft. long. In other words, a board foot is equivalent to one-twelfth cubic foot. The number of board feet in one or more pieces of lumber may be determined, therefore, by first finding the number of cubic feet contained in the lumber and then multiplying by 12. This method at first may seem to be tedious, but in practice it is easy to use. Simply apply the following simplified formula:

$$\text{Board feet} = \frac{\text{number of pieces} \times \text{inches thick} \times \text{inches wide} \times \text{feet long}}{12}$$

Cancellation will nearly always simplify the figuring so that it may be done without tedious multiplication and division. For example, to find the number of board feet in three 2 by 4's, 10 ft. long

$$\text{Board feet} = \frac{3 \times 2 \times 4 \times 10}{12} = 20$$

Lumber less than 1 in. thick is commonly measured in square feet instead of board feet.

Mill-surfaced lumber is never full width or full thickness, owing to the waste removed when the boards are surfaced or planed. In figuring board measure, however, it is always figured as if the boards were full width and thickness. For example, a 2 by 4 actually measures about 1⅝ by 3⅝ in., but it is always figured as full 2 by 4 in.

Determining Amount of Lumber to Cover a Surface. Since boards are not full width, they will not completely cover the area indicated by their nominal size. In figuring bills of materials, therefore, a certain amount over and above the indicated number of board feet must be allowed. A general rule is to add about one-twelfth when using 1 by 12 boards, about one-tenth for 1 by 10's, about one-eighth for 1 by 8's, etc. For tongue-and-groove or matched lumber like shiplap, flooring, siding, and ceiling, an amount varying from one-eighth to one-third or more must be added, depending upon the width of the boards and the amount of lapping in the grooves.

8. Writing Specifications to Accompany Drawings

When a building or a large piece of equipment is to be constructed, remodeled, or repaired, it is customary to write a set of specifications to amplify and supplement the drawings. Specifications are essentially a set of statements of kinds and grades of materials to be used, methods of construction, qualities of workmanship, finish, etc. Specifications when used together with the drawings should leave no doubt as to the materials to be used or how any part of the work is to be done. The test of any set of specifications and drawings is whether or not they answer questions that arise as the work proceeds. In the case of large buildings, the specifications may be long and involved, and they should be written or checked by an experienced builder.

Jobs and Projects

1. Sketch top, side, and end views of a workbench in the shop. Indicate the principal dimensions.

2. Sketch top, side, or end views of one or two other appliances or pieces of equipment in the shop, showing the main dimensions. Use the right kind of lines to show hidden parts, extension lines, dimension lines, etc.

3. Sketch a sectional view of a nail or tool box that may be available in the shop.

4. On a sheet of paper, write the letters A, B, C, D, E, and F corresponding to the blocks shown at the top of page 36. Beside each of these letters, write T, F, and E for top, front, and end views. Under each of these letters, write the figures that represent the proper top, front, and end views. Example:

A—T F E.
  1 4 4

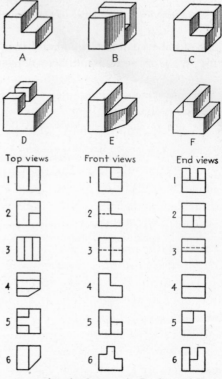

Top views   Front views   End views

5. On a sheet of paper, write the letters A, B, C, and D corresponding to the four sets of drawings shown below. Beside each of the letters, write the number of the block that the drawings represent.

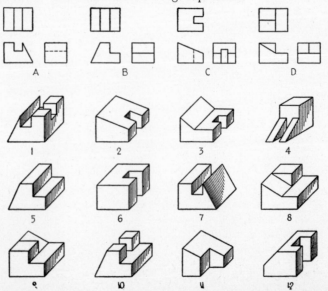

6. Sketch top, front, and end views of several of the blocks shown in No. 5 above.

7. Sketch the floor plan of some small farm building (shop, garage, machine shed, poultry house, etc.) on your home farm or some farm in your community.

Use cross-ruled paper and sketch the floor plan to scale, using an appropriate scale, such as 1 division equals 1 ft. or 2 divisions equal 1 ft. Show the location and the size of doors, windows, and permanently placed equipment.

8. On cross-ruled paper, sketch the floor plan for a combination farm workshop and machine shed which you believe would be practical for your home farm or some other farm in your community.

9. Design and make dimensioned sketches or drawings of some small shop or farm appliances (such as workbench, sawhorses, poultry feeder, tool cabinet) you would like to make in the shop.

3. Woodwork and Farm Carpentry

THERE will always be need for the farmer to make repairs and construct appliances involving the use of wood. It is easy to become reasonably proficient in the use of woods and woodworking tools, because woodworking, like most other kinds of mechanical work, is based on a comparatively few fundamental tool processes or operations, such as measuring, sawing, and planing. Once these processes are mastered, one is well on his way toward becoming a proficient woodworker.

MAJOR ACTIVITIES

1. Selecting Kinds and Grades of Lumber for a Job

2. Measuring and Marking Wood

3. Sawing Wood

4. Planing and Smoothing Wood

5. Cutting with Wood Chisels

6. Boring and Drilling Holes in Wood

7. Fastening Wood

8. Shaping Curved and Irregular Surfaces

9. Cutting Common Rafters

10. Building Stairs and Steps

11. Laying Out and Erecting a Small Building

1. Selecting Kinds and Grades of Lumber for a Job

Kinds and grades of lumber differ in such properties as strength, stiffness, hardness, toughness, freedom from warping, ease of working, nail-holding power, wear-resistance, decay-resistance, paint-holding power, and appearance. In selecting lumber for a particular job, give consideration to the requirements of the job and to the properties of the lumber available, as well as costs. Often the practice or custom in the locality is a good guide as to the best kind and grade of lumber to use. Sometimes lumber sawed from locally grown trees is more practical, and sometimes lumber shipped from a distance and sold through local lumberyards is better.

Lumber is classified technically into two general classes: softwoods, or lumber cut from needle-leaf evergreen trees, like pine, fir, and cypress; and hardwoods, or lumber cut from broadleaf trees which shed their leaves, like oak, hickory, and maple. Softwoods are in more general use for building construction and hardwoods for factory work. Southern yellow pine and Douglas fir are most widely used for construction work. Other lumber like white pine, cypress, and redwood are used where they are available. Oak is the most commonly used hardwood. Hardwoods are generally better for work like tool and implement handles, floors, and furniture.

There are two general grades of softwoods—select and common. The select grade is subdivided into four subgrades A, B, C, and D; and the common grade into five subgrades No. 1, No. 2, No. 3, No. 4, and No. 5. The two highest grades of hardwoods are called *firsts* and *seconds*. They are commonly sold together and designated as "FAS" (firsts and seconds). The lower grades of hardwoods are the same as for softwoods.

Select grades are generally used only where good finish is important, such as for inside trim and window and door casings, and for parts that are to be varnished or given a very smooth finish. Some of the lower grades of lumber can often be economically used for a job by cutting out and discarding knotty or otherwise unsuitable parts.

2. Measuring and Marking Wood

Accurate measuring and marking is the first requirement for success in shopwork. The 2-ft. folding rule (Fig. 36) and the zigzag-

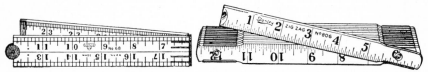

Fig. 36.—A 2-ft. folding rule.　　Fig. 37.—A zigzag folding rule.
(Courtesy of Stanley Tools.)　　(Courtesy of Stanley Tools.)

type folding rule (Fig. 37), which is available in various lengths, are common measuring tools used in woodwork. The try square and the carpenter's steel square are also used for measuring as well as for squaring.

Reading a Rule. The graduation lines on a rule are varied in length to facilitate quick and accurate reading. The 1-in. lines are longest, the ½-in. lines a little shorter, the ¼-in. lines still shorter, etc. (see Fig. 38). The smallest division on most rules used in wood-working is ¹⁄₁₆ in.

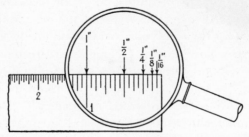

Fig. 38.—The graduation lines on a rule are varied in length to facilitate reading.

In reading a fractional measurement with a rule, think of the measurement as a major fraction plus or minus a small fraction. For example, ¹¹⁄₁₆ in. is ¾ in. minus ¹⁄₁₆; ⁵⁄₁₆ in. is ¼ in. plus ¹⁄₁₆; etc.

Measuring with a Rule. To measure a certain distance between two points, place the end of the rule exactly on or even with one point, and read the rule at the graduation line on or nearest the other point

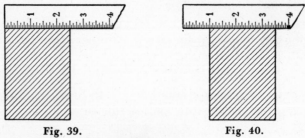

Fig. 39.   Fig. 40.

Fig. 39.—Measuring with a rule.

Fig. 40.—A common method of measuring with a rule when the end is worn. Place the 1-in. mark even with one edge of the work and read the rule at the other edge. The true measurement is then the reading minus 1 in.

(see Fig. 39). If the end of the rule is worn, start at the 1-in. mark and subtract 1 in. from the rule reading (see Fig. 40).

To lay off measurements with a rule, place the end of the rule (or the 1-in. graduation) carefully at one end of the measurement, and then make a fine mark with a pencil or knife exactly even with the desired graduation line on the rule. A knife gives a more accurate marking, but a pencil line is more easily seen and is used for all except the most accurate measurements.

To lay off several measurements in a straight line, it is best to mark off all measurements without raising the rule. If the rule is

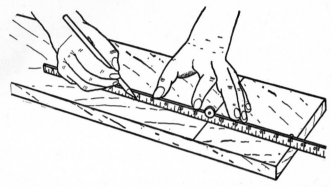

Fig. 41.—For extremely accurate measuring, lay the rule on edge so that the graduations touch the work being measured. When making several measurements in a straight line, do not raise the rule until all are marked.

raised and each measurement made separately, then there is a much greater possibility of errors.

For extremely accurate measuring, lay the rule on edge so that the graduations touch the work being marked (see Fig. 41).

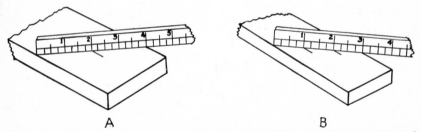

A   B

Fig. 42.—A. To mark the middle of a board, lay the rule across it at an angle with two even inch marks coinciding with the edges and mark the mid-point. B. In a similar manner, a board is easily divided into three (or more) equal widths.

Marking the Middle of a Board. To locate and mark the middle of a board, place the rule across the board at an angle so that major divisions, like inch marks, coincide with the edges of the board (see Fig. 42), and mark midway between these two major divisions. In a similar manner, a board may be divided into several equal widths.

Using the Try Square. The try square is used mostly at the bench for (1) measuring short distances, (2) laying out lines perpendicular to an edge or side of a board, (3) checking edges and ends of

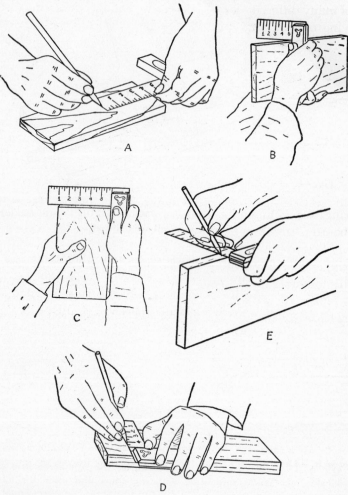

Fig. 43.—Common uses of the try square: A, marking off short measurements; B, checking an edge for squareness with a surface; C, checking an end for squareness with an edge; D, squaring across a surface; E, squaring across an edge.

boards to see if they are square with adjacent surfaces, and (4) checking the width or thickness of narrow boards (see Fig. 43). To keep a try square accurate and true, be careful not to drop it, and never use it for prying or hammering.

In squaring with a try square, always hold the handle firmly against the working edge or working surface (the main edge or surface from which other surfaces are measured or squared).

Laying Off Angles with the Steel Square.   The steel square is a tool of many different uses.   It is easily used for measuring distances and laying off and checking right angles.   It is also easily used for

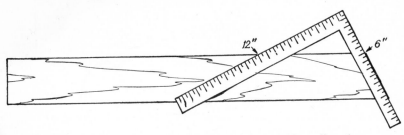

Fig. 44.—The square is a valuable tool for marking off angles.

measuring and duplicating various angles.   For example, suppose that a board is cut off at a certain angle and it is desired to cut another board at exactly the same angle.   Place the square with the tongue along the end of the board (see Fig. 44).   Note the readings where both the tongue and the body of the square touch the edge of the board.   Place the square on the second board with these same two readings along one edge of the board.   A mark along the tongue will

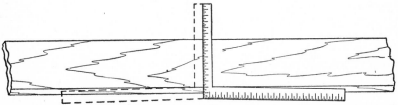

Fig. 45.—To test a square, select a board with a straight edge and square a line across it.   Then turn the square over and see if the line checks square.

give exactly the same angle as on the first board.   Take the readings as large as convenient in order to ensure accuracy.   For instance, a setting of 12 and 4 would be preferable to 6 and 2 or 3 and 1.

Testing a Square.   If a square is suspected of not being true, it is easily tested.   Use a board that has a perfectly straight edge and mark a line across the board (presumably at right angles to the edge).   Then turn the square around (see Fig. 45) and see if the line still checks square.   If not, the square is not true.

If a square is found not true, it may be adjusted by placing the corner of the square flat on an anvil and hammering carefully to stretch the outer or the inner part of the corner as may be required.

Fig. 46.—The bevel is used for laying out and checking angles and bevels.

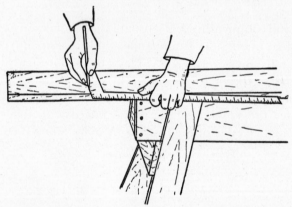

Fig. 47.—When squaring across a board, hold the body of the square firmly against the edge and mark close to the tongue of the square.

Using the Bevel. The bevel is used for laying out and checking angles and bevels (see Fig. 46). The blade is adjustable and is held in place by a thumb screw. After it is set to the desired angle, it is used in much the same manner as a try square. A good way to set it is to mark off the desired angle on a board, and then adjust the blade to fit the angle.

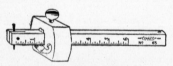

Fig. 48.—A marking gage. (Courtesy of Stanley Tools.)

Marking with a Pencil or Knife. To mark with a pencil or knife, first place the square or rule very carefully, and then make a fine narrow line very close to the edge of the square or rule. A knife makes a finer line and is recommended for very accurate work. For most farm woodwork and carpentry, however, a sharp pencil is

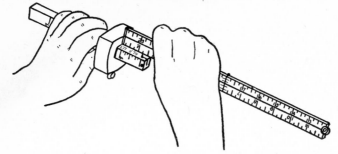

Fig. 49.—For accurate marking with the marking gage, always check the setting of the gage with a rule.

accurate enough. A hard pencil makes a finer line and stays sharp longer than a soft one. A soft pencil, however, makes a line that is more easily seen and is therefore generally preferred for rough work.

Fig. 50.—Using the marking gage. Be sure to hold the head of the gage firmly against the edge of the board and to roll the beam slightly forward so that the spur drags at a slight angle.

Using the Marking Gage. The marking gage (Fig. 48) is used for marking lines parallel to the sides, edges, or ends of a board. The spur should protrude through the beam about ⅛ in. and should be kept sharp so as to make a very fine line. For accurate marking, always check the setting of the gage with a rule (see Fig. 49).

To use the gage, grasp it with the fingers around the head and with the thumb behind the spur in position to push. Then push

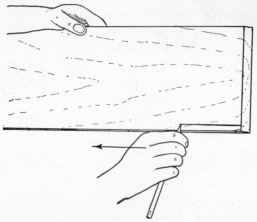

Fig. 51.—An easy method of gaging a line close to the edge of a board.

the gage forward, holding the head firmly against the surface from which the line is to be gaged, and with the gage rolled slightly forward so that the spur drags at a slight angle (see Fig. 50). See page 70 for gaging chamfers with a marking gage.

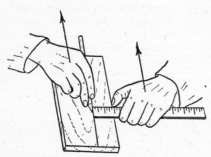

Fig. 52.—An easy method of gaging a line a few inches from the edge of a board.

Gaging with a Pencil. When a marking gage is not at hand or when the spur of the marking gage would mar the work, gaging may be done with an ordinary pencil as follows: Grasp the pencil loosely in the closed fist with the point protruding the desired distance. Then draw the pencil along with the thumbnail firmly against the edge of the board (see Fig. 51).

For gaging lines somewhat farther from the edge of the board, use a rule and a pencil. Grasp the rule in one hand with the thumbnail firmly against the edge of the rule at the desired distance from the end. Then draw the rule along with the thumbnail against the edge of the board while a pencil is held in the other hand at the end of the rule (see Fig. 52).

Setting and Using Dividers.   Dividers are used (1) for marking out circles or parts of circles, (2) for transferring or duplicating short measurements, and (3) for dividing distances into a number of equal parts.

To set a pair of dividers, loosen the thumbscrew and spread the legs to an approximate setting, and then tighten the screw.   Finally make the fine, or close, adjustment with the thumb nut at the end of the arc (see Fig. 53).

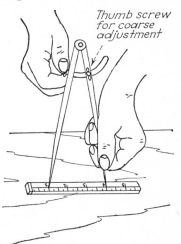

To divide a line into a number of equal parts, say three, set the dividers by guess to one-third the total length. Then check the setting by stepping off three steps to see if they equal the total length.   If the setting is too large, then set the dividers closer by guess, by one-third the distance overstepped on the last step.   If the setting is too short, then set the dividers wider.   Always check a new setting by stepping off the line again, continuing until the proper setting is obtained.

Fig. 53.—To set a pair of dividers, make an approximate adjustment with the thumbscrew and then make the fine adjustment with the nut at the end of the arc.   Steadying the knuckles against the bench top helps to hold the points accurately.

Laying Out Duplicate Parts; Superposition.   In marking out two or more pieces that are to be alike in all or part of their dimensions, much time may be saved and more accurate work ensured by marking all pieces at the same time, or from a pattern (see Fig. 54).

Frequently a piece can be marked for the required sawing, cutting, or boring by superposition, that is, by properly placing the parts together and marking them.

In marking out a number of pieces that are to be alike, rafters for example, the same piece should always be used as a pattern to ensure uniformity.

Laying Out Irregular Designs with Patterns.   Patterns are very convenient for laying out irregularly shaped pieces.   Patterns, also called *templates*, are sometimes furnished with plans.   When patterns are not available, designs may be sketched on paper, cardboard, presswood, or thin wood boards and then cut out with appro-

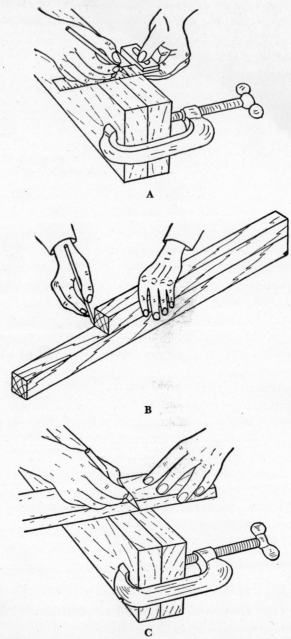

Fig. 54.—Forethought and planning in laying out duplicate parts save time and ensure accuracy. A, marking duplicate parts at the same time with a square. B, marking duplicate parts with a pattern. (Always use the same piece for the pattern.) C, marking the width of a notch by superposition.

priate tools.   In case of pieces that are symmetrical about a center line, a pattern for only one side, say the right or left half, is sufficient (see Fig. 55).

To use such a pattern, simply put it on the material to be cut, hold it firmly in place, and make a mark around it.

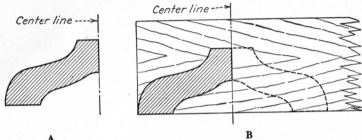

A   B

Fig. 55.—Patterns, or templates, are convenient and ensure accuracy in laying out irregular designs. A, a pattern; B, the pattern in place for marking the left half. To mark the right half, simply reverse the pattern.

Laying Out with Squares.   Many drawings and plans for irregularly shaped pieces are made on squared paper   To use such plans, rule off squares of the size indicated, usually 1 in. on a side, either on material for a pattern, or directly on the material to be used for the article.   Then locate and mark the points where the curves intersect the sides of the squares, and sketch the curves smoothly through the points (see Fig. 56).

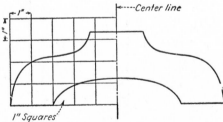

Fig. 56.—Designs drawn on squares are easily reproduced.   First rule squares on the material and then sketch in the curves so that they intersect the sides of the squares as on the drawing or plan.

Snapping a Chalk Line.   A quick and simple method of marking off a straight line on a floor, wall, ceiling, or piece of lumber is to use a chalk line.   It is simply a piece of string or cord that has been rubbed with chalk and coated with chalk dust.   To use a chalk line, stretch it between the two points that are to be joined by a straight line.   (It may be held in place by tying or wrapping around nails,

Fig. 57.—Snapping a chalk line.

Fig. 58.—Using the plumb bob to locate a point directly beneath another.

or with the assistance of a helper.)   Then lift the line off the surface at a point somewhere between the two ends, and allow it to snap back into place (see Fig. 57).   A straight chalk mark is the result.

Using the Plumb Bob.   A plumb bob is a pointed weight that may be suspended by means of a cord or string.   It is used for locating a point directly beneath another (see Fig. 58).   To use it, tie the string to a nail or other suitable support, and allow the bob to come to rest.   Then mark at the point of the bob.   If a plumb bob is not

Fig. 59.—Establishing a level line by means of a plumb bob and square.

available, a symmetrical weight, like a nut, may be used by carefully suspending it from a string.

The plumb bob may also be used, in connection with a square, for establishing a level line (see Fig. 59).   Simply allow the plumb bob to come to rest, and align the body of the square with the string Then mark along the tongue of the square.

Using and Testing a Level.   The carpenter's level is commonly used for marking level lines, placing the surface of a board in a level plane, and for such work as leveling foundation forms or a part of a machine that must be level for best operation.   The bubble vial or tube has a very gentle curvature and is almost filled with a nonfreezing liquid.   The tube is mounted so that when the base of the level is horizontal or level the bubble will be in the middle of the tube.

To use a level, simply place it on the surface or part to be leveled, and then raise or lower one end of the surface until the bubble stands in the middle of the bubble tube.

Most levels are equipped with a second bubble tube located near one end. This second tube is mounted perpendicular to the long edges of the level and is used for checking vertical pieces for plumb.

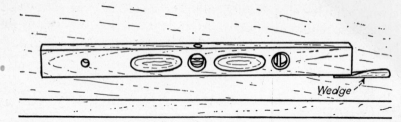

Fig. 60.—Testing a level for accuracy. Wedge up one end until the bubble is in the center. Then turn the level end for end and see if the bubble still is in the center.

To test a level for accuracy, place it on a bench or some other convenient surface, and wedge up under one end until the bubble comes to the middle of the tube (see Fig. 60). Then turn the level end for end. The bubble should return to the middle. If it does

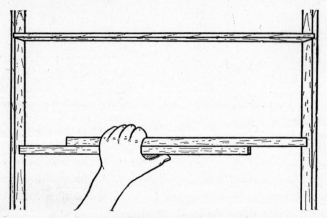

Fig. 61.—Measuring the width of an opening.

not, the level is not in adjustment. Most good levels can be adjusted by turning a screw in one end of the bubble-tube mounting.

Measuring the Width of Openings. A convenient way to measure a door opening, or similar openings, is to extend two yardsticks or pieces of scrap lumber until they just span the opening, as shown in Fig. 61. The measurement may then be read directly from the yardsticks or transferred to a single board or piece and measured accurately with a rule, square, or other suitable measuring tool

3. Sawing Wood

The proper method of using the saw is not difficult to learn, and everyone studying shopwork should early master the art of sawing.

The first requirement for satisfactory work with the saw is that it be in good condition. Creditable work cannot be done with a saw

Fig. 62.—The saw, arm, elbow, shoulder, and right eye should all be in the same vertical plane.

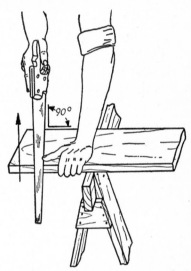

Fig. 63.—Start the saw cut with two or three backstrokes. Guide the saw with the thumb and hold the blade square with the board.

that is dull or poorly filed or set (see pages 177 to 185 on saw sharpening). If the workman cannot or does not wish to sharpen his own saw, he should have it sharpened by a competent mechanic.

Holding the Saw; Sawing Position. Grasp the handle of the saw firmly, yet not tightly. Let the forefinger extend along the side of the handle, and not through the handle with the other fingers. This enables one to better guide the saw.

Stand back from the work a little and in a position so that a line across the chest and shoulders is about 45 to 60 deg. with the line of sawing. Place the saw, the arm, elbow, shoulder, and right eye (for a right-handed workman) all in the same vertical plane (see

Fig. 62). In this position, the saw can be more easily controlled and made to follow a straight line and cut perpendicular to the surface of the board.

Starting the Hand Crosscut Saw. To saw off a board, clamp it in a vise, or hold it firmly on a box or sawhorse with the left knee (in

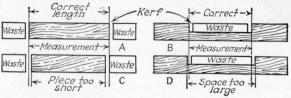

Fig. 64.—Be sure to saw carefully on the waste side of the line, as at **A** and **B**. Sawing on the line or on the wrong side of the line makes the stock too short, as at **C**, or the opening too large, as at **D**.

the case of a right-handed workman). Grasp the far edge of the board with the left hand, using the thumb to guide the saw while starting the cut (see Fig. 63). Make two or three backstrokes, lifting the saw on the forward strokes. Draw the saw back slowly and carefully just where the cut is to be made.

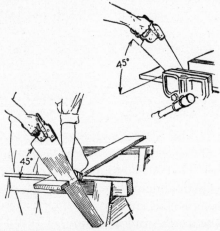

Fig. 65.—About 45 deg. is the correct angle between the saw and the work for crosscut sawing. (Courtesy of Stanley Tools.)

Do not start the saw cut or kerf on the line, *but beside the line in the waste material* (see Fig. 64), leaving the line itself. In case the piece is to be finished with a plane, make allowance for this and saw a *little* farther from the line. Removing excess waste with a plane, however, is tedious work and should be kept to a minimum.

Sawing Off a Board. After the saw is started, push it forward and pull it back, using *long, easy* strokes and *light pressure*. Do not work too fast. Short, fast, choppy strokes are signs of an amateur or careless workman.

Hold the saw at an angle of about 45 deg. with the board (see Fig. 65). If the saw tends to go to one side of the line, twist the handle slightly and gently to make it come back to the line gradually as the sawing proceeds (see Fig. 66). If it cannot be made to follow a straight line, the set may not be enough, or it may be uneven, the teeth on one side being bent out more than those on the other.

Fig. 66.—If the saw leaves the line, twist the handle slightly and gradually draw it back to the line.

Keep the blade square with the surface of the board. Testing occasionally with the square may be advisable (see Fig. 67). If the saw is getting off square, bend it a little (do not twist) to straighten it gradually as the sawing proceeds (see Fig. 67A).

If heavy pressure is used, or if short, quick strokes are made, there is danger of catching the saw and bending or kinking the blade. It will also be much more difficult to saw a straight line. If heavy pressure is required, the saw needs sharpening.

Finishing the Cut. In order to prevent splintering just as the saw is about to finish the cut, hold up the outer end of the board (see Fig. 68) and use short, easy strokes.

Using the Ripsaw. The ripsaw is used for cutting lengthwise of the grain. It is used in practically the same manner as a hand crosscut saw, except that the cutting edge should make a steeper angle with the surface of the board—about 60 deg. instead of 45 (see Fig. 69).

A good method of holding a board for ripping is to place it lengthwise on a sawhorse, or between two sawhorses, and hold it in place with the knee. Use a small wedge in the saw kerf if the blade binds.

If a ripsaw is not available, a crosscut saw may be used for an occasional job of ripping, but it will be slower and require more work than a ripsaw.

Sawing Curves. The *compass saw* is useful in sawing curves, especially inside curves where the cut must be started in a hole bored with an auger bit (see Fig. 70). In sawing sharp curves, use short strokes and do most of the sawing near the end of the blade where it is narrow. Be careful not to catch the blade and bend it. Saw with the cutting edge perpendicular to the surface of the board and not at an angle as with other saws. For accurate work or where a smooth surface must be left, do not saw too close to the line, but leave about $\frac{1}{16}$ in. or somewhat less to be removed with other tools, such as the

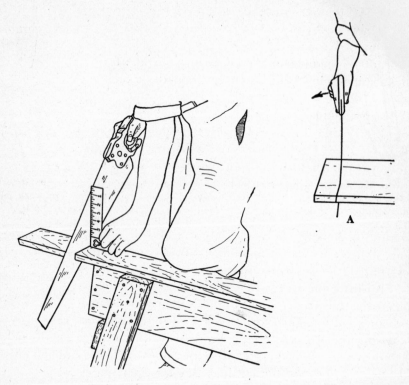

Fig. 67.—It is best for the beginner to check his work occasionally to see that he is making a square cut. If he is not, he should bend the saw a little, as shown at A, to straighten it as the sawing proceeds.

spokeshave or wood file. It is difficult to saw exactly to the line and leave a smooth cut.

The *coping saw* has a light, thin, short blade held in a frame and is used for sawing curves in thin material. The blade may be inserted to cut either on the pull stroke or on the push stroke. Cutting on the pull stroke is less apt to kink or break the blade. Long, steady, moderately slow strokes should be used. Short, fast strokes are apt to overheat the blade.

Fig. 68.—Support the outer end of the board as the saw finishes the cut. Splintering is thus avoided.

Fig. 69.—About 60 deg. is the correct angle between the saw and the work for ripsawing. (Courtesy of Stanley Tools.)

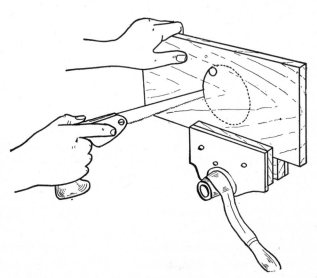

Fig. 70.—The compass saw is used for sawing curves.

Sawing with a coping saw can usually best be done by holding the work level, allowing it to project over the bench top or supporting it in a "saddle" or V-shaped bracket held in a vise as shown in Fig. 71. The sawing is then done with the handle below the work and the blade inserted to cut on the pull or downstroke.

When sawing has progressed as far as the frame of the saw will permit, it is usually possible to turn the blade a quarter turn in the frame and saw farther.

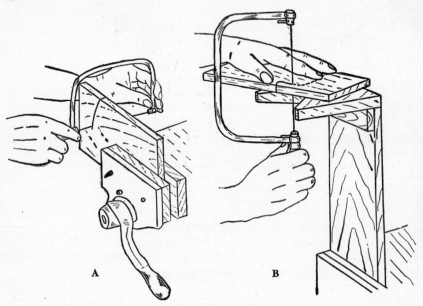

Fig. 71.—The coping saw is useful for sawing curves in light work. Although the work may be held in a vise, as at A, it is usually better to use a bracket or saddle, as at B, with the blade inserted in the saw frame to cut on the downstroke.

4. Planing and Smoothing Wood

In order to do good work with a plane, it must be sharp and properly assembled and adjusted. (See pages 155 to 161 for information on sharpening.)

Assembling the Standard Plane. Fasten the plane-iron cap to the flat side of the plane iron, allowing the cutting edge to project about $\frac{1}{16}$ in. beyond the plane-iron cap (see Fig. 73). Tighten the screw tightly to hold these two pieces firmly together and thus prevent shavings from wedging between them and possibly causing the plane to choke.

Place the assembled plane iron and plane-iron cap into the throat of the plane, with the plane iron down (see Fig. 74). Put the lever

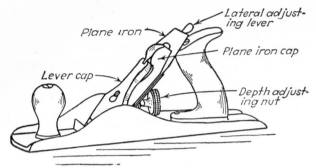

Fig. 72.—Parts of the standard plane.

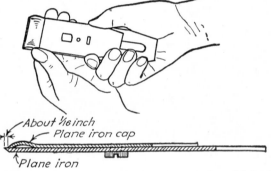

Fig. 73.—For average work, set the plane-iron cap back about $\frac{1}{16}$ in. from the cutting edge of the plane iron.

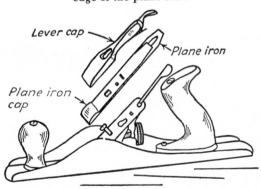

Fig. 74.—Assemble the plane with the beveled edge of the plane iron down.

cap in place on top of the plane-iron cap, and clamp it down. Be sure the lever cap fits down securely. If it does not, tighten the lever cap screw a little.

Adjusting the Plane. There are two main adjustments on the standard plane. The knurled nut just in front of the handle is to regulate the depth of cut; and the lateral adjusting lever just under the back end of the blade is to straighten the blade in the plane to make it cut the same depth on both sides.

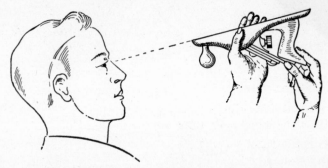

Fig. 75.—To make a trial adjustment, turn the plane upside down and sight along the bottom. The blade should project through evenly and just about the thickness of a sheet of paper.

To make a preliminary or trial adjustment, turn the plane upside down, holding the front end toward you, and sight along the bottom (see Fig. 75). Turn the depth-adjusting nut until the blade projects through about the thickness of a sheet of writing paper, and move the adjusting lever until the blade projects through the throat evenly on

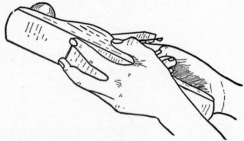

Fig. 76.—A good way to check the setting of a plane is to feel it with the fingers.

both sides. It is well also to check the adjustment by feeling the corners of the bit with the first two fingers of one hand (see Fig. 76). If one corner projects through the throat farther than the other, it can be easily detected by this method.

Using the Plane. Grasp the handle of the plane with the right hand and the knob with the left hand, palm on top (see Fig. 77).

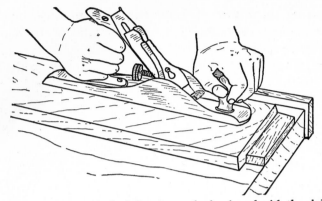

Fig. 77.—Hold the plane with the left palm on the knob and with the right forearm pushing straight in line behind the plane.

Stand with the right side to the bench, feet apart, and with the left foot slightly ahead (see Fig. 78). As the plane is pushed forward, gradually shift weight to the left foot. Keep the forearm straight in line behind the plane. In this manner, a workman can best control the plane and work with least fatigue.

Plane with the Grain. Before starting to plane, always examine the board to see which way the grain runs, and then plane with the grain. If there is doubt as to which way the grain runs, a stroke with the plane will quickly indicate the direction. An attempt to plane against the grain will result in rough work and possibly in choking the plane. Sometimes, because of irregular grain, it may be necessary to plane part of the board in one direction and the remainder in another.

Hold the Board Properly. Hold the board being planed by clamping it securely in a vise if possible, or by placing one end against a stop or a block on the top of the bench. A thin strip of wood may be nailed to the bench top to serve as a stop when planing

Fig. 78.—In planing, stand with the right side to the bench, feet apart and with the left foot slightly ahead of the right. As the plane is pushed forward gradually shift more weight to the left foot.

a wide board (see Fig. 77).   A V block, like that shown in Fig. 83, is convenient for holding a board for edge planing when a vise is not available.

Use the Planing Stroke.   To start the plane at the end of a board, press down firmly on the knob of the plane and push forward on the

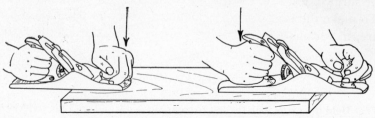

Fig. 79.—Press hard on the knob at the beginning of the stroke and hard on the back of the plane at the end of the stroke.   This makes the plane cut straight all the way across.

handle.   As the plane goes over the other end of the work, finishing the stroke, gradually release pressure on the knob and be sure to hold the back of the plane down firmly (see Fig. 79).   Thus the board can be planed straight all the way across.

Do Not Plane Too Deep.   Keep the plane set to cut a thin shaving, except in smoothing rough lumber or in removing considerable

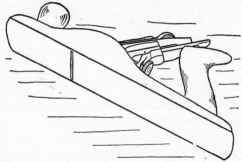

Fi   80.—To prevent the dulling of the cutting edge, lay the plane on its side when not in use.

waste.   Even in such cases, it is best to set the plane shallow for the finishing cuts.   A common mistake among beginners is to set the plane too deep, which results in gouging and rough, uneven work.

Lay the plane on its side when not in use (see Fig. 80).   This prevents the cutting edge from being dulled by contact with a gritty bench top.   When putting the plane away, place a thin strip of wood under the front end to keep the cutting edge off the tool chest or case; or else turn the depth-adjusting screw to draw the plane iron well up into the throat of the plane and thus protect it.

Planing a Surface.   Begin at one edge of the board and plane with full-length strokes, working to the other edge.   When the plane takes a thin shaving all over the board, and has touched all points of the surface, test it to see if it is true.

To make the test, place a straightedge, such as the edge of a steel square, in various positions on the surface, sighting under it to locate

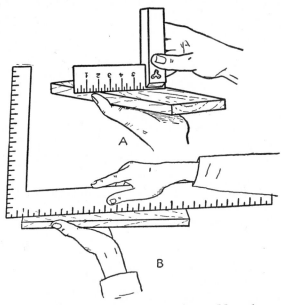

Fig. 81.—Testing to see if the surface is a true plane.   Move the straightedge from end to end, as at A, and from edge to edge, as at B, then place it on one diagonal and then on the other.   About the same amount of light should be seen under the straightedge in all positions.

the high and low places.   First place the straightedge crosswise on the board and move it slowly from one end to the other while sighting (see Fig. 81A).   Then place the straightedge lengthwise and move it slowly from one edge to the other (see Fig. 81B).   Finally place the straightedge on one diagonal and then on the other.

When the amount of light that can be seen under the straightedge in all positions is about the same, the surface may be considered true.

Remove the Planer or Mill Marks.   When planing a piece that is to be varnished or finished by staining and waxing, be sure to remove all traces of the planer marks left by the planing mill. Any such marks can be detected by holding the board up to the light and moving it about slowly. These marks will appear as a series of small hollows and ridges and, if not removed by planing, will be magnified by varnishing or polishing. Such marks left on a finished piece suggest careless workmanship.

Planing an Edge Straight and Square with an Adjoining Surface. It is frequently necessary in woodworking to plane the edge of a board to make it (1) straight and (2) square with an adjoining

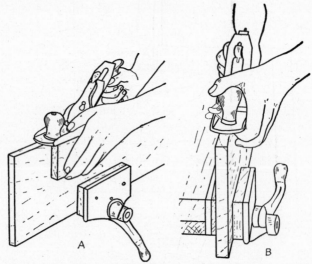

Fig. 82.—Two good methods of holding a plane while planing an edge square with a surface.

surface. Beginners should therefore early master this simple but important operation.

Before starting to plane an edge to straighten it, sight along it to note the location of any high spots. Plane down such high spots first, and then take long strokes extending the full length of the board if possible. Be sure to keep the front end of the plane down firmly at the beginning of the stroke and the back end down firmly at the finish (see Fig. 79).

In planing an edge, be sure to *keep the bottom, or sole, of the plane square with the side of the board.* A simple way of doing this is to hold a small square-edged block under the front of the plane and against the side of the board (see Fig. 82A). Another good way, commonly used by experienced workmen, is to allow the fingers of the left hand to project down under the plane and rub along the board (see Fig. 82B). This helps to steady the plane and keep it square with the side of the board.

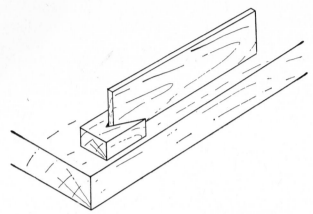

Fig. 83.—A good way to hold a board for edge planing.

A block with a V-shaped notch in the end when nailed to the bench top (see Fig. 83) is excellent for holding boards for edge planing.

Fig. 84.—Planing end grain. The pushing of a plane at an angle gives an oblique cutting action and makes it cut better. Set the plane very shallow and keep it sharp.

As the planing proceeds, check the edge frequently for straightness by sighting or using a straightedge, and for squareness by using a try square or steel square.

Planing End Grain. To plane end grain, be sure the plane is very sharp and set extremely shallow. If it is dull, or if it is set too deep, it will gouge and jump, causing rough, uneven work.

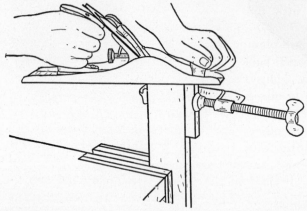

Fig. 85.—Clamping a block of scrap material on the far edge prevents splintering when planing end grain. Another method is to plane part from one edge and part from the other.

Hold the plane at an angle of about 45 deg. to the board (see Fig. 84), and push it along sidewise and parallel to the end—not in the direction the plane points. This gives an oblique or draw-cutting action and better enables the workman to control the plane.

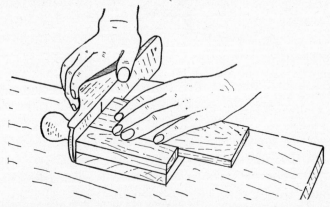

Fig. 86.—The ends of small pieces may be planed with the aid of a bench hook.

To avoid splintering the edge of a board, clamp a small block of scrap material on the edge as shown in Fig. 85. If such a method cannot be used, plane the end of the board partly from one edge and the remainder from the other edge.

Check the work frequently as the planing proceeds, sighting or using a straightedge in checking for straightness and using a square in checking for squareness. It is the mark of a careful workman to remove as little material as possible in straightening and squaring his work.

For square planing the ends of small pieces, the homemade miter box, or a bench hook (see Fig. 86), may be used as an aid. Hold the piece firmly against the backstop, allowing the end to project a little—almost not at all—beyond the edge of the bench hook. Then push the plane entirely across the end.

The block plane is excellent for planing end grain (see page 72).

Smoothing End Grain with a File. The use of the file for smoothing and squaring the ends of boards should usually be discouraged. If the end has been carefully marked and sawed, very little smoothing and squaring will be required; and such as is needed can usually be much better done with a plane. When a file is used, however, it should be used properly. Proper use ensures not only better work, but also faster and easier work.

In using a file, *use long, steady strokes*—not short, quick, jerky ones. Also, lift the file slightly, or release the pressure, on the backstroke. When filing narrow surfaces, put more pressure on the front end of the file as it starts the stroke, and gradually shift the pressure until more is on the handle at the end of the stroke. In this manner it is much easier to control the file and work the end down straight and square.

Where considerable waste is to be removed, use the file at an angle across the board as shown in Fig. 87A. For light finishing cuts, push the file in line with the edge as shown in Fig. 87B.

Squaring Up a Board. By squaring up a board is meant making all sides, ends, and edges smooth, true planes and at right angles to adjoining surfaces. (A true plane is one that has all points in the same plane. A surface may be smooth, yet not true. See page 63 for methods of testing a surface for trueness.)

For most woodwork jobs on the farm, mill-planed lumber will be near enough true and square and will be smooth enough without planing. Some jobs, however, require greater accuracy and smoother work than can be done with the lumber at hand, and the pieces will need to be partly if not completely squared up.

The method of squaring up a board is as follows:

1. *Plane one broad surface smooth and true.* This surface is then known as the *working surface.* Mark it with one short line somewhere

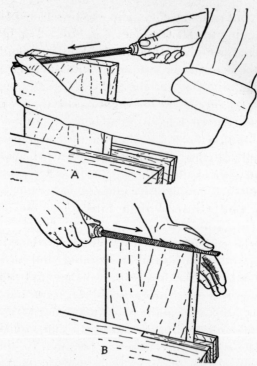

Fig. 87.—Smoothing and squaring of ends may be done by filing if only a little waste is to be removed. Planing is usually better, however. A, making a fast roughing cut; B, making a light finishing cut.

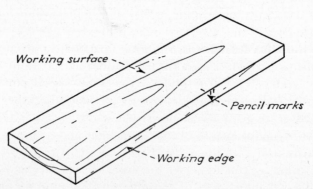

Fig. 88.—A board properly marked to designate (1) the working surface, and (2) the working edge. With the marking done as indicated, one may tell which is the working surface and which is the working edge by a glance at either.

on the surface, but near the edge that is to be selected for the working edge and extending to this edge (see Fig. 88). The first step should not be considered complete until the marking is done. If inspection or test shows the board to be true and smooth enough for its purpose without planing, then the working surface is simply marked.

2. *Select the best edge for the working edge, and plane it* (a) *straight and* (b) *square with the working surface.* Test for straightness with a straight-edge or by sighting, and test for squareness with a square. This edge is called the *working edge*. Mark it with *two* short lines extending to the working surface. If the edge is already straight and square with the working surface, simply mark it. (With the marking done as indicated, one may tell which is the working surface and which the working edge by a glance at *either*.)

3. *Make the second edge parallel to the working edge.* It will then be (a) straight and (b) square with the working surface. Probably the easiest way to perform this third step is to gage (or otherwise mark) for the desired width, marking on both the working surface and the opposite surface, and then plane to the gage lines or marks.

4. *Mark and cut one end* (a) *straight,* (b) *square with the working surface, and* (c) *square with the working edge.* Always hold the handle of the try square firmly against either the working edge or the working surface in marking around a board. Saw very carefully and very close to the lines.

5. *Mark the piece for the desired length, and cut the second end like the first one,* making it (a) straight, (b) square with the working surface, and (c) square with the working edge.

6. *Gage for thickness and plane to the gage lines,* making the second surface parallel to the working surface. This step is usually omitted except when working with very rough lumber.

Many workmen prefer to perform the operations of squaring up a board in the order given above. After the working surface and the working edge are established, however, the remaining steps may be performed in any order.

Marking and Planing a Chamfer. A chamfer is a straight flat surface formed by cutting away the arris, or sharp edge, formed by the meeting of two surfaces. A chamfer is used to improve the appearance of a piece or to lessen the danger of splintering when in use, or both. To make a chamfer, first mark it out by gaging lines back from the arris a uniform distance, usually about $\frac{3}{16}$ to $\frac{1}{4}$ in. (see Fig. 89). The marking can be done with a pencil or with a pencil and rule (see

Figs. 51 and 52). The marking gage with the regular steel point or spur is not suitable for marking out chamfers, because it leaves a scratch or mark that is difficult to remove. A marking gage may be

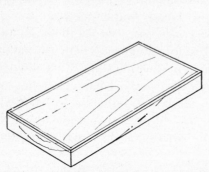

Fig. 89.—A piece marked preparatory to cutting a chamfer.

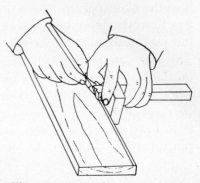

Fig. 90.—A good way to mark a chamfer. Hold the pencil point in a small notch filed in the end of the beam of a marking gage.

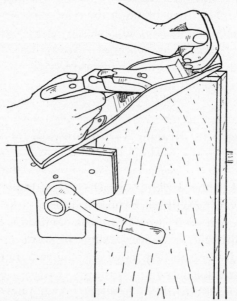

Fig. 91.—In planing chamfers around a board, plane the edges first and the ends last. In planing end chamfers, hold the plane at an angle and push it parallel to the end. This gives an oblique cutting action.

used, however, by filing a small notch in the end of the beam and holding the point of a pencil in the notch as it is moved along (see Fig. 90).

Plane those chamfers which are on the sides of a piece first, and plane the chamfers across the ends last. This avoids splintering when planing across the end grain. In planing the side chamfers, hold the plane parallel to the side or edge. In planing the end chamfers, however, hold the plane at about 45 deg. to the end (see Fig. 91), but push it parallel to the end—not in the direction the plane points. This gives an oblique cutting action and makes the plane cut better on the end grain. (Be sure to keep the plane sharp and set shallow when planing end grain.)

As a chamfer nears completion, work carefully, and try to reach both lines on the last cut. A chamfer should be straight and true, not rounded. It may be tested by sighting and careful observation, or with a straightedge as in testing a true surface (see page 63).

Fig. 92.—The block plane is a very good tool for planing end grain.

Using Different Types of Planes. The *jack plane*, which is about 14 in. long, is a general-purpose plane. There are other kinds of planes especially adapted to certain kinds of work.

The *smooth plane* is about 6 to 10 in. long and, as its name implies, is used for smoothing boards. Being short, it can follow into slight depressions in a board better than the longer planes. The smooth plane is normally used after the main straightening of the surface has been done with a jack plane. The smooth plane is sometimes selected for the farm shop when only one plane can be bought, although the jack plane is usually preferred as a general-purpose tool.

The *jointer plane* is 22 to 24 in. long and is used primarily for straightening the edges of long pieces.

The *block plane* is a small plane about 6 in. long. It is used mostly for planing across end grain and for planing small pieces where it is not convenient to hold them in a vise. The plane, being small, can be used with one hand while the other hand holds the work. The plane bit is mounted in the body of the plane at a much lower angle than in the standard or jack plane. This makes it better for cutting across end grain (see Fig. 92).

Scraping Wood Surfaces. The wood scraper is a thin, flat piece of steel with fine scraping burrs turned on its edges. The scraper is used after planing and before sandpapering. When properly sharpened it will take off a very fine, silky shaving, leaving a much smoother surface than would be possible with a plane. It is also valuable in smoothing wood that would be difficult to plane on

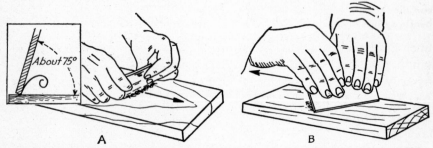

Fig. 93.—Spring the wood scraper to a slight curve and push it, as at A, or pull it, as at B, with one end slightly ahead of the other. Keep it sharp. Dust instead of shavings indicates a dull scraper.

account of irregular or gnarly grain. Scraping with a dull scraper is exceedingly slow, tedious, and discouraging work. Only an inexperienced or poor workman would try to use a dull scraper. (See pages 161 to 164 for instructions on sharpening scrapers.)

To use a scraper, hold it at an angle of about 75 deg. to the surface of the wood and push or pull it along with one end slightly ahead of the other (see Fig. 93). Hold the scraper firmly and keep it sprung to a slight curve. Stop and reburnish the edge (see page 162) as often as required to keep it sharp.

Sandpapering Unfinished Wood Surfaces. Do not use sandpaper until all work with cutting tools and scrapers is finished. The beginner usually wants to use sandpaper before he should. There is generally no advantage in using sandpaper on wood that has not previously been planed or scraped. It is practically impossible to remove the planer marks (small hollows and ridges) left by the mill planer by hand sandpapering. In fact, sandpapering such surfaces generally magnifies the mill marks and actually detracts from the appearance, rather than improving it.

Select a grade of sandpaper suitable to the kind of work to be done, using coarser grades for rougher surfaces or the first sanding and

Fig. 94.—For economy, tear sheets of sandpaper into four quarters by creasing and then tearing along the sharp edge of a rule, as at A, or over the edge of the bench, as at B.

finer grades for the final sanding. The commonly used grades of sandpaper range from No. 00 (fine) to No. 2 (coarse). Usually No. ½ or No. 1 is satisfactory for the first or coarse sanding on wood, and No. 0 for the final or finish sanding.

For ordinary use, tear a sheet of sandpaper into four quarters by creasing it firmly and then tearing over the sharp edge of a rule or the

Fig. 95.—To sandpaper flat pieces, wrap the sandpaper part way around a flat block. Always sand back and forth with the grain—never across it.

edge of the bench (see Fig. 94). Wrap one of the small pieces part way around a flat block. For economy, the block should be of such a size that the paper will come only part way up on each edge and not around on top.

Always sandpaper back and forth *with the grain* (see Fig. 95), and never with a circular motion or across the grain, as this would roughen and scratch the work instead of smoothing it. Keep the block flat against the surface, particularly on narrow surfaces, and be careful not to round the edges. If desired, to prevent splintering, a sharp edge, or arris, may be removed by running a plane over it lightly before sanding, or by running the sandpaper block over it carefully once or twice after the other sandpapering is finished.

Use only moderate pressure on the sandpaper block. Too much pressure may cause the paper to wrinkle or tear. Keep the sandpaper free of dust by knocking and shaking it out frequently.

Sandpapering Round and Irregular Surfaces. For sandpapering inside or concave curved surfaces, the sandpaper may be wrapped around a round rod or stick. For very irregular work, use smaller pieces—about half of a quarter sheet—with the fingers or hand only and without a block.

5. Cutting with Wood Chisels

Choosing Wood Chisels. Wood chisels are made in various widths of blade, ranging from ⅛ to 2 in. They are also made in different lengths of blade, the longer ones being known as *firmer chisels*, the medium length ones as *pocket chisels*, and the shorter ones as *butt chisels*. Wood chisels may also be classified as socket type or

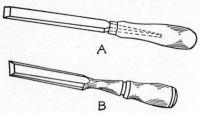

tang type, according to the method of attaching the handle. The socket type has a socket on the driving end into which the handle fits. The tang type has a steel tang, much like the tang on the end of a file, which fits into the wooden handle (see Fig. 96). Tang chisels are preferred by some workmen for paring, because of their light weight

Fig. 96.—Types of wood chisels: A, tang type; B, socket type. The socket type is usually preferred for the farm shop.

and better balance, but they are not adapted to heavy chiseling with a mallet. For general farm shopwork, medium-weight socket chisels are preferred. Such chisels can be handled easily for paring, and they can also be used for the heavier work where a mallet is needed.

Keeping the Chisel Sharp. The first requirement for good work with a chisel is to keep it very sharp. A dull chisel not only requires extra effort to force it through the wood, but, what is more serious, it cannot be easily guided and controlled. Consequently, rough, inaccurate work is almost certain to result with a dull chisel.

The chisel is very easily sharpened (see page 155). Whenever it becomes dull, stop and sharpen it. The time required will soon be gained back in faster and better work with the sharpened tool.

Right   Wrong

Fig. 97.—When laying a chisel on the bench, always place the cutting edge up (bevel side down). This prevents dulling the edge.

In order to prevent dulling the chisel, do not allow the cutting edge to touch other tools or pieces of metal or even the bench top. *Always lay the chisel on the bench with the bevel side down*—not up (see Fig. 97).

Chiseling with the Grain. In chiseling with the grain, as on the surface or edge of a board, observe the following points:

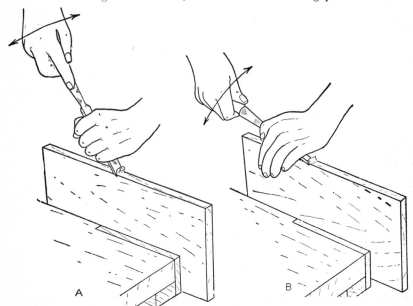

A   B

Fig. 98.—Chiseling with the grain. Use the chisel with the bevel down, as at A, for deep roughing cuts, and the bevel up, as at B, for light finishing cuts. Move the handle from side to side slowly as the chisel is pushed forward in order to give an oblique cutting action.

1. Always cut with the grain, as in planing, to avoid splitting or splintering.
2. Fasten the work in a vise whenever possible, so as to leave both hands free to use the chisel.

3. Always push the chisel from you, *keeping both hands behind the cutting edge.*
4. Use the left hand to guide the chisel and the right hand to push the handle forward.
5. Use the chisel with the bevel down for roughing cuts and with the bevel up for fine paring or finishing cuts (see Fig. 98).
6. Hold the handle slightly to one side, or move it back and forth slightly, as the chisel is pushed forward. This gives a sliding or oblique cutting action, which makes the chisel cut better and easier.

Chiseling across a Board. This kind of work is done mostly in making notches, gains, or dadoes (see pages 79 to 81). In chiseling across the grain, observe the following points:

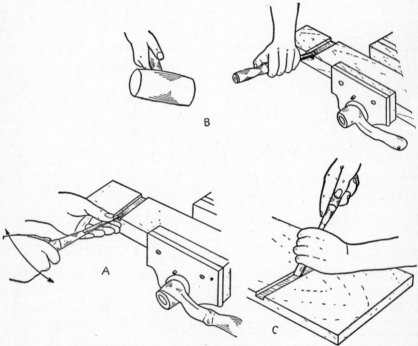

Fig. 99.—Chiseling across a board. A. Work with the bevel up (except for wide boards). Raise the handle just enough to make the chisel cut and move the handle from side to side slowly. Guide the front of the chisel with thumb and fingers of left hand. B. For heavier chiseling or roughing cuts, the mallet may be used. Do not use a hammer. C. For chiseling across wide boards where the chisel will not reach to the center, work with the bevel down.

1. Grasp the blade of the chisel between the thumb and the first two fingers of the left hand, to guide it and to act as a brake, while the pushing is done with the right hand (see Fig. 99*A*).

2. Do not cut all the way across a board from one side, but cut part way from one edge and part way from the other to avoid splintering.
3. Move the handle from side to side slightly as the chisel is pushed forward to give a sliding or oblique cutting action.
4. Cut with the bevel side up, raising the handle just enough to make the chisel cut. In chiseling across wide boards, however, where the chisel cannot reach the middle of the board, work with the bevel side down (see Fig. 99C).

Chiseling across End Grain. Chiseling across end grain is difficult work. By careful marking and sawing, however, chiseling of end grain can be kept to a minimum and sometimes eliminated altogether. In chiseling end grain, observe the following points:

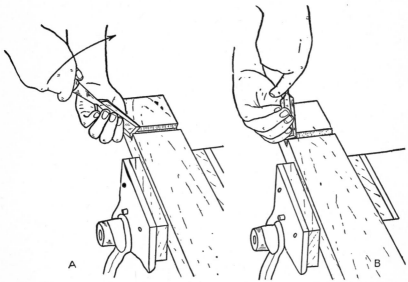

Fig. 100.—Chiseling down across end grain. Guide the chisel with the left hand pushing forward and downward at the start, and gradually raise the handle. A beginning a stroke; B, finishing a stroke.

1. If much waste is to be removed, take a roughing cut first, leaving about $\frac{1}{16}$ in. to be removed with a finishing cut.
2. Start on the near edge of the board, and push forward at an angle and downward (see Fig. 100). As the stroke proceeds, straighten the handle up until it is about vertical at the end of the stroke.
3. Guide the chisel with the left hand, and apply force with the right.
4. Use about half the width of the chisel for cutting on each new stroke. Keep the back half of the blade flat against the surface left by the previous

stroke. Thus the work of cutting is made easier, and the line of cutting is more easily kept straight.

5. If the chisel is to cut entirely through or across a piece, place the work on a cutting board or piece of scrap lumber to keep the chisel from cutting into the bench, thus marring its surface and possibly dulling the chisel.

Cutting Curves with the Chisel. Outside, or convex, curves can be cut easily by first sawing two or three straight cuts tangent to the curve, and then working with the chisel as shown in Fig. 101. Use the chisel with the bevel side up, the left hand holding it down and guid-

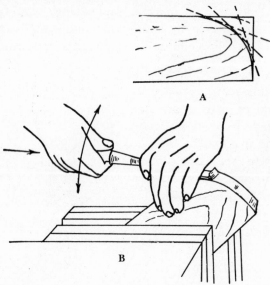

Fig. 101.—An easy method of cutting outside, or convex curves. A. First saw two or three lines tangent to the curve. B. Then finish with the chisel, moving the handle from side to side while pushing it forward.

ing it while the right hand pushes forward and moves the handle back and forth sidewise slightly at the same time.

The chisel is also a good tool for finishing inside, or concave, curves (see Fig. 102). For such work, use the chisel with the bevel side down. Guide the chisel with the left hand, while the right hand pushes down and pulls backward at the same time.

Using the Mallet. Use a mallet to drive the chisel where considerable force is required as in making deep rough cuts. Never use a steel hammer, because this would soon ruin the chisel handle. A series of light taps with a mallet is better than heavy blows, because the chisel can thus be better controlled.

Paring Chamfers. The chisel may be used satisfactorily in paring chamfers, either with the grain or across end grain. In paring chamfers, keep the bevel side of the chisel up and the flat side down. Hold the handle slightly to one side, or move it from side to side, as it is pushed forward in order to give a sliding or oblique cutting action (see Fig. 103). To prevent splintering when cutting end chamfers, work part way from one edge of the board and part way from the other.

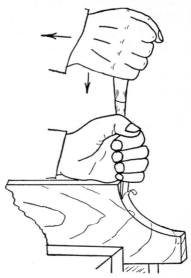

Fig. 102.—A good method of finishing inside curves. Use the chisel bevel side down. With the right hand, push down and pull back at the same time.

Making a Dado. A dado is a groove that runs across a board to receive the end or edge of another board. Dadoes are commonly made in shelving and in cabinetwork.

The first step in making a dado is to mark it out accurately for depth and for width—the same as the thickness of the piece that is to fit into the dado. The piece itself may be used to mark the width of

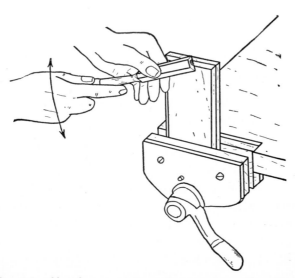

Fig. 103.—Chamfers are easily made by paring with a chisel.

the dado by superposition (see Fig. 104). Use a square to ensure marking the dado square with the edges of a board. A knife is best

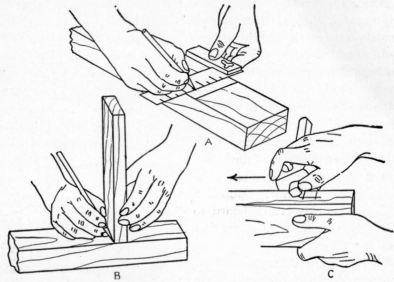

Fig. 104.—The first step in making a dado is to mark it out accurately. A, marking the width with a try square; B, marking the width by superposition; C, marking the depth with a marking gage.

Fig. 105.—A good way to saw the sides of a dado accurately. Clamp a straight-edged block in place to guide the saw. Thus little or no chiseling of end grain will be required. An extra saw cut or two between the sides of a dado will facilitate chiseling out the waste.

ιor marking, although a sharp pencil can be used. The depth of the dado is easily marked on the edges of the board with a marking gage (see Fig. 104C).

After the dado is accurately marked out, saw just inside the lines in the waste material. Be careful not to saw too deep. To do a good job, a straight square-edged block may be clamped in place temporarily to guide the saw (see Fig. 105). In case of a wide dado, saw an extra kerf or two in the waste to facilitate its removal with the chisel. Then chisel out the waste, working carefully to depth and observing the suggestions for chiseling across a board as listed on page 76.

If the sawing has been done carefully and accurately, no paring of the sides of the dado will be required. If the sawing has not been done accurately to the lines, however, the sides of the dado may be finished to width by vertical paring with a chisel (see Fig. 100), or by filing or sandpapering.

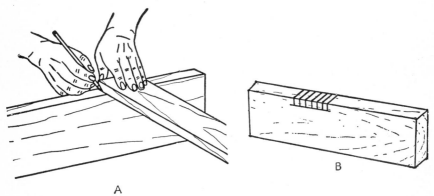

Fig. 106.—Gaining in. A, marking for a gain by superposition; B, making several saw cuts lessens the work of chiseling out waste.

Gaining. It is frequently desirable to gain into, or notch into, the edge of a piece to fasten a second piece securely. Typical examples are fastening the lower crosspieces to the legs of a workbench or table and fastening the steps to the side rails of a ladder.

To make a gain, first mark it out accurately to the exact width and depth, using a knife or sharp pencil and square and possibly a marking gage. Then saw accurately and chisel out the waste in a manner similar to that described for dadoes in the preceding article. The chiseling will be easier if several saw kerfs are first made in the waste as shown in Fig. 106B.

Marking for a gain may often be simplified by using the piece that is to fit into the gain and marking by superposition (see Fig. 106A).

Attaching Butt Hinges.   To attach a butt hinge, first put it in place and mark around it carefully with a sharp pencil or knife (see Fig. 107*A*).   Then remove the hinge and gage a line on the side of the piece to indicate the depth the hinge is to be set or gained in.   Then carefully cut out the waste with a chisel (see Fig. 107*B*), trying the hinge in place for fit as the work nears completion.

After the gain is finished, fasten the hinge in place with screws, first making holes for the screws with an awl or a drill.

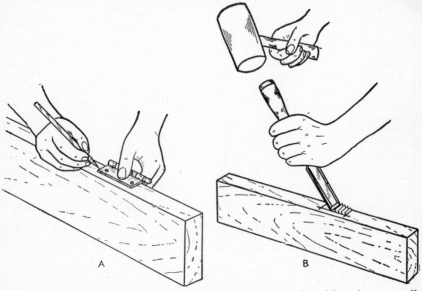

Fig. 107.—Attaching a butt hinge.   A, marking for the gain with a sharp pencil. Some prefer a sharp knife instead of a pencil.   B, chiseling out the gain.

Making Mortise-and-tenon Joints.   A mortise is a hole cut into or through one piece and into which, or through which, another piece fits.   A tenon is an end of a piece especially shaped (usually with a shoulder) to fit into a mortise.   Figure 108 illustrates several mortise-and-tenon joints as well as other joints sometimes used in fastening pieces of wood together.

To make a mortise, first mark it out accurately, using a square and sharp pencil or knife and marking gage.   Then remove the waste with a chisel of appropriate width (see Fig. 109*A*, *B*, and *C*) or by boring first with an auger bit to remove most of the waste and then finishing with a chisel (see Fig. 109*D*).   If the waste is to be removed partly by boring, mark a center line (preferably with a marking gage) to ensure that all auger holes will be exactly in line.   Use an auger bit that has the same diameter as the width of the mortise.

If a mortise is to go entirely through a piece, mark its location accurately on both sides, and cut the mortise partly through from one side and partly from the other.

A tenon, although easily made, requires careful work.   To make a tenon, first mark it out completely and accurately, and then work it to size with the saw and chisel.

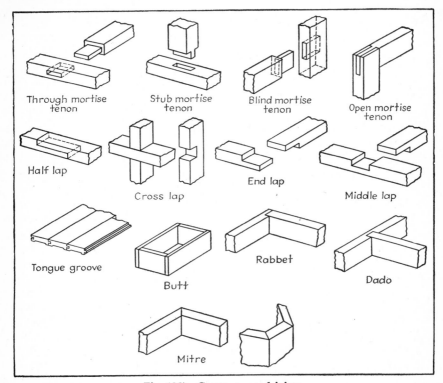

Fig. 108.—Common wood joints.

Rabbeting.   A rabbet is a groove cut in the edge or end of a piece to receive a second piece like a panel.   Rabbeting is commonly done in making frames to hold glass and frequently, also, in constructing drawers and other cabinetwork.   A rabbet may be made with a chisel if it is first accurately marked out and the workman is careful. Marking is best done with a marking gage.   It is much easier to cut a rabbet with a power saw or jointer, or with a special grooving or rabbeting plane, if such tools are available.

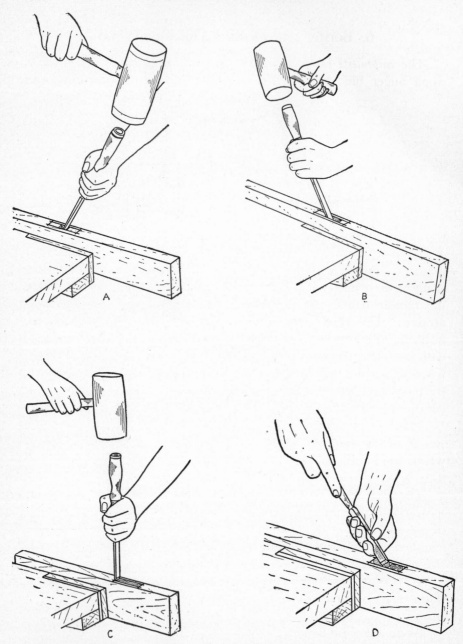

Fig. 109.—Making a mortise. The first step is to mark it out accurately. A, mortise started; B, mortise partly done; C, finishing the mortise. D. If desired, most of the waste may be removed with a wood auger and the mortise finished with a chisel.

6. Boring and Drilling Holes in Wood

The *carpenter's brace* (Fig. 110) is used for turning such tools as wood auger bits, twist drills, screw-driver bits, countersinks, and

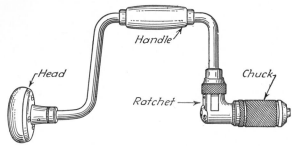

Fig. 110.—A carpenter's brace.

reamers. Braces are made either with or without the ratchet device. A *ratchet brace* makes it possible to bore holes in close quarters where the handle cannot be turned all the way around. This type is also more convenient in boring in hardwood and in driving screws with a screw-driver bit. For such work, it is frequently easier and better to turn the brace by part turns rather than by making full continuous turns.

The size of a brace is designated by its sweep, or the diameter of the circle through which the handle swings. A brace with an 8-in. sweep is suitable for average work.

The *auger bit* (Fig. 111) is the most common tool for boring holes in wood. The size of an auger bit is designated by a number stamped on the shank, the number being the size of the bit in sixteenths of an inch. Thus a bit marked 7 bores a hole $\frac{7}{16}$ in. in diameter, a bit marked 11 bores a hole $1\frac{1}{16}$ in., etc.

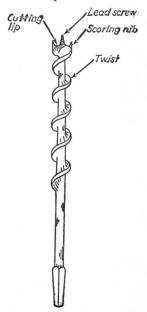

Fig. 111.—An auger bit.

As an auger bit is turned, the lead screw guides the bit and draws the cutting parts into the wood, so that only moderate pressure is required on the brace. The spurs or scoring nibs cut off the wood fibers, and the cutting lips cut out the waste inside the circle scored by the nibs. The twists on the bit carry the waste to the surface. To bore a clean, straight hole, the bit must be in good condition. See page 164 on how to sharpen auger bits.

Starting the Auger Bit. For accurate boring, first mark the location for the center of the hole, by the intersection of two cross lines, or by a small hole made with an awl or other sharp-pointed tool. Then, with one hand, guide the point of the bit carefully into place, while the other hand exerts a slight pressure on the head of the brace (see Fig. 112).

As the auger starts boring, be careful to keep it perpendicular to the surface (unless it is desired to bore the hole at an angle). To check

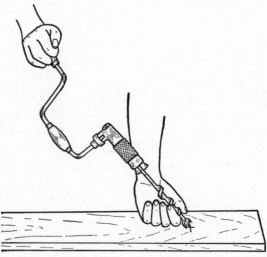

Fig. 112.—Steadying the hand, knuckles down against the board, helps to place the point of the bit accurately.

to see that the auger is boring square with the surface, step back a little, steadying the brace with one hand, and sight; then move around and sight in another direction about at right angles to the first direction of sighting (see Fig. 113*A*). The try square may also be used to see if the bit is going straight (see Fig. 113*B*). It is better for a learner not to depend too much on the square, however, but to develop his ability in sighting. Leaning the top of the brace slightly one way or another will change the direction of boring.

Boring Through. If a hole is to be bored entirely through a board, bore until the point of the bit can be felt on the back side of the board (see Fig. 114). Then turn the board over and bore from the other side. This prevents splintering around the edge of the hole.

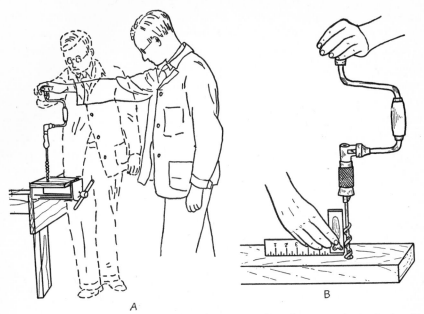

Fig. 113.—To ensure boring straight, sight from two directions, as at A; or check with a square, as at B.

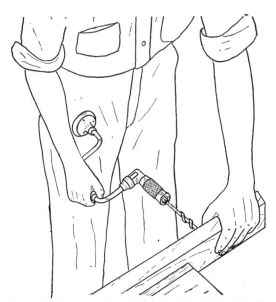

Fig. 114.—To prevent splintering when boring a hole, stop when the point of the bit can be felt and finish by boring back through from the other side.

Another method that may be used, especially on pieces that can be held in a vise, is to clamp a block of scrap wood behind the piece through which the hole is to be bored. The boring can then be done from one side without danger of splintering.

Boring to Depth. A good way to bore a hole to a definite depth is as follows: Stop turning as soon as the cutting lips touch the wood, and measure the distance from the end of the chuck to the surface of the piece being bored. Then bore until the measurement on the rule

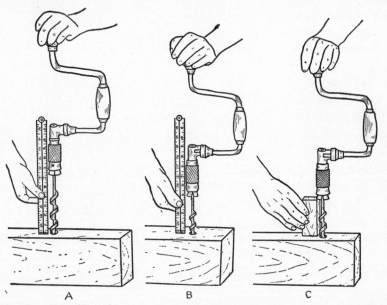

Fig. 115.—In order to bore to an exact depth, measure the distance from the surface to the chuck just as the lips start cutting, as at A; then bore until the measurement is decreased an amount equal to the desired depth, as at B. When several holes are to be bored the same depth, a block may be cut to fit under the chuck and used as a gage, as at C.

is decreased by an amount equal to the desired depth of hole (see Fig. 115*A* and *B*).

If a number of holes are to be bored to the same depth, considerable time may be saved by cutting a block to the correct length and using it as a gage as shown in Fig. 115*C*, or by boring a hole through the block and using it on the auger in a manner similar to that shown in Fig. 119.

Counterboring. Counterboring is making a hole larger at the mouth than deeper down. Where pieces of wood are to be fastened together with bolts or some kinds of screws, counterboring is frequently done to sink the boltheads or screwheads below the surface (see Fig. 137).

To make a counterbored hole, first bore with a bit of a diameter suitable for the counterbore. After the desired depth of counterbore is reached, finish with a smaller bit. Do not bore the small hole first, for then there would be no center to guide the large bit.

Preventing Splitting While Boring. Boring large holes in thin or narrow pieces sometimes causes splitting, owing to the wedging action of the lead screw on the bit. Such splitting can be avoided by drilling through first with a small twist drill, or by clamping the piece in a vise with the vise jaws against the edges—not the sides—of the piece.

Drilling with Twist Drills. These drills (Fig. 116*A* and *B*) can be used for drilling holes in either wood or metal, and their use is recommended where there is danger of striking a nail or other piece of metal that would dull an auger bit. The smaller sizes of twist drills are very good for drilling holes to receive wood screws.

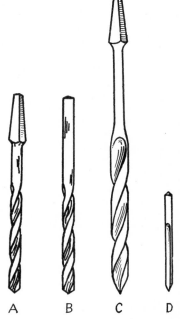

To drill a hole in wood with a twist drill or other blunt-pointed drill, start the drill in a mark or depression made with an awl or nail. Otherwise, the point may "drift" from the proper location when the drill starts turning, and the hole will not be drilled exactly where it is wanted.

Fig. 116.—Wood-drilling bits. A, bit-stock twist drill; B, straight round-shank twist drill; C, wood-boring drill; D, wood-drill point.

In drilling holes in wood with a twist drill, remove the drill frequently to clean the cuttings from the twists or flutes. This prevents overheating the drill and also speeds up the work of drilling. See pages 165 to 171 for suggestions on sharpening twist drills.

Wood-boring drills (Fig. 116*C*) are similar to twist drills, but are usually longer and have sharper points.

Wood-drill points (Fig. 116*D*) are used for drilling small holes in wood. They are very much like small twist drills except that they are made of softer steel and have straight instead of spiral grooves or flutes. They are commonly sold in sets ranging in size from $\frac{1}{16}$ to $1\frac{1}{64}$ in. They are used in hand drills or automatic push-type drills.

Using the Hand Drill. The hand drill (Fig. 117) is one of the

most useful tools for drilling small holes either in wood or in metal. It is small and light and is much faster and more convenient than the carpenter's brace. Also, there is less danger of breaking small drill bits when using them in the hand drill.

Fig. 117.—The hand drill is a very useful tool for drilling small holes in either wood or metal.

In using a hand drill, hold it straight and steady, push with an even pressure against the handle, and turn the crank with a steady,

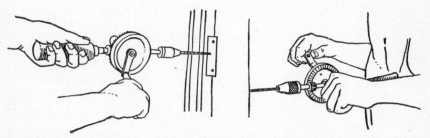

Fig. 118.—In drilling with a hand drill, hold it straight and steady and turn with a moderate, even speed. It is sometimes more convenient to hold the drill by the side handle and lean against the end handle. (Courtesy of Stanley Tools.)

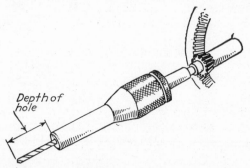

Fig. 119.—A convenient depth gage can be made by cutting a piece of wood to correct length, drilling a hole through it, and slipping it over the bit. (Courtesy of Stanley Tools.)

moderate speed (see Fig. 118). Either too slow or too high a speed or too heavy pressure is likely to bend or break small drill bits.

Using the Automatic Push Drill. The automatic push drill (Fig. 120) is sometimes used for drilling small holes in wood. By pushing the handle down and letting it come back up, a forward and backward rotary motion is imparted to the drill bit. Drilling with the push drill is a little slower than with the hand drill. The push

Fig. 120.—The automatic push drill.

drill can be operated with one hand, however, leaving the other free to hold the work.

7. Fastening Wood

Fastening with Nails

There are two general kinds of nail hammers: *bell-face* hammers and flat or *plain-face* hammers. Bell-face hammers have striking surfaces that are slightly round or convex, and by careful use nails can be driven up tight with their heads flush with the surface of a board, or even slightly below, without leaving hammer marks. Plain-face hammers, having flat faces, are a little easier to learn to use.

Hammers with straight claws, like those shown in Fig. 128, are called *ripping hammers* and are especially good for ripping off old boards.

The size of a hammer is designated by the weight of the head exclusive of the handle, the most common sizes ranging from 12 to 16 oz.

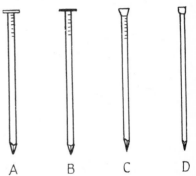

Fig. 121.—Some of the more common kinds of nails: A, common nail; B, box nail; C, casing nail; D, finishing nail.

Nails are made from steel wire by special machines that receive the wire from large rolls, cut it into the desired lengths, and form the points and heads automatically. Various kinds of nails are used for different purposes. *Common nails* (see Fig. 121) are of large diameter, have large flat heads, and are most commonly used for rough carpentry work. The larger size of common nails are called *spikes*. *Box nails* are similar to common nails, but are more slender and are used on wooden boxes and crates. Being smaller in diameter, there is less danger of splitting. They are available either plain or cement coated to give greater holding power. *Casing nails* are of the same size as box nails, but have small tapered heads instead of flat heads. They are used where large nailheads would be objectionable, as in cabinetwork and in blind nailing of tongue-and-groove flooring. *Finishing nails* are somewhat more slender than casing and box nails and have small heads. They are used mainly in inside trim or finish carpentry and in cabinetwork. The heads are driven flush, or set below the surface with a nail set. Other commonly used nails are lathing nails, shingle nails, roofing nails, fence nails, and plasterboard nails, used for the special purposes indicated by their names. Very small nails called *brads* are also available for fastening small pieces.

The size of nails is designated by the term *penny*, 2-penny nails being small, and 4-, 6-, and 8-penny nails being larger. All kinds of nails of the same penny size are of the same length. For example, 8-penny nails are all $2\frac{1}{2}$ in. long; all 6-penny nails, 2 in. long; etc. All nails of a given penny size are not the same diameter, however.

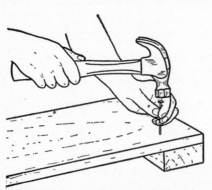

As indicated in the preceding paragraph, common nails are of larger diameter, box nails and casing nails somewhat smaller, and finishing nails still smaller in diameter.

Driving Nails. To start a nail, hold it steady with the thumb and fingers of one hand, and strike one or two light blows with the hammer (see Fig. 122). After the nail is well started, drive it up tight with firm well-directed blows. Hold the hammer handle near the end, and strike squarely on top of the nailhead. For light driving, use mostly wrist motion; for heavier hammering, use wrist motion and elbow motion; and for very heavy hammering, use shoulder action, as well as wrist and elbow motion.

Fig. 122.—Steady the nail with one hand while striking one or two light taps of the hammer with the other. Then follow with firm, well-directed blows. Grasp the hammer handle near the end.

In driving a nail flush with the surface of a board, make the final blow with care so as not to leave a hammer mark on the surface.

Keep the striking face of the hammer clean to prevent it from slipping off the nailhead.

Preventing Splitting. If there is danger of splitting a board when a nail is driven, a smaller nail should be used, as a smaller nail will have more holding power than a larger one that splits the board. Some nails have chisel-shaped points, owing to the method of manu-

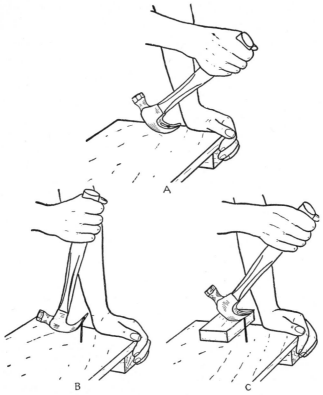

Fig. 123.—Steps in pulling a nail. A, slip the claws under the nailhead; B, pull until the hammer handle is nearly vertical; and C, put a block under the hammer head to increase the leverage and relieve the strain on the handle.

facture. Drive such nails with the long way of the point across the grain, so that they cut the fibers of the wood instead of wedging them apart and causing splitting. Splitting may be prevented in thin boards and where short nails are used by cutting the nails off square or chisel-shaped with nippers or pliers.

Pulling Nails. To pull a nail, slip the claws of the hammer under the nailhead, and pull up and back on the hammer handle. If the nail is not pulled out by the time the handle is about straight up, stop and place a block under the head of the hammer, and then proceed (see Fig. 123). Pulling the hammer handle too far back may overstrain it, or possibly break it. The block increases the leverage and reduces the strain on the hammer handle.

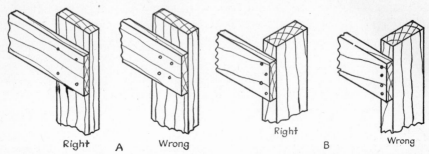

Fig. 124.—Right and wrong methods of nailing.

If the nailhead is down in the wood so far that the claws cannot be slipped under it, a pair of pincers may be used to start pulling the nail, or a small cold chisel and hammer may be used. Such methods, however, are almost certain to mar the wood.

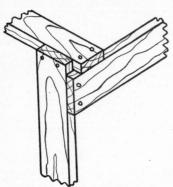

Fig. 125.—A strong nailed corner used in crating.

Locating Nails in Nailed Joints. The strength of a nailed joint depends largely upon the distribution and location of the nails. Where possible, stagger the nails and do not drive them too close together or in line with the grain. Examples of right and wrong methods are shown in Fig. 124. It is always good practice to nail through a thin piece into a thick one, and not through a thick one into a thin one. It is also good practice to drive nails across the grain of the wood rather than into end grain (into the ends of boards) wherever possible. Figure 125 shows a well-nailed crate corner with none of the nails driven into ends of boards.

Clinching Nails. Nails are clinched by bending the ends that protrude after they are driven. Clinching makes nails hold better. For greatest holding power, bend the end of the nail opposite to the direction the head would move in case the nail should draw under load. For example, if strain on the nailed parts tends to pull the head down, then bend the point of the nail up; if the strain tends to move the head to the right, bend the point to the left (see **Fig. 126**).

Toenailing. This is done to fasten a piece that butts against another (see Fig. 127). Good judgment must be used in selecting

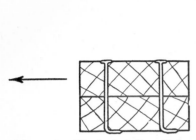

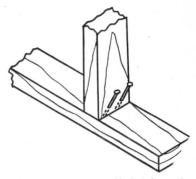

Fig. 126.—Clinched nails. If forces tend to move the top board to the left, the points of the nails should be bent over to the right.

Fig. 127.—A toenailed joint. Start the nails high enough to prevent splitting and low enough to give deep penetration into the second piece.

the point to start the nail and the angle to drive it. The nail should get a good hold in the first piece without danger of splitting, and yet go deep enough into the second piece to ensure good holding.

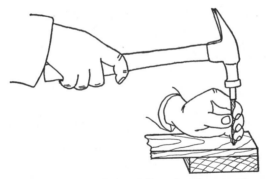

Fig. 128.—Setting a nail. To hold the nail set in place on the nail head, steady the hand on the board and hold the tip of the little finger against the set.

Setting Nails. To set the head of a finishing nail or a casing nail slightly below the surface, drive it in the usual manner until the head is almost, but not quite, flush. Then finish with a nail set. A nail set resembles a small punch and has a cup-shaped point. If possible, select a nail set with a cupped point slightly larger than the nailhead. Hold the set in the left hand, supporting the top part with the thumb and first fingers, and holding the point on the nailhead with the tip of the little finger (see Fig. 128). Set the nailhead about $\frac{1}{16}$ in. below the surface.

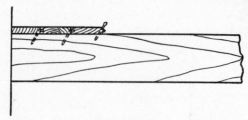

Fig. 129.—Draw nailing. Driving the nails at an angle helps to draw the boards tightly together.

Draw Nailing. Where it is desired to make a tight joint between two boards, as between pieces of tongue-and-groove flooring, drive

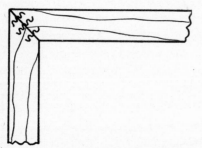

the nails at an angle with the surface, as shown in Fig. 129. The nails then have a drawing effect as they are driven up, making the joint tight.

Using Corrugated Fasteners. Corrugated fasteners can frequently be used to advantage in reinforcing joints like miter joints (see Fig. 130) and butt joints. They are especially good when used in end

Fig. 130.—Corrugated fasteners used to strengthen a miter joint.

grain. In driving them, be careful to keep them straight, and do not drive one end faster than the other.

Fastening with Screws

Pieces of wood can be fastened together more securely with screws than with nails. Yet pieces fastened with screws can be taken apart more easily and with less damage. Fastening with screws, if done in a systematic manner, can be done quickly and easily and with excellent results. On the other hand, if careless, slovenly methods are used, much time will be wasted, and the work will most likely be disappointing. The main points to observe are the following:

1. Select screws of suitable size and length.
2. Drill holes of the proper size and depth for the screws.
3. Use screw drivers that are in good condition and that fit the screws.

There are various kinds and sizes of screws. The *flathead* screw is most commonly used in woodwork, although *oval-head* and *round-head* screws are sometimes used, mainly for their ornamental effect.

The size of wood screws is designated by (1) the gage or size of wire from which they are made and (2) their length, measured as shown in Fig. 131.

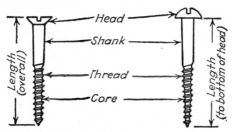

Fig. 131.—Parts of a wood screw.

Finish. Steel screws without any special finish, designated simply as *bright*, were once commonly used in woodwork. Cadmium-plated rustproof screws are much better and are now more commonly used. Other finishes are nickel plated and blued. Screws made of brass are also sometimes used. Brass screws are not so strong as steel screws, however, and more care must be used in driving them to prevent twisting them in two or marring the screwheads.

In ordering screws, always specify the size, kind, and finish, as 1½-in. No. 8 flathead cadmium-plated wood screws.

Using Lag Screws. A lag screw might be described as a square headed bolt, but pointed and with coarse threads to screw into the wood instead of fine threads to receive a nut. Lag screws are used where ordinary wood screws would not be strong enough. The size of a lag screw is de-

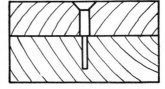

Fig. 132.—Always drill holes to receive wood screws in fastening two pieces of wood together. Make the hole in the first piece the size of the screw shank or a little larger. Make the hole in the second piece the size of the core of the screw and drill it almost as deep as the screw will go.

signated by the diameter of the unthreaded part near the head and by the length.

Determining Sizes of Holes to Drill for Wood Screws. When-ever pieces of wood are to be fastened together, always drill holes as shown in Fig. 132. Drill the hole through the first piece the size of the shank of the screw, *or a little larger;* and drill the pilot hole in the second piece the size of the core or body of the screw under the threads, *or slightly smaller*. In case of hardwood, or of large screws, or of screws made of soft metal like brass, drill the pilot holes nearly as deep as the screws will go. In case of softwood, or medium to small screws, drill the pilot holes about half as deep as the screws will go. When very short screws are used and the wood is soft, the pilot holes in the second piece may be made with an awl or nail.

Table 1. Sizes of Holes to Drill for Wood Screws

Screw size	Size of first hole (shank), 32d in.	Size second hole (thread), 32d in.
2	3	2
3	4	2
4	4	2
5	4	3
6	5	3
7	5	4
8	6	4
9	6	4
10	6	4
11	7	5
12	7	5
14	8	6
16	9	7

A table like Table 1 may be used as a guide in selecting the proper sizes of drills for the different sizes of screws. If such a table is not at hand, it is a good plan to drill holes in scrap material and try the screws in them before drilling holes in the work. The hole through the first piece should never be so small as to require a screw driver to force the screw into it.

Locating and Drilling the Holes. Mark the locations for screw holes accurately, usually first by the intersection of cross lines, and then with a deep mark or depression made with an awl. Make the depression deep enough to keep the drill from "wandering" when it starts to turn.

Drill the holes through the first piece before marking the locations for holes in the second piece. Then, using the first piece as a guide or template, mark the locations for holes in the second piece with an awl, a pencil, or a nail (see Fig. 133). Or, if the two pieces can be held in alignment in a vise or a clamp, simply drill the holes in the second piece without marking, using the holes in the first piece to guide the drill bit.

Fig. 133.—Marking locations of drill holes in the second board by using the top one as a template for guiding the awl.

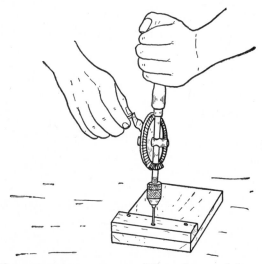

Fig. 134.—After one or two screws are driven, the remaining pilot holes in the bottom piece are easily drilled by using the holes in the top piece to guide the drill.

Sometimes when several screws are to hold two pieces together, it is possible to drill one or two pilot holes in the second piece and set the screws in these holes to hold the pieces together. The remaining

pilot holes can then be drilled easily by using the holes in the first
board as guides (see Fig. 134).

Countersinking. When flathead screws are used, always countersink the holes to allow the heads to draw down flush with the sur-

**Fig. 135.—A common type
of countersink used for countersinking flathead screws.**

face, or in some cases slightly below.
Countersink all holes to the same depth,
and be careful not to countersink them
too deep. It is a common mistake to
countersink too deep. Trying a screw
upside down in a countersunk hole is a
good way to test for depth (see Fig. 136). The *diameter* of the top
of the hole should be about the same as the diameter of the top of
the screwhead.

Sometimes when round-headed screws are used, the holes are
counterbored to sink the heads below the surface, as shown in Fig. 137.
In making a counterbored hole, be sure to bore with the large auger
or bit first, and the small one last. Otherwise, it will be difficult or
impossible to bore or drill the large section of the hole clean and
smooth.

Using the Screw Driver. *First select a screw driver of suitable size.*
A screw driver that is too wide or too narrow may mar the work or
the screwhead or both. Also there is danger of damaging a screw
driver that is too small. A good mechanic keeps a set of screw drivers
of different sizes and selects one that fits the screw.

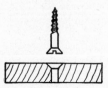

**Fig. 136.—Checking the counter-
sunk hole for depth by trying the screw
upside down in the hole.**

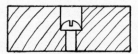

**Fig. 137.—Counterbored holes may
be used to sink round-headed screws
below the surface.**

Be sure that the screw-driver blade is properly shaped and in good condition.
The end should fit the screw slot. The two flat faces should be
straight and parallel, or slightly concave, near the tip where it fits
into the screw slot. The tip should be uniform in width or thickness
and square with the broad surfaces. It should never be ground or
filed rounded or to a sharp edge like a knife. The shank should be

straight, and the tip should be in line with the shank. (See page 173 on fitting screw drivers.)

In using a screw driver, grasp the handle firmly in the right hand with the palm resting on the end of the handle and the thumb and first finger extending along the handle. While the right hand gets a new grip for the next turn, use the left hand to steady the screw driver and keep it in the screw slot (see Fig. 138).

If the screw turns too hard, the pilot hole in the second piece may be too small or not deep enough, or the shank hole through the first piece may be too small. The screw should be removed and the

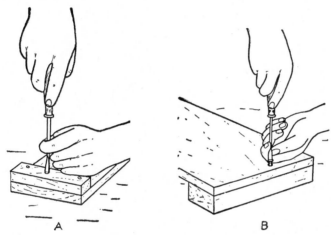

A   B

Fig. 138.—Two good methods of using a screw driver. While the right hand gets a new grip on the handle for the next turn, the left hand holds the bit steady and keeps it in the screw slot.

trouble determined and remedied. Otherwise, the screw may twist off or split the board, or the screw driver may slip from the screw slot and mar the work or the screwhead, or both.

A little soap, wax, or oil applied to the threads of the screw will make it go into the hole more easily.

Points on the Care and Use of Screw Drivers; Safety Rules

1. Keep the end free from grease or oil while using it.
2. Do not hold the work in the hand while tightening or loosening a screw. If the blade should slip, a bad cut might result. Hold the work in a vise or on a solid surface.
3. Never use the screw driver as a chisel,

4. Do not strike the handle with a hammer.

5. Do not use the screw driver for a pry bar.

6. Be sure the screw driver is in line with the screw.

7. Do not carry a screw driver in a pocket. The point might cause an injury.

8. Do not use a screw driver with a bent blade. Straighten it or discard it.

9. Hold the work so that if the screw driver should slip there would be no injury to the hands, face, or other parts of the body.

Using the Screw-driver Bit. Where several screws are to be driven, particularly large screws, a screw-driver bit in a brace makes for faster and easier work (see Figs. 139 and 140*B*). Such a bit is simply a short screw driver that has a square tapered end to fit into a carpenter's brace instead of a handle. Use care with the screw-driver bit to keep it from slipping from the screw slot. This is more easily done if the ratchet on the brace is used and the crank is backed up slightly every quarter or half turn.

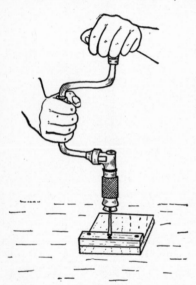

Fig. 139.—Driving screws with the screw-driver bit. By backing the crank a little every quarter or half turn the bit is kept in the screw slot.

Using Other Types of Screw Drivers. The *offset screw driver* (Fig. 140*C*) is for use in close quarters where a standard screw driver cannot be used. It has a short blade on each end, one in the same plane as the handle and the other crosswise to the handle. To use such a screw driver, place one end in the screw slot and turn the screw as far as possible; then reverse the screw driver and turn the screw as far as possible with the other end. The *Phillips screw driver* (Fig. 140*D*) has a special cross-shaped blunt-pointed blade to fit into the recessed head of Phillips type screws. With this type screw and screw driver, there is little danger of the blade slipping from the screwhead. Phillips screws are widely used in the automotive industry. A *spiral-ratchet screw driver* (Fig. 140*E*) is one that can be used to impart a rotary motion to the screw by pushing on the handle and, when properly set, to impart an intermittent rotary motion by simply twisting the handle back and forth.   When using such a screw driver by pushing on the handle, be especially careful not to allow the blade to slip from the screw slot.

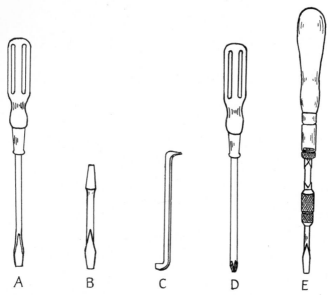

Fig. 140.—Types of screw drivers: A, standard screw driver; B, screw-driver bit; C, offset screw driver; D, Phillips screw driver; E, spiral-racket screw driver.

Fastening With Glue

Several different kinds of glue are available on the market.   *Animal glue, casein glue,* and *cold liquid glue* are the kinds most generally used in the average shop.   Each kind has its advantages.   Animal glue is strong, casein glue is more water-resistant, and cold liquid glue is always ready for use.   Animal glue, which comes in flakes or sheets, is prepared by breaking into small pieces, soaking in a definite amount of water, and then carefully heating in a gluepot.   It is applied hot. Casein glue, which is made from one of the ingredients of milk, is sold as a dry powder.   It is prepared for use by stirring it in water in accordance with the directions on the container.   Since it is reasonably strong, water-resistant, and easily and quickly prepared, it is favored by many for general shop use.   Cold liquid glue is always ready for use and is commonly used for general repairwork and odd jobs, although it may not be so strong nor so water-resistant as some other kinds of glue.

Success in the use of glue depends largely upon the care used in following directions. Warming of cold liquid glue is important in cold weather, and thinning with alcohol or other appropriate thinners as may be indicated on the container is important in case the glue is too thick, owing to age and to evaporation from a container that was not tightly sealed.

Applying Glue. Be sure that the parts to be glued fit properly and that all clamps, material, and equipment are in readiness before applying the glue. Then apply the glue thoroughly to all parts, and brush it or otherwise work it well into the pores of the wood, but do not apply more glue than is needed. (It is a common tendency of beginners to apply too much glue.) If the joint is movable, rub one piece back and forth over the other to distribute the glue thoroughly and work it into the pores.

Clamping Pieces Together. Once the glue is applied, clamp the pieces together and allow them to stand until the glue has hardened. If regular cabinetmaker's clamps are not available, clamps may be improvised by using the vise, or the bench top and wedges, or by twisting wire or rope.

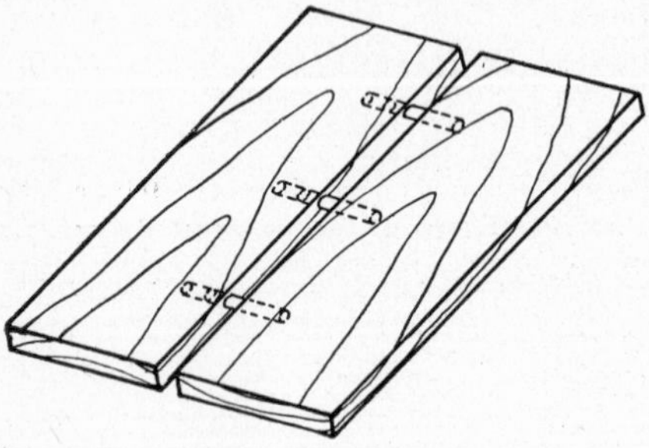

If two or more boards are to be glued edge to edge to form a wider piece, the edges must be carefully jointed, that is, made straight and square with the working surfaces of the boards. If the boards are to be planed after gluing, they must be so placed that the grain will run the same way in all of them. If the grain should run one way in one board and the opposite in the adjoining one, it would be impossible to plane one smooth without roughing the other.

Fig. 141.—A doweled joint ready to be glued and assembled.

Glued joints are generally reinforced by the use of dowels or corrugated fasteners, or both.

Doweling. A doweled joint is one in which the pieces are held together by round wooden pins called *dowels*. Figure 141 illustrates a typical doweled joint. Dowels are also often used instead of mortise-and-tenon joints for fastening boards together in cabinetwork. Dowels may be bought, or they may be made by splitting some straight-grained wood and planing or whittling roughly to size, and then driving the pieces through a round hole in a piece of steel called a *dowel plate*. The holes in a dowel plate are tapered slightly, and the wooden pieces are driven through the small end of the hole first to prevent binding. It is a good practice to drive the piece of wood through an oversize hole in the dowel plate first and finally through a hole of the required size.

A dowel plate can be easily made by drilling holes of the desired size, usually $\frac{5}{16}$ to $\frac{1}{2}$ in., in a piece of steel and reaming them slightly with a tapered reamer or with a round file if the work is done carefully. Care must be used in reaming not to enlarge the hole on one side of the plate.

Locating and Boring Dowel Holes. In using dowels, it is important that the holes be accurately located and bored so as to match and that the holes be bored perpendicular to the surface. If two boards are to be fastened edge to edge by doweling, clamp them together in a vise with the edges to be joined up and even with each other and with the working surfaces out. Locate the holes by squaring across the edges with a try square and a knife or sharp pencil and by gaging about half the thickness of the pieces from the working surfaces.

Bore dowel holes the same size as the dowels. They generally need not be over 1 in. deep. Countersink the holes slightly, and point the ends of the dowels slightly. Also cut a small groove lengthwise in each dowel with a knife or saw to allow the air and excess glue to escape when it is forced into place.

8. Shaping Curved and Irregular Surfaces

Using the Drawknife. The drawknife is very useful in shaping curved or tapered surfaces, especially where considerable waste stock is to be removed. To use a drawknife, first clamp the work in a vise or otherwise hold it securely. Use both hands to pull the knife, and be careful to trim or cut with the grain. Keep the bevel side of the knife up for ordinary work, and move one end of the blade slightly ahead of the other to give an oblique or sliding cut (see Fig. 142). This gives better control of the tool and enables the workman more easily to cut to a line.

Hewing. Hewing with a hand ax or hatchet, or even with a chopping ax, is often the fastest way of removing a large amount of excess stock. A higher degree of skill is required, however, to do good work with an ax than with a drawknife.

In hewing, first deeply cut or hack the surface every inch or two with the ax, striking the surface at an angle of 45 to 60 deg., as shown

Fig. 142.—Trimming with a drawknife. Move the blade with one handle slightly ahead of the other. This gives an oblique cutting action that enables the workman to control the tool better.

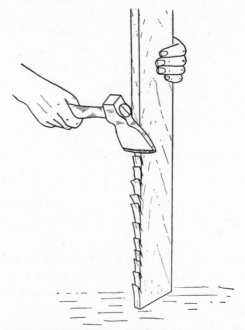

Fig. 143.—In hewing, first hack the surface deeply and then remove the roughened waste with light well-directed strokes.

in Fig. 143. Then remove the roughened waste by striking with the blade making a very small angle with the surface. A hand ax with a single bevel, being flat on one side, is best for hewing. As in all other cutting operations on wood, hewing should be done with the grain and not against it.

Using the Spokeshave. The spokeshave is used for planing curved and irregular surfaces (see Fig. 144). Its action is very much like that of a plane, and the blade is sharpened and set in much the same manner. Best work can be done with the blade set to give a thin shaving. Being very short, the spokeshave can follow rather abrupt turns or curves. It is generally pushed, but may be pulled. Keeping one handle slightly ahead of the other gives an oblique or sliding cut and thus gives better control of the tool. In planing abrupt curves, use a wrist motion.

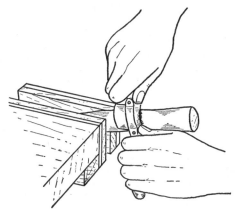

Fig. 144.—The spokeshave acts like a very short plane and is used for planing curved surfaces.

Whittling with the Pocketknife. The pocketknife is an invaluable tool for cutting or whittling on curved or irregular surfaces. It can frequently be used in places where it would not be possible to use other cutting tools. For best work it should be kept sharp and in good condition. (See page 148 for information on sharpening pocketknives.)

The same principles apply to the use of the pocketknife as to other cutting tools, the main ones being to cut with the grain wherever possible, and not against it, and to move the blade obliquely with one end slightly ahead of the other to enable the workman to cut better to a line. In doing heavy cutting, one should always cut away from himself. Another safety precaution is never to exert end pressure on the blade in such a manner that the blade could snap shut while in use.

Using Rasps and Files.   The wood rasp is valuable for removing waste in forming curved and irregular surfaces.   The "half-round" rasp, with a flat surface on one side and a curved surface on the other, is usually preferred.   Rasps cut faster than files, but leave rougher surfaces.   After using a rasp, the surface may be smoothed with a file or a spokeshave.   A reasonably coarse file, such as a flat bastard file or a woodworker's file, is usually preferred for filing wood.

In using a file or a rasp, take full, long, steady strokes and use only a moderate speed.   Lift the file or rasp slightly, or release the pressure, on the backstroke.   (Fast, short, choppy strokes are the mark of an amateur or a careless workman.)   Keep the file teeth clean by brushing frequently with a file card (file brush).

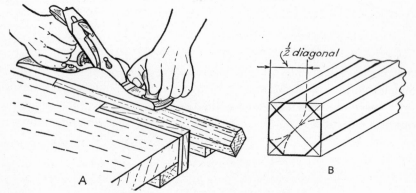

Fig. 145.—A. By careful work a cylinder may easily be made with a plane.   First make it square, then eight-sided, sixteen-sided, and finally round.   B. Method of marking an octagon on the end of a square stock preparatory to planing.

Sawing Curves.   The compass saw and the coping saw are useful for sawing curves.   The method of using them is explained on pages 55 and 56.

Cutting curves with the chisel is explained on page 78.

Planing a Piece Round.   To make a cylinder, such as a round handle for a tool box, first square up a piece of stock of the desired length, making it square in cross section.   Then mark out an octagon on the end of the piece, marking back from each corner a distance equal to one-half the diagonal (see Fig. 145B).   Next gage lines along the sides of the piece and plane off the corners, making the piece eight-sided.   Then without further marking, plane off each one of the eight corners about the same amount, making the piece sixteen-sided, or practically round.   Final smoothing may be done with sandpaper, rubbing the paper lengthwise of the piece.

Making a Curved Object. In making a curved object like a hammer handle, the workman will need to rely upon his judgment much more than in making objects with flat surfaces. With curved

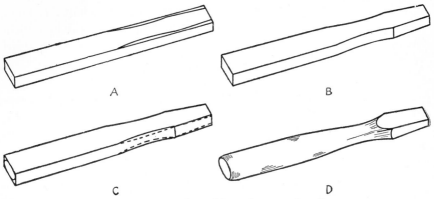

Fig. 146.—Steps in making a hammer handle.

surfaces it is difficult or impossible always to work to lines. Most articles with curved surfaces can best be made, however, by squaring up a piece of stock just large enough to make the article. Then lay out and cut curves and tapers on two edges, say top and bottom, using

Fig. 147.—Marking a board to fit against an irregular rock wall. By means of the dividers, a mark is made on the board parallel to the surface of the wall.

such tools as the drawknife, spokeshave, rasp, or file (see Fig. 146*A* and *B*). These curved surfaces should be made square with the sides.

Next lay out and cut curves and tapers on the other two sides; then round the corners and finish the work with such tools as the spokeshave, pocketknife, scraper, and sandpaper. If a steel scraper is not available, the sharp edge of a piece of broken glass can be used satisfactorily for scraping irregular surfaces. In sandpapering an irregular surface, use a small piece of sandpaper, and rub it back and forth with the grain. The paper can best be held in contact with the wood with the hand or fingers, no block being needed as in sandpapering a flat surface.

Fitting a Board against an Irregular Surface. To mark the edge of a board to be fitted against an irregular surface, like a stone wall, hold the board firmly beside the wall and mark it with a compass or a pair of dividers, as shown in Fig. 147. As one leg of the compass or dividers is moved along the surface of the wall, the other leg marks off a line parallel to the surface. The legs of the compass or dividers should be set apart a distance a little greater than the width of the largest space between the board and the wall. The edge of the board can then be trimmed to the line with such tools as the saw, chisel, and drawknife.

9. Cutting Common Rafters

Laying out a rafter is not a difficult problem if the basic principles are understood. The student should, therefore, strive to understand

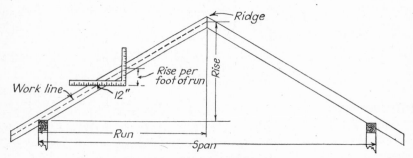

Fig. 148.—Common gable rafters.

the principles rather than to rely upon certain rules and figures that may be memorized.

The main jobs in laying out a rafter are: (1) determining the angles of cut at the ends of the rafter and at the bird's mount or seat, and (2) determining the length. This is easily done by means of the carpenter's steel square and is made possible by the fact that a rafter may be considered as the hypotenuse of a right triangle, the rise and run of the rafter being the other two sides or legs of the triangle.

The *work line* is a straight line laid off about midway between the edges of the rafter and parallel to them. It is used as a base line in measuring and marking out the rafters. (In case of rafter stock that is straight, it is not necessary to lay off a work line, the marking and measuring being done along one of the edges instead of on the work line.)

The *run* of a rafter is the horizontal distance measured from the outside edge of the plate to a point directly below the top end of the rafter (see Fig. 148). In the case of a plain gable roof with the ridge in the middle of the building, the run is half the width or *span* of the building.

The *rise* is the vertical distance from the plate to the upper end of the rafter (upper end of the work line).

The *pitch* of a rafter is a measure of its slope and is defined as the ratio of the rise of the rafter to twice its run, or as a fraction having the rise as the numerator and two times the run as the denominator. Expressed as a formula,

$$\text{Pitch} = \frac{\text{rise}}{2 \times \text{run}}$$

In the case of a gable rafter, the pitch is then the ratio of the rise of the rafter to the span of the building. For a rafter having a rise of 4 ft. and a run of 8 ft., the pitch is

$$\frac{4}{2 \times 8} = \frac{4}{16} = \frac{1}{4}$$

If the rise were 6 ft. and the run 9 ft., the pitch would be

$$\frac{6}{2 \times 9} = \frac{6}{18} = \frac{1}{3}$$

The relationship between rise, run, and pitch may also be expressed by a modification of this formula, as follows:

$$\text{Rise} = \text{pitch} \times 2 \times \text{run}$$

For example, if it is desired to find the rise of a rafter having a pitch of one-third and a run of 15 ft., substitute the values into the formula thus:

$$\text{Rise} = \tfrac{1}{3} \times 2 \times 15 = 10 \text{ ft.}$$

The *rise per foot of run* is a term used frequently in rafter work, and should, therefore, be thoroughly understood.   It may be determined by dividing the rise of the rafter by the feet of run; or it may be determined by use of the formula for rise given in the preceding paragraph. For example, if the pitch is $\frac{1}{3}$, substitute into the formula, using 12 in. as the run, and we have

$$\text{Rise} = \frac{1}{3} \times 2 \times 12 = 8 \text{ in.}$$

If the pitch is $\frac{1}{4}$, then

$$\text{Rise} = \frac{1}{4} \times 2 \times 12 = 6 \text{ in.}$$

Table 2 gives the rise per foot of run for the common pitches.

Table 2.   Rise per Foot of Run and Square Settings for Common Rafter Pitches

Pitch	Rise per foot of run	Square setting
$\frac{1}{8}$	3	3 and 12
$\frac{1}{6}$	4	4 and 12
$\frac{1}{4}$	6	6 and 12
$\frac{1}{3}$	8	8 and 12
$\frac{1}{2}$	12	12 and 12

Laying Out a Gable Rafter.   The laying out of a gable rafter usually involves the following steps:

1. Laying out the work line.
2. Marking the ridge cut or upper plumb cut.
3. Determining and marking the length of the main part of the rafter.
4. Marking the bird's mouth.
5. Marking off the rafter tail.
6. Shortening for the ridge board (when ridge board is used).

Laying Out the Work Line.   This is best done with a chalk line. If the rafter is bowed, place the bow or crown up.   The work line may be marked in the middle of the rafter, or up about 2 in. from the bottom edge in the case of a 2 by 4.   The marking of the work line may be omitted if the rafter has a good straight edge from which to measure and mark.

Marking the Ridge or Upper Plumb Cut.   To mark the plumb cut, place the square near one end of the stock with 12 on the body (see Fig. 149) and a number on the tongue corresponding to the rise per foot of run, coinciding with the work line (or one edge of the rafter in case no work line is used). Then mark along the tongue.

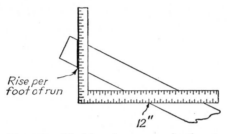

Rise per foot of run

12"

Fig. 149.—Marking the upper plumb cut.

Determining and Marking the Length of Main Part of Rafter.
Method 1. *Scaling.* Place the square on a piece of stock having a straight edge, with the side up which has the inches divided into twelfths. Measure the rise on one leg and the run on the other to the scale of 1 in. equals 1 ft. Make the figures on the legs of the square coincide exactly with the edge of the stock. Mark along the legs of the square with a knife or a very sharp pencil. Measure the distance between the intersections of these marks with the edge of the stock. This is the length of the rafter to the scale of 1 in. equals 1 ft. The inches and twelfths of inches on the square are simply read as feet and inches.

For example, to determine the length of a rafter having a run of 10 ft. 6 in. and a rise of 5 ft. 3 in., set the square with $10\frac{6}{12}$ in. on the body and $5\frac{3}{12}$ in. on the tongue coinciding with the edge of the stock (see Fig. 150*A*). Measuring the distance between the points of intersection of these marks with the edge of the stock, we find it to be $11\frac{10}{12}$ in. (see Fig. 150*B*). The rafter length is therefore 11 ft. 10 in.

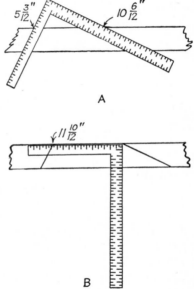

Fig. 150.—Steps in using the scaling method to determine the length of a rafter whose run is 10 ft. 6 in. and whose rise is 5 ft. 3 in.

Method 2.   *Using a Rafter Table.*   Probably the simplest method of determining the length of a rafter is to use a rafter table.   Figure 151 shows such a table on a rafter or framing square.   To use it, proceed as follows: Determine the rise per foot of run for the particular rafter to be marked out.   Find this figure on the body of the square.   Under this figure will be found another figure which is the length of common 'rafter per foot of run.   Multiplying this figure by the total feet of run in the rafter gives the total length.

For example, to determine the length of a quarter-pitch rafter having a run of 10 ft., look under 6 on the body of the square (6 being the rise per foot of run for quarter-pitch rafters) and find the figure 13.42.   Multiplying this by 10, we get 134.2 in.   Reducing to feet, we get 11 ft. 2.2 in., or for practical purposes a little less than 11 ft. and 2¼ in.

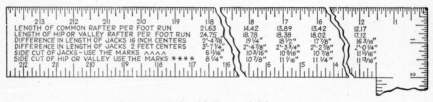

Fig. 151.—Rafter table on a rafter square.

On many squares the inside scale on one side of the tongue is graduated in tenths of inches to facilitate measuring off distances where decimal fractions are involved.   If such a square is available, 2.2 in. may be measured off directly.   Most squares also have a hundredths scale, which is 1 in. divided into 100 parts and from which decimal fractions of an inch may be measured off with a pair of dividers.   The hundredths scale is located on the back of the square near the junction of the body and the tongue.

Method 3.   *Stepping with Square.*   To use this method, take 12 on one leg of the square and the rise per foot of run on the other and place the square with these figures coinciding with the work line (or one edge of the rafter).   Mark carefully along the body and the tongue of the square, using a very sharp pencil (see Fig. 152).   This marks off one step on the work line (or edge).   Then move the square along the stock and repeat, taking as many steps as there are feet of run in the rafter.

Another variation of this method is to use a number on one leg corresponding to the total feet of run, and a number on the other corresponding to the total rise, and to take 12 steps.

Unless the work is done carefully, there is a chance for errors in marking when the stepping method is used. For this reason, many good carpenters do not use it, or if they do, they check the work by some other method.

Method 4.   Using Square Root.   The length of a rafter may be determined by extracting the square root of the sum of the squares of the rise and the run.   Stating this as a formula

$$\text{Length of rafter} = \sqrt{\text{rise}^2 + \text{run}^2}$$

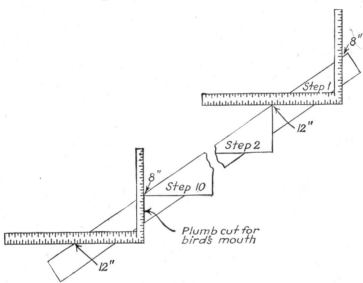

Fig. 152.—**Determining the length of a one-third-pitch rafter whose run is 10 ft., using the stepping method.**

This method gives accurate results but requires considerable involved computation with the possibility of errors in arithmetic, or the use of tables of square roots, which are usually not available.   This method therefore is seldom used by practical carpenters.

Marking the Bird's Mouth.   After the length of the main part of the rafter is determined and measured off on the work line (or one edge of the rafter), a plumb line is marked at the lower end just the same as at the ridge or upper end of the rafter.   The horizontal cut of the bird's mouth is made to pass through the intersection of this lower plumb line and the work line (or, in case the work line is not used, through the mid-point of the plumb line).   This horizontal cut may be made by placing the square with the usual figures of 12 on the body and the rise per foot of run on the tongue, coinciding with the work line (or upper edge of rafter), and moving the square up or down the stock until the edge of the body passes through the mid-point

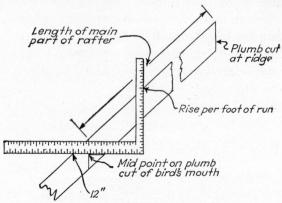

Fig. 153.—Marking the bird's mouth.

of the lower plumb line (or through the intersection of the plumb line and the work line) (see Fig. 153).

If the bird's-mouth notch as marked out is so deep as to weaken the rafter too much, or if it is so shallow as to give inadequate bearing

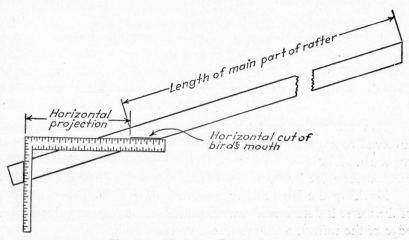

Fig. 154.—Marking off the rafter tail.

on the plate, the depth should be varied somewhat to give stronger construction.

Marking Off the Rafter Tail. The length of the rafter tail, or overhanging part, and the angle of cut at the lower end of the rafter are determined as follows: place the square with the edge of the body along the horizontal cut of the bird's mouth, and measure out from the plumb line of the bird's mouth, the horizontal projection (see Fig. 154). A mark along the tongue gives the end cut.

Shortening for the Ridge Board. If a ridge board is used, the rafter will have to be shortened by an amount equal to half the thickness of the ridge board, this amount to be measured back from the end of the rafter perpendicular to the plumb cut, and not parallel to the edges of the rafter (see Fig. 155).

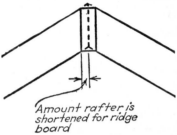

Amount rafter is shortened for ridge board

Fig. 155.—If a ridge board is used, shorten each rafter one-half the thickness of the ridge board. Measure back perpendicularly to the plumb cut.

10. Building Stairs and Steps

Building stairs and steps is often considered a job for only experienced carpenters. A knowledge of a few simple principles, however, will enable most ordinary mechanics to build porch stairs, stairs in barns, and stairs in other buildings where fine carpentry is not required. Laying out and cutting stair stringers or supports is very similar to laying out and cutting common rafters.

The first step in building stairs is to plan the work carefully by determining the total rise and run of the flight of stairs and the rise and run, or tread, of the individual steps. A rise of about 7 in. per step and a run or tread of about 10 in. will usually be about right. Usually the rise should not be less than 6½ in. nor more than 7½ in., and the width of tread not less than 9½ in.

Determining the Rise and Run. Measure the total rise of the flight of stairs by measuring vertically from the ground, or lower, floor level to the floor level above. Then divide this total rise into a number of equal parts so that each part is as near 7 in. as possible. It is important that all steps in a flight, including both the bottom step and the top step, have the same rise.

One way of determining the rise per step is to measure off the total rise on a board and then set a pair of dividers to 7 in. and step off the marked measurement on the board. If there is a fraction of a step left over, increase or decrease the setting of the dividers and step off the measurement again; repeat until the total rise is divided into a number of exactly equal parts. This will give the number of risers on the stairs. The setting of the dividers will be the rise of each individual step. (A riser is a vertical board between two treads of a staircase.)

The total run of a staircase can usually be varied within certain limits, and where this is the case, allow a tread or run of about 10 in. for each step. If the total run of the stairs is exact and cannot be varied, then determine the width of tread or run for each step by dividing the total run by the number of risers less one. (There is always one more riser than treads.)

Cutting a Stair Stringer. Before marking a stair stringer for sawing, be sure the top edge is straight and square with the sides of the stringer. Plane the top edge if necessary. Then place the square on the side of the stringer and mark out the notches for the individual steps. Use a method similar to that for stepping off the length of a rafter (see Fig. 152). Use the exact rise and run of the individual steps (as previously determined) as square settings, make these square settings on the legs of the square coincide with the top edge of the stringer, and then mark along both legs of the square. After the stringer is marked out, saw out the notches. Be careful to saw square to the sides of the stringer.

11. Laying Out and Erecting a Small Building

Before starting a building, one should have a complete working plan and bill of materials. The plan may be taken from some bulletin, a farm paper, or it may be drawn by the builder himself. The important thing is that it be well worked out and in detail so as to avoid delay and waste of materials in construction. In deciding upon a plan, it is well to study all available plans and select one that best suits the particular needs. Often a more or less standard plan may be used, with perhaps only minor, if any, alterations.

Having a plan, it should be carefully studied and the various details of construction understood before actual work of construction starts. All required materials and tools should be on hand, or there should be assurance that they will be available as needed.

Laying Out the Foundation. To lay out a foundation for a building, first mark out one side or one end (see line *AB*, Fig. 156). It is usually desirable to make this end or side parallel or perpendicular to the side of some other building or a road or fence line. Set stakes *A* and *B* on this line, locating two corners of the building. Then drive nails in the top of each stake to locate the corners exactly.

Next drive another stake *F* on line *AB* 6 ft. from *A*.   Drive a nail in the top of stake *F* exactly 6 ft. from the nail in stake *A*.   Then drive stake *E* and put a nail in the top exactly 8 ft. from the nail in stake *A* and exactly 10 ft. from the nail in stake *F*.   The corner *EAF* is then a right angle.   This method of laying out a right angle is known as the 6-8-10 method.   Extend line *AE* to form the second boundary line of the building, and establish the other corners in a similar manner.   It is a good plan to check the accuracy of the work by measuring the diagonals *AC* and *BD*.   They should be equal in length.

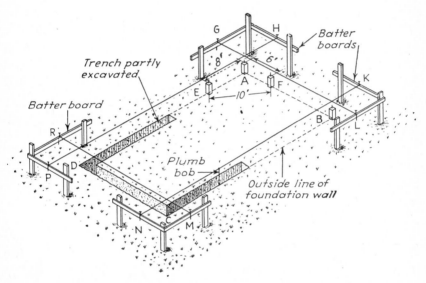

Fig. 156.—A method of laying out square corners and of using batter boards to establish building lines.   (Courtesy of Portland Cement Association.)

After the corners of the building are located, drive stakes and put up batter boards, outside the building lines, as shown in Fig. 156.   Then stretch strings between the batter boards, so they intersect exactly over the nails in the corner stakes.   The batter boards should be set level and at the height of the top of the foundation or at some other convenient height.

Digging the Trench and Constructing the Foundation.   A foundation wall should have a footing or broadened base to carry the load without appreciable settling.   The size of the foundation and the size and depth of the footing depend upon the size of the building to be supported and the type of soil.   For small buildings like poultry

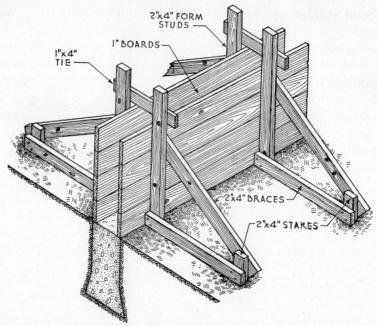

Fig. 157.—Foundation wall forms may be made in this manner provided the soil does not crumble or cave and a wide footing is not required. (Courtesy of Portland Cement Association.)

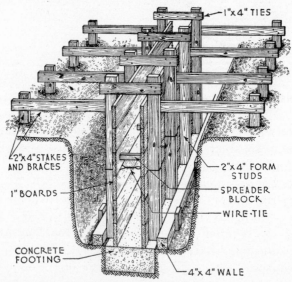

Fig. 158.—A good method of making foundation forms when footings are required. (Courtesy of Portland Cement Association.)

houses on average soil, footings 12 in. wide and 8 in. deep are large
enough.

If the soil is firm and the digging is done carefully, the trench
may be simply widened at the bottom as shown in Fig. 157.   If the
soil is crumbly, however, or if an especially thorough job is to be done,
dig the trench wider and make the footing and foundation as shown
in Fig. 158.   The foundation can be built of brick, tile, or stone if
desired instead of concrete.

While the footing should extend below the frost line to avoid
heaving from freezing in the case of major buildings, a depth of
about 2 ft. is usually considered adequate for small buildings like
poultry houses, garages, and machine sheds.   Always extend footings

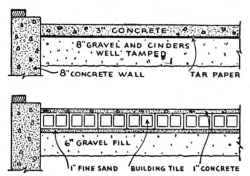

Fig. 159.—Two styles of laying-house floor construction.   (Courtesy of Everybodys
Poultry Magazine.)

to firm well-settled soil, however.   The foundation walls should
usually extend at least 12 in. above the ground line.   They should
be at least 6 in. thick for the smaller buildings, and up to 10 or 12 in.
for large ones.

When the foundation has .been built to within 6 in. of the top,
embed ½-in. bolts about 10 in. long in the foundation about every
8 ft. to hold the sills in place.

See pages 256 to 273 for methods of making and handling concrete.

Building the Floor.   The kind of floor used will depend upon
the kind of building, the desires of the owner, and the amount of
money to be spent on it.   Concrete, wooden, or even earth floors
are variously used in farm buildings.   Concrete is usually preferred
for poultry houses, except for small portable ones.

For a poultry house, use a fill of 6 to 8 in. of well-tamped cinders,
gravel, or crushed stone placed inside the foundation walls, and

place the floor on top of the fill. This is to avoid dampness working up through the floor by capillary action. Two styles of poultry-house floor construction are shown in Fig. 159.

For a small-colony brooder house or portable poultry house, build the floor as shown in Fig. 160. The floor joists are nailed in place crosswise on the main runners, and the flooring is nailed onto the floor joists.

Erecting the Walls. Fasten the sills to the foundation walls, or on top of the floor in case of a portable poultry house, and erect the framework according to the plan being used. Nail or otherwise securely fasten all pieces. Determine the locations for the windows and doors, and cut and nail in the framework for them.

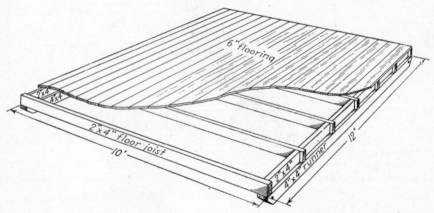

Fig. 160.—A method of building the floor for a portable brooder or poultry house.

After the framework for the walls is in place, install the siding or other material to be used on the outside walls. In the case of horizontal siding, start at the bottom and work up; and in the case of vertical siding, start at one corner and work around the building.

Putting On the Roof. Cut the rafters, and erect and nail them securely in place. Be sure to mark all of them from the same pattern and to saw carefully to the lines. See page 110 for information on laying out rafters. Next put on the roof sheathing, using solid sheathing for composition shingles or roll roofing and 1 by 4's spaced about 1¼ in. apart for wooden shingles. For a tight draftproof roof, it may be desirable to use solid sheathing even for wooden shingles. Start laying the shingles or other roofing along the bottom edge of the roof, and work toward the top. Roll roofing should be carefully

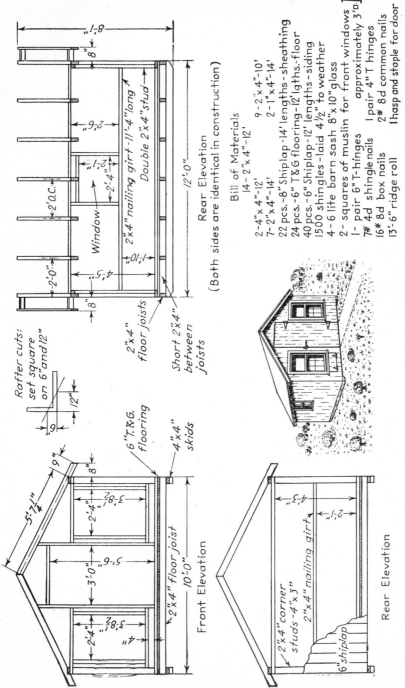

Rafter cuts:
set square
on 6" and 12"

6" T.&G.
flooring

4"x4"
skids

Rear Elevation
(Both sides are identical in construction)

2"x4" nailing girt - 11'-4" long
Double 2"x4" stud

Window

2"x4"
floor joists

Short 2"x4"
between
joists

Bill of Materials
14 - 2"x4" - 12'

2 - 4"x4" - 12'   9 - 2"x4" - 10'
7 - 2"x4" - 14'   2 - 1"x4" - 14'
22 pcs. - 8" Shiplap - 14' lengths - sheathing
24 pcs. - 6" T&G flooring - 12' lgths. - floor
40 pcs. - 6" Shiplap - 12' lengths - siding
1500 shingles - laid 4½" to weather
4 - 6 lite barn sash 8"x10" glass
2 - squares of muslin for front windows
1 - pair 6" T-hinges   approximately 3'□.]
7# 4d shingle nails   1 pair 4" T hinges
16# 8d box nails   2# 8d common nails
13'-6" ridge roll   1 hasp and staple for door

Front Elevation

2"x4" floor joist

10'-0"

Rear Elevation

2"x4" corner
studs - 4"x 3"

2"x4" nailing girt

6" shiplap

Fig. 161.—Plans for a 10 ft. by 12 ft. brooder house.

sealed at the joints with roofing cement. Any special instructions of the maker of the roofing should be carefully observed.

Finishing the Building. After the walls are up and the roof is on, install the doors and the windows. Put on the window and door trim. To ensure good appearance, as well as a draʿtproof building, use care in marking, sawing, and fitting the doors and windows. Inside fixtures, such as nests, feeders, and roosts for a poultry house, are next made and installed. The location of fixtures and equipment, as well as the details of their construction, should receive considerable thought and attention.

Jobs and Projects

1. *a.* Look through shop books, manuals, and bulletins that include plans for small appliances made of wood, such as nail and tool boxes, bench hooks, sawhorses, bag holders, wood floats for concrete, book shelves, lawn chairs, workbenches, and poultry feeders. (See check list below.)

   b. List three or four that you would like to make and that you believe you could make in the shop. Select some of the simpler jobs and some that are larger and more involved.

   c. Study the plans for these jobs carefully, and make sure you understand them. If the plans do not exactly suit you, sketch changes in the plans to make them suit your purposes better.

   d. Make a list of the common woodworking operations, such as measuring and marking, sawing, planing, squaring up, boring, and chiseling. List them in a vertical column down the left side of a sheet of paper. Across the top of the paper, write the names of the jobs you propose to do, ruling a column for each job. Analyze each job, and check the operations involved. Would the jobs you have selected give considerable practice in all the more important operations?

2. As you make various appliances or do different jobs in the shop, be sure to practice right methods of doing the basic processes or operations. It will ensure the turning out of better jobs, and what is more important, it will enable you to acquire speed and skill in shopwork more quickly and easily.

3. Examine some woodwork jobs that have been done in the shop, possibly yours and some others. Use a grading or scoring system, such as E for excellent, S for superior, M for medium, and I for inferior, and rate the workmanship. Note, in particular, if corners and ends are square, if saw cuts are straight and smooth; if nails are well spaced, of suitable size and kind, and well driven; if screws are uniformly and neatly driven with the heads flush with the surface; if planing, chiseling, and boring have been well done; if parts have been made to dimensions called for on plans.

4. A suggested check list of things to make:

Tool box	Lawn chair or seat
Nail and tool box	Sawhorse
Wood float for concrete	Workbench
Miter box	Kitchen stool
Bench hook	Tool cabinet
Tool rack for chisels and files	Saw filing clamp
Single tree	Wash bench
Evener	Flower box
Harness-sewing clamp	Doghouse
Bag holder	Poultry water stand
Footstool	Chicken feeder
Book shelves	Nests for poultry
Ladder	Poultry crate and coop
Self-feeder for hogs	Hog crate
Hog trough	Loading chute

4. Painting, Finishing, and Window Glazing

IT IS generally practical for a farmer to paint his smaller buildings and many pieces of farm equipment himself and, under some conditions, his larger buildings also. Painting prolongs the serviceable life of materials, improves appearance, and, on the inside of buildings, promotes cleanliness and sanitation.

Paint is composed essentially of a pigment, such as white lead, and a vehicle, such as linseed oil. Under most conditions, a thinner like gum turpentine is added to make the paint spread more easily and penetrate better, and a drier is added to make it dry more rapidly. Upon drying, the vehicle forms a tough, somewhat elastic film that binds the particles of pigment together and to the surface being painted. Some house paints, as they weather, become slightly chalky on the surface and wear down gradually, leaving a protective, although somewhat dull, coat on the wood for many years. Other paints, because of their composition, dry to a harder and more glossy surface, but in time check and crack.

MAJOR ACTIVITIES

1. Inspecting Buildings for Painting Failures

2. Selecting Factory-prepared Paints

3. Mixing Paint at Home

4. Preparing Outside Wood Surfaces for Painting

5. Applying Exterior Paint

6. Using Stains, Varnishes, Enamels, and Lacquers

7. Painting Metal Surfaces

8. Selecting, Cleaning, and Caring for Brushes

9. Storing and Handling Paints Safely

10. Whitewashing

11. Cutting Glass to Size

12. Replacing a Broken Window Glass

13. Repairing a Defective Window Sash

1. Inspecting Buildings for Painting Failures

A knowledge of painting troubles and failures is valuable to anyone planning to paint a building. It will enable him better to select paint and to avoid or remedy common paint troubles.

Chalking. High-quality house paints wear away through a gradual chalking of the paint surface caused by a natural deterioration of the oil in the paint. Premature or excessive chalking may be due to the use of a low-grade paint.

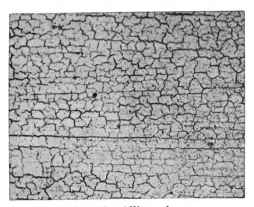

Fig. 162.—Alligatoring.

Alligatoring (see Fig. 162) is usually caused by applying a faster drying paint over a soft or slow-drying paint. The harder outer coat, being unable to contract and expand with the softer undercoat, cracks and forms an "alligator-skin" appearance. Allowing insufficient drying between coats, or the use of a paint with inferior or slow-drying oils as a substitute for linseed oil, may cause alligatoring.

Blistering and peeling (see Figs. 163 and 164) are caused by moisture behind the paint film. It may be from moisture in the wood at the time of painting or from moisture, possibly in the form of vapor, coming through the wall and getting behind the paint. Expansion of this moisture under the paint film, due to a rise in temperature, may cause blisters to form. When such blisters break, the paint will peel.

Checking is the formation of a network of fine hair lines in the outer coat. It is the result of natural shrinking of the oil which has changed from a liquid to a semisolid as the paint ages. It usually is not considered a serious defect and can be avoided by properly formu-

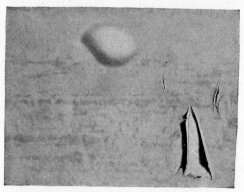

Fig. 163.—Blistering and peeling.

lating the paint for the various coats (adding suitable amounts of thinner and oil) and by allowing plenty of time for drying between coats.

Cracking and scaling (see Fig. 165) results when a paint becomes too brittle as it ages. Wood expands and contracts, and if the paint

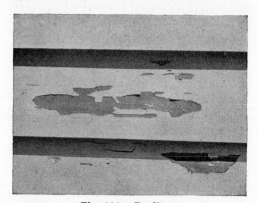

Fig. 164.—Peeling.

film is not elastic enough to expand and contract with the wood, cracks develop in the paint and it eventually scales off. Cracking and scaling may be avoided by using a higher grade, more elastic paint. When repainting a cracked or scaled surface, it is necessary to remove

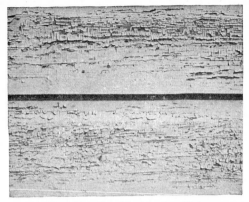

Fig. 165.—Cracking and scaling.

Fig. 166.—Running and sagging.

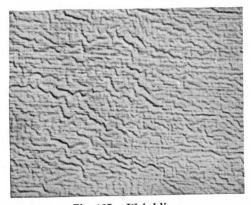

Fig. 167.—Wrinkling.

all the old paint, by using scrapers, sandpaper, blowtorch, or liquid paint remover; otherwise, a rough job will result.

Running and sagging (see Fig. 166) is caused by using a paint with too much oil and applying it too thick. Applying paint over a glossy surface may also cause sagging. A glossy surface should be sandpapered or wiped with a cloth soaked in benzene before painting.

Wrinkling (see Fig. 167) is caused by improper drying of the paint film. The surface dries quickly, leaving undried paint beneath it. Using too much oil in the paint, failure to brush the paint out to a thin even coat, or a sudden drop in temperature during the drying period may cause improper drying, resulting in wrinkling.

Incompatibility. Poor results similar to alligatoring may occur if two paints of distinctly different composition are used, one over the other. Such paints are apparently incapable of being used together and are said to be incompatible. This type of failure is more apt to occur where white or light-colored paint is applied over dark paints.

2. Selecting Factory-prepared Paints

Most paint manufacturers make two or more grades of paint. It is nearly always more satisfactory and cheaper in the end to buy the best grade, although for some purposes, such as temporary protection, a second-grade paint might be advisable. Third-grade paints should usually be avoided.

The biggest item of painting expense is generally labor. Therefore, even if the cost of paint is reduced considerably, only a little saving may be made in the total cost of painting. Inferior paints do not cover as well, and considerably more is required, offsetting to a large degree the lower price per gallon. Furthermore, the period of protection afforded by inferior paints is much shorter. All things considered, the use of inferior paints is generally false economy.

Selecting House Paints. There are three different types of outside house paints commonly available on the market: (1) *white lead paint,* (2) *white lead and zinc oxide paint,* and (3) *white lead, zinc oxide, and titanium dioxide paint* (white only). A white lead paint gives a surface that can be neglected for long periods, yet be easily repainted. White lead and zinc oxide paints give harder, more glossy surfaces which do not chalk off like lead and oil paint. They may be subject to cracking and peeling if long neglected and may, therefore, be harder to paint over. Titanium dioxide has great hiding power and is often used in white paints.

The quality of any one of the above three types of paints depends upon the proportions of the various ingredients used and the amount of inert materials, extenders, or substitute fillers used. The following standards, used by some agencies of the Federal government in buying paints for their own use, may be used in judging paints.

White Lead and Oil Paint

Pigment
    White lead, not less than 98 per cent
    Tinting pigments, not more than 2 per cent
Vehicle
    Linseed oil, not less than 87 per cent
    Thinner and drier, not more than 13 per cent
    Weight per gallon, not less than 19½ lb., pigment to be at least 71 per cent of total weight

White Lead and Zinc Oxide Paint

Pigment
    White lead, not less than 70 per cent
    Zinc oxide, not less than 15%, nor more than 20 per cent
    Tinting colors and siliceous mineral pigments, not more than 10 per cent
Vehicle
    Linseed oil, not less than 87 per cent
    Thinner and drier, not more than 13 per cent (thinner to be turpentine, volatile mineral spirits, or any mixture thereof)
    Weight per gallon, not less than 17½ lb., pigment to be at least 66 per cent of the total weight

White Lead, Zinc Oxide, and Titanium Dioxide Paint

Pigment
    White lead, not less than 58 per cent
    Zinc Oxide, not more than 25 per cent
    Titanium dioxide, at least 7 per cent
    Extenders (white silicate pigments, barium sulfate or mixture thereof), not over 10 per cent
Vehicle
    Linseed oil, not less than 85 per cent
    Thinner and drier, not more than 15 per cent
    Weight per gallon, not less than 17 lb., pigment to be at least 68 per cent of the total weight

Selecting Barn Paints. Although any good house paint can be used on barns, they are most commonly painted red because a good-quality red paint costs less and lasts longer than good-quality white or light-colored paints. Iron oxide pigment, which gives red barn paint its color, is quite durable. There should be at least 30 per cent of iron oxide in the pigment of such a paint, and not more than 70 per cent of siliceous materials. The vehicle should contain at least 75 per cent linseed oil, the remainder being drier and thinner. A red barn paint should weigh at least 12 lb. per gal. and at least 53 per cent of the total weight should be pigment.

Estimating Amount of Paint Required. To estimate the amount of paint required for a job, first determine approximately the number of square feet to be covered, and divide by the number of square feet a gallon of paint is estimated to cover. The covering capacity of a paint will depend principally upon the paint, the condition of the surface, and how well it is brushed out when applied. Under average conditions, 1 gal. of outside house paint will cover about 600 sq. ft. on the first coat on new wood, and about 700 sq. ft. on the second and third coats. On repaint jobs with the surface in good condition, 1 gal. of good paint may be expected to cover 400 to 500 sq. ft. with two coats.

3. Mixing Paint at Home

There are certain advantages to mixing paint at home, the chief ones being a lower cost per gallon and assurance of high-grade materials used in the paint. There is more work to mixing the materials, however, and it may be difficult to add the correct amount of tinting pigment to match colors.

TABLE 3. PROPORTIONS OF INGREDIENTS TO USE IN MAKING EXTERIOR HOUSE PAINTS*

	New wood			Repainting (2 coats)		Repainting (1 coat)
	Primer	Body	Finish	Body	Finish	
Soft paste white lead †	100 lb.	100 lb.	100 lb.	100 lb.	100 lb.	100 lb.
Pure raw linseed oil	4 gal.	1½ gal.	3 gal.	2 gal.	3 gal.	2 gal.
Pure gum spirits of turpentine	1¾ gal.	1¼ gal.	0	1¾ gal.	0	⅓ gal.
Liquid drier ‡	1 pt.	1 pt.	1 pt.	1 pt.	1 pt.	1 pt.
Approximate yield in gallons	9 gal.	6 gal.	6¼ gal.	7 gal.	6¼ gal.	5⅔ gal.

* From *Minnesota Agricultural Extension Bulletin* 233.
† If a heavy paste white lead is used, add 1 qt. of turpentine for each 100 lb. of paste.
‡ The drier should be omitted if boiled linseed oil is used.

Soft paste lead should be used for home-mixed paints.   Pour the paste white lead into a clean mixing pail or tub that will hold somewhat more than the total amount of paint to be made.   Add about one-third of the oil, and stir until the mixture becomes a thin paste. Then add the remainder of the oil and the other materials according to the formula used (see Table 3), and stir thoroughly.   If a tinted paint is to be made, use colors ground in oil.   Put them in a little

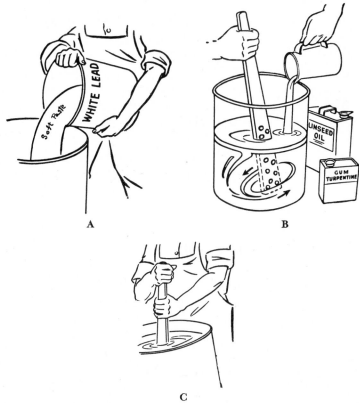

A   B

C

Fig. 168.—Steps in mixing white lead paint.

at a time while stirring as soon as the paint is thin enough to stir well and before all the liquids are added.   Use just enough color to give the desired shade.   It is best to mix up enough paint for the whole job, as it might be difficult to match the color exactly should a second batch be required.   Always strain the paint through a fine screen or cheesecloth folded double just before using.   This will remove lumps or foreign material that may have been introduced, as well as improve the brushing qualities of the paint.

4. Preparing Outside Wood Surfaces for Painting

Surfaces to be painted should be dry and clean, that is, free from mud, dust, grease, plaster, smoke, rust, or old loose, scaly paint. Usually a wire brush, a putty knife, and a dusting cloth or brush are the only tools needed for cleaning. Sometimes a surface will need to be washed, or, if greasy, wiped with a cloth moistened with turpentine or gasoline, and then allowed to dry. Sandpaper can sometimes be used to advantage on a rough surface. If old cracked scaly paint must be removed, it may be necessary to use a blow torch or a liquid paint and varnish remover.

Any broken or rotten boards should be replaced, and any loose ones should be nailed down. The surface should be carefully inspected for indications of paint failure due to moisture getting under the paint. Any leaks or defects in the building that might admit moisture should be repaired.

Be sure the surface is dry. If there has been a recent rain, allow several days of drying weather before applying a priming coat.

5. Applying Exterior Paint

Choose Painting Weather Carefully. Almost any time from April to November when it is warm and dry and not too windy or dusty is suitable for painting outside work. A temperature between 65 and 80° is best, and painting should not be done when the temperature is below 50°. Not only the surface to be painted, but the wood clear through should be thoroughly dry.

Thinning Factory-prepared Paint. When using a ready-mixed paint, all coats except the last require thinning. Carefully follow the directions for each coat as shown on the label. Linseed oil is added to compensate for the oil absorbed by the dry wood or old painted surface. Turpentine is added to aid penetration into the wood and to make the paint easier to brush on.

It is very important that the paint be thoroughly mixed. Pour the liquid off the top of the can into a clean container, and add any required thinners to it. Stir the thick pigment into a smooth paste, using a stiff flat paddle. As stirring proceeds, slowly add the liquids. "Boxing," or pouring the paint back and forth from one container to another, ensures better mixing.

Painting with a Brush. Hold the brush lightly with the long part of the handle resting in the hollow between the thumb and first

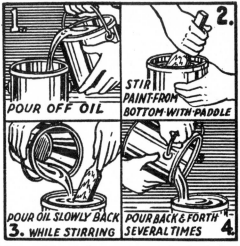

Fig. 169.—Thoroughly mix factory-prepared paint before applying it.

finger and with the ends of the fingers well up on the metal ferrule (see Fig. 170). Do not allow the fingers to extend down on the bristles.

Dip the brush into the paint about one-third the length of the bristles, and then remove the excess paint by gently tapping the brush against the inside of the pail, or wiping it over the edge. Use long sweeping strokes, generally with the grain. *Feather* the strokes by bringing the brush down against the surface gradually at the beginning of each stroke and lifting it gradually at the end. Brush the paint out well to form a thin even coating. Start at the top of a surface and at one edge, working across and down. Stir the paint occasionally to make sure it is thoroughly mixed.

Applying Paint with a Spray Gun. Spray painting of buildings saves a large amount of labor, especially when large surfaces with few windows and doors are to be painted. A spray-painting outfit consists essentially of an air compressor with an engine or motor to drive it, suitable controls and regulators, and a paint gun. For painting buildings, it is important that good equipment of

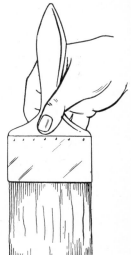

Fig. 170.—A good way to hold a paint-brush. Note that the ends of the fingers are well up on the ferrule and do not touch the bristles.

ample capacity be used and that it be properly adjusted and operated. Small outfits, although they may be satisfactory for painting small appliances or machines or apparatus, are not usually satisfactory for painting buildings. Spray painting, properly done, will give as good a job as brush painting, and while it will waste some paint, possibly as much as 5 to 10 per cent, a painter can spray up to ten times as much in a day as he could paint with a brush.

In using a paint gun, care should be taken not to apply too thick a coat. The gun should be moved at a steady, moderate speed and at a uniform distance from the surface, usually about 8 to 10 in. With some practice, one can soon make feathered strokes and satisfactory laps. It is important that paint to be applied with a spray gun be strained through a very fine screen or cloth, such as a piece of rayon stocking.

A paint gun should be thoroughly cleaned after use. This may be done by spraying a suitable solvent, such as benzene or painters' naphtha, through it and then blowing it clean with compressed air. The nozzle may then be taken apart and cleaned. When painting is to be resumed in an hour or two, the gun need not be thoroughly cleaned if it can be set in a can of solvent. In using a hose-fed gun, be sure to run the cleaning liquid through the paint hose at the end of a day's work; otherwise, dried paint will fill it up and reduce feed capacity.

6. Using Stains, Varnishes, Enamels, and Lacquers

Staining. Stains are used for coloring wood. They are available in a wide variety of colors and are of three general types, based upon the vehicle or carrier used, namely, water stains, alcohol or spirit stains, and oil stains. Oil stains, although usually more expensive, are probably the most satisfactory for general use. They are easily applied and do not raise the grain like water or spirit stains. An application of linseed oil alone makes a very desirable light-color stain.

Varnishing. A varnished finish makes a very attractive appearance and has good wearing qualities. The best varnishes are made of copal gum dissolved in linseed oil and turpentine. There are various grades and kinds of varnishes on the market, and, to ensure best results, a varnish made for a particular purpose should not be used for other purposes. Interior varnishes cannot be expected to give good results when exposed to the weather. Varnishes should be thinned and applied in accordance with the directions on their containers.

Varnish should always be applied with a high-grade clean brush that has never been used for anything except varnish. Varnish is sticky and slow in drying. Dust should therefore be kept down to a minimum around freshly varnished surfaces.

Using Enamels and Lacquers. There is on the market a wide variety of enamels and lacquers that are especially suited to finishing inside woodwork and furniture. They are available in various colors, and, although the better grades are somewhat expensive, they produce finishes that are attractive and easily cleaned and wear well unless subjected to unusual wear and abuse. Enamels are made by grinding pigments in varnish and should therefore be handled and applied like varnish. Lacquers and quick-drying enamels have more volatile vehicles or solvents and, therefore, dry more quickly.

7. Painting Metal Surfaces

In painting metal surfaces, it is most important to use a priming coat of rust-inhibiting paint, such as red lead and oil. In general, the priming coat may be followed by any paint that will give the desired finish and color. Regular house paints are commonly used on gutters and downspouts after they have been primed.

For painting machinery and implements, inside metal barn equipment, etc., it is best to use a special implement or metal paint or enamel which will dry harder and give a more durable wearing surface. Such metal paints usually have some varnish in them to produce the desired hard-wearing surfaces.

It is important that metal surfaces be thoroughly cleaned of grease, old loose paint, or loose rust before paint is applied.

8. Selecting, Cleaning, and Caring for Brushes

Selecting Brushes. A 3½- or 4-in. brush is commonly used for painting large surfaces. The bristles should not be too long, not much over 4 in., for inexperienced painters. A flat brush 3 in. wide is a good size for painting trim, and a sash brush 1 or 2 in. wide is good for painting windows. For varnishing, use a good-quality brush that has never been dipped in paint.

Taking Care of Brushes While in Use. Never allow a brush to rest upright on its bristles. If work is stopped for a few minutes, remove the surplus paint from the brush by wiping it on the edge of the pail, and then lay it flat, across the top of the pail or on some smooth, clean surface. If the work is stopped for a longer time—overnight or a few days—suspend the brush in a can of turpentine and raw linseed oil in the case of paintbrushes, or in turpentine or paint thinner in the case of varnish brushes. This can best be done by drilling a small hole through the handle and hanging it on a small wire hook on the side of the can, or on a wire laid across the top of the can, so that the bristles are covered by the liquid and yet do not touch the bottom of the can (see Fig. 171).

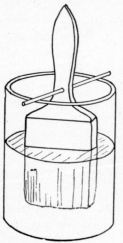

Cleaning and Storing Brushes. When you have finished painting or varnishing, clean the brush out thoroughly with turpentine, benzene, gasoline, or kerosene, and then wash with warm soapsuds. Then shake the brush well, and while still damp, wrap in heavy paper and lay it away or hang it in a dry cool place.

Old neglected brushes can generally be reclaimed by soaking them in paint remover or special brush cleaners available at paint stores, and then washing in turpentine, alcohol, gasoline, or benzene.

Fig. 171.—A good way to take care of paintbrushes when work is stopped overnight or for a few days. Note that the bristles are covered with the liquid but do not rest on the bottom of the can.

9. Storing and Handling Paints Safely

Paints are inflammable, as are turpentine, oils, and thinners commonly used for thinning paint and cleaning brushes. These materials should not be used or handled near an open flame. Oil-soaked and paint-soaked rags sometimes catch fire spontaneously. They should therefore not be left scattered about, but should be burned in a safe place or kept in a metal container with a tight cover until they can be burned safely.

Paint, as well as turpentine, paint oils, and thinners, should be stored in sealed containers, preferably of metal. Glass containers are subject to breakage, and unsealed cans are subject to evaporation and spilling. A definite, safe place for storage of paint and painting materials contributes not only to safety, but also to orderliness and system about the shop or premises. Left-over paint can be saved for future use if properly sealed and stored.

10. Whitewashing

Whitewash affords an inexpensive means of improving the appearance and lighting of basements, barns, poultry houses, and similar places.   It is sometimes used for painting fences and other outside surfaces, but for such purposes its poor wearing qualities make it much inferior to oil paints.

Common whitewash is composed of lime and water and may be applied either with a brush or a sprayer.   Many different formulas are used for making whitewash.   A good general-purpose whitewash may be made as follows: Make a lime paste by soaking 25 lb. of hydrated lime in about 3 gal. of water.   Dissolve 2½ lb. of casein glue in water according to manufacturer's instructions, dilute to a thin consistency, and mix thoroughly with the lime paste just before using.   Thin to the desired consistency.

Another good whitewash may be made by mixing 2½ lb. of salt, 4½ gal. of skim milk, and 50 lb. of hydrated lime.

11. Cutting Glass to Size

Glass is "cut" by first scratching the surface with a tool called a *glass cutter* and then broken by applying pressure along the scratch.

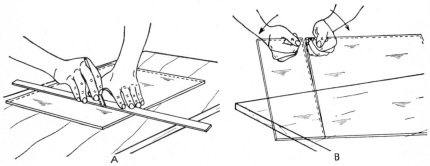

Fig. 172.—Cutting glass: A, draw the glass cutter along the straightedge, using a firm but moderate pressure—do not let the cutter lean sidewise; B, break the glass by applying pressure with the two hands.

To cut glass, clean it and put it on a flat table or bench.   Hold a straightedge firmly on top of the glass to guide the glass cutter, and draw it along slowly with an even, *moderate* pressure (see Fig. 172*A*). Begin at the far edge of the glass, and make a clean scratch all the way across in one stroke.   Do not use too much pressure, or the glass may splinter.   Do not go over a scratch a second time as this injures the cutter and may cause the glass to break unevenly. The cutter may be leaned slightly in the direction of cutting, but do not let it lean sidewise.

After scratching, break the glass by applying pressure up from beneath the scratch (see Fig. 172*B*). Tapping the glass first underneath the scratch with the end of the glass cutter may make the breaking easier and cleaner. If any small jagged projections are left, break them off with one of the square notches in the end of the glass cutter or with a pair of pliers.

Another method of breaking the glass is to place it, scratched side up, on a flat table with the scratch along one edge and then apply pressure downward on the overhanging part (see Fig. 173).

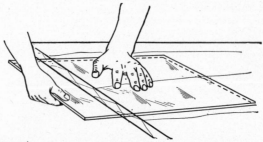

Fig. 173.—An alternate method of breaking a glass after it is scratched with a glass cutter. Align the scratch over the edge of the bench and press downward on the overhanging portion.

A good way to get a piece of glass cut to exact size is to mark out the size on paper and then place the paper under the glass. The straightedge is then easily placed to guide the cutter exactly where it should go. It is a good plan to cut glass $\frac{1}{16}$ to $\frac{1}{8}$ in. shorter and narrower than the frame into which it is to fit.

12. Replacing a Broken Window Glass

The first step in replacing a broken window glass is to remove all the old pieces of putty and glass and glaziers' points (small flat triangular or diamond-shaped pieces of metal used to hold the glass in place). Next clean the rabbet (groove) and coat it with linseed oil or thin paint, so that when the new putty is applied it will not dry out too rapidly. Then put a thin layer of putty in the rabbet and press the glass firmly in place. Fasten the glass with glazier's points driven 6 to 8 in. apart. A good way to drive glaziers' points is to use a wood chisel. Hold the chisel with the bevel flat against the glass, and tap the points with the edge of the chisel.   If desired, the edge of the chisel may be tapped lightly with a hammer.   If glazier's points are not available, use small brads instead.

13. Repairing a Defective Window Sash

Window sashes often fail at the corners.   By reinforcing a loose corner with thin corner irons, as shown in Fig. 174*A* or *B*, a window sash may be strengthened and kept in service much longer.   If an iron is installed around the outside corner (Fig. 174*B*), set it in flush, and countersink and drive the screws carefully so their heads will be flush or slightly below the surface.

Usually the first step in repairing a window is to remove it from the frame and take it to the shop where it can be worked on more

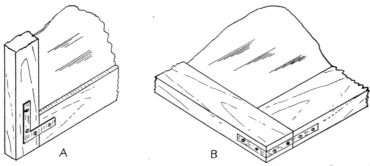

Fig. 174.—Methods of reinforcing a defective window sash by means of corner irons.

conveniently.   Windows are commonly held in place by window stops lightly nailed with small finish nails.   To remove a window, carefully remove the stop at one edge of the window.   An old wood chisel or a small pry bar may be used, but be careful not to mar the stop.   Removing only one stop will allow the bottom sash of a double-sash window to be removed.   If the upper sash is to be removed, the parting strip must be removed from one side of the window frame or jamb.   It is usually held in place with small finish nails and can easily be pried out.

Jobs and Projects

1. Examine a few buildings for painting defects or troubles, and determine the causes of the troubles, whether moisture in the wood, faulty building construction, improper thinning or mixing of the paint, poor methods of applying the paint, improper selection, etc.

2. Inspect some building, such as a poultry house, garage, barn, or house, and make a list of repair work that should be made to windows, siding, roof or other parts before painting.

3. Estimate the amount of paint and other painting materials that would be required to paint the building you inspected.

4. Make all arrangements for paints, materials, supplies, and equipment you will need for painting some small building, on your home farm, and paint it. Make a statement of all costs of doing the job, allowing yourself a reasonable wage.

5. Thoroughly clean and paint some machine or implement that has been repaired in the shop.

6. Repair some window sash in need of repairs. Remove it and take it to the shop, if feasible, replacing any broken panes, strengthening corners with corner irons or by other means, or doing such other work as may be needed.

5. Sharpening and Fitting Hand Tools

THE cutting edge of a sharp knife or similar tool is composed of a series of small microscopic teeth. Such tools cut best when used with a diagonal or sawlike motion. The size of the teeth determines whether the edge is known as coarse or fine. For certain kinds of work, coarse edges are best, and for other kinds of work, medium, fine, or very fine edges are required. Bread knives and paring knives work better when sharpened to coarse edges, whereas plane bits and wood chisels need fine edges, and razors still finer edges.

Sharp tools are the mark of a good workman. Only a poor workman or an amateur will struggle along with a dull tool rather than take time to sharpen it because the time required is soon regained in faster and better work. Furthermore, sharpening of most common tools is simple and easy. Those interested in shopwork, therefore, should strive for an early mastery of sharpening the tools commonly used.

MAJOR ACTIVITIES

1. Selecting and Using Grinders and Sharpening Stones

2. Sharpening Knives

3. Sharpening Axes and Hatchets

4. Sharpening Plane Bits and Wood Chisels

5. Sharpening Wood Scrapers

6. Sharpening Auger Bits

7. Sharpening Twist Drills

8. Sharpening Cold Chisels and Punches

9. Fitting Screw Drivers

10. Sharpening Scissors and Snips

11. Sharpening Hoes, Spades, and Shovels

143

1. Selecting and Using Grinders and Sharpening Stones

Every farm shop should have a grinder of some kind for sharpening tools and for the many odd grinding jobs that arise, such as grinding miscellaneous parts to fit and grinding hard materials that cannot well be filed or otherwise shaped. A hand-operated grinder that clamps to the workbench will serve if a power-driven grinder cannot be afforded.

Where electricity is available, a motor-driven bench grinder with the grinding wheels mounted on the ends of the motor shaft is an excellent tool for general grinding. Where cost must be kept to a minimum, a bench grinder carrying two wheels (one perhaps a mower sickle grinding wheel) and driven with a V belt from a $\frac{1}{4}$-hp. motor is a practical unit.

In selecting a grinder of any kind, it is important to get one that is sturdy and well-built, with good bearings that can be adequately lubricated and that are well protected from grit and dirt. Good, sturdy, adjustable work rests are also important.

Selecting Grinding Wheels. The secret of success with a grinder is to select good wheels that are suited to the kind of grinding to be done. Emery, a mineral found in nature, was at one time used extensively in grinding wheels. Abrasives made in the electric furnace are much superior and are now used exclusively in better grinding wheels. Wheels containing emery tend to become slick or glazed and cause much more heating while grinding than wheels made of electric-furnace abrasives. In buying grinding wheels, it is usually false economy to buy any but the best.

Choosing Grain and Grade. The coarseness or fineness of a grinding wheel is designated by a number representing the size of grains or particles used in making the wheel. A *grain* of 36, for example, means that the finest screen through which the particles will pass has 36 meshes to the linear inch. A 36 grain size is approximately $\frac{1}{36}$ in. across.

For grinding tools like plane bits and knives, a medium-fine grain of about 80 is best. For fast cutting where a highly polished surface is not necessary, a grain of about 30 may be used. The speed with which a grinder runs, as well as the grain size, affects the smoothness of grinding. The faster a wheel runs, the smoother it will grind.

By the *grade* of a wheel is meant its softness or hardness, or the ease with which the dulled particles of grit are shed or pulled from the wheel in grinding. A good grinding wheel that is suited to the work being done gradually but slowly wears away, shedding the particles from the surface as they become dull and exposing sharp particles. If the particles shed too fast, the wheel does not hold its shape well and soon wears out. On the other hand, if the particles do not shed fast enough, the surface of the wheel becomes glazed and rubs instead of cutting, thereby causing excessive heat, which would draw the temper of tools. Grinding wheels vary in grade from very hard **to**

Fig. 175.—Dressing a grinding wheel. The cutting surface of the wheel is easily and quickly trued, cleaned, and renewed by simply holding the tool firmly against the wheel while it is turning.

very soft. The grade of a grinding wheel is determined largely by the strength and amount of the binding material used to hold the particles of grit together. Soft-grade wheels are preferred for tool grinding, as there is less danger of drawing temper with such wheels.

Testing and Mounting a New Wheel. Sometimes wheels are cracked in shipment or in handling. Before mounting a new wheel, it is a good plan, therefore, to test it for hidden cracks or flaws. This may be done by striking a light blow with a small hammer. If the wheel is sound, it will ring; if there are flaws, it will give a dull thud.

A wheel is commonly fastened to the grinder spindle or shaft by a nut that clamps the wheel between two flanges or disks. Draw the nut up only moderately tight, and be sure there are washers of heavy paper, rubber, or leather between the flanges and the wheel. This is to prevent undue strain on the wheel, which might cause it to crack.

Dressing a Grinding Wheel. A grinding wheel must be dressed occasionally if it is to continue to give good service. It should be dressed if any of the following conditions exist: (1) if the surface is glazed and the dull particles need to be removed; (2) if the pores of the wheel are clogged with dirt, grease, brass, lead, etc.; (3) if the grinding surface is grooved or otherwise out of shape.

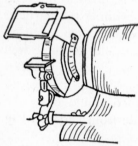

To dress a wheel, simply hold the dressing tool firmly against the wheel while it is turning at normal speed (see Fig. 175). If a wheel requires dressing too often, it is an indication that it is too soft, too hard, or otherwise unsuited to the kind of grinding being done.

Fig. 176.—A shatterproof glass shield affords protection against flying particles of grit and metal.

A dressing tool is inexpensive and should be included in the shop equipment along with the grinder.

Observing Safety Precautions in Grinding. *Wear goggles or a face shield, or equip the grinder with safety-glass shields.* Always protect the eyes and face while grinding. It is best to wear goggles or a face shield. When goggles are not worn, glass shields made of shatterproof glass and mounted on the grinder are recommended (see Fig.

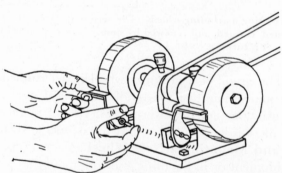

Fig. 177.—Keep the work rests adjusted so that they just clear the wheels.

176). In any event, it is advisable for the operator to stand slightly to one side so his face will not be in line with the grinding wheel and there will be the least danger from flying sparks, grit, and bits of metal.

Use wheel guards on high-speed power-driven grinders. They keep most of the particles of grit and steel from flying outward toward the operator, and they also provide a certain measure of protection in case a wheel should break while in use.

Adjust Work Rests. The work rest should be set as close to the wheel as possible without touching (see Fig. 177). This is very important. If the rest is too far from the wheel, the piece being ground may catch and wedge between the wheel and the rest and possibly chip or break the wheel or spring the grinder spindle.

Adjust Bearings. Bearings should be kept tight and well lubricated. Loose bearings not only allow vibration and cause inferior grinding, but also introduce an element of danger, especially on high-speed grinders.

Selecting and Using Oilstones. An oilstone for sharpening keen-edged tools is indispensable in a good shop. Two kinds of stones are available, natural stones and those made of artificial or electric-furnace abrasives. The artificial abrasive stones are generally preferred as they are more uniform. A combination stone with one side made of coarse- or medium-grain abrasive and the other made of fine-grain abrasive is recommended. The coarse side is used for faster cutting during the first part of the sharpening, and the fine side for finishing to a keen smooth edge.

Use a light oil, such as kerosene and motor oil mixed in equal parts, on an oilstone to float off the small cuttings of steel and to prevent the surface from becoming clogged with dirt. If a stone is used dry, it soon becomes slick and will not cut fast. A dirty stone may be cleaned by placing it in a pan and heating it in an oven or over a fire, or by washing it in gasoline or kerosene.

Points on Selecting and Using Grinders and Oilstones

1. Select a good-quality wheel that is suited to the kind of grinding to be done.
2. Well-built sturdy grinders with good bearings are worth the small additional cost over cheap grinders.
3. With a hand grinder, turn with a moderately fast, steady speed.
4. Keep the work rest adjusted as close to the wheel as possible without rubbing.
5. Hold the tool against the wheel with a light to medium pressure.
6. Move the tool from side to side while grinding to distribute the wear evenly on the wheel and prevent grooving and also to ensure even grinding of the tool and prevent overheating.
7. Keep the bearings well lubricated.
8. Keep the bearings tight.

9. Dress the wheel whenever it becomes dull (surface glazed), whenever the pores become clogged, or whenever it becomes worn out of shape.

10. It is best to wear goggles or a face shield when grinding or to have the grinder equipped with safety-glass eye shields. The operator should stand slightly to one side with his face out of line with the wheel to lessen the danger from flying sparks, grit, and bits of metal.

11. A wheel may be tested for hidden flaws by striking lightly with a small hammer. A clear ring indicates a sound wheel; a dull thud, a wheel that is cracked.

12. The nut that holds the wheel in place should be drawn up only moderately tight.

13. Use washers of heavy paper or similar material between the mounting flanges and the wheel.

14. An oilstone for sharpening keen-edged tools is indispensable in any good shop. A combination stone, with one side of coarse or medium grit and the other of fine grit, is generally preferred.

15. Always use a light oil, such as kerosene and motor oil mixed in equal parts, on an oilstone.

2. Sharpening Knives

The general method of sharpening keen-edged tools is first to produce a coarse edge by grinding with a grinding wheel, or by

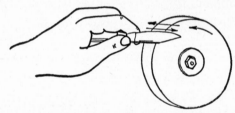

Fig. 178.—In grinding knives, turn the wheel against the cutting edge. Hold the knife at an angle to the wheel, use moderate pressure, and move the knife slowly from side to side.

whetting on the coarse side of an oilstone, and then to finish by whetting on the fine side of an oilstone. If a very fine edge is needed, the process is carried a step further and the tool is stropped on leather.

Grinding a Knife. If the knife blade is nicked or if it is extremely dull, first grind it on a medium or fine grinding wheel. Place the blade flat against the wheel, raising the back edge just enough to barely grind on the cutting edge. Turn the wheel at a moderate speed and *against the cutting edge*—not away from it (see Fig. 178). Be careful not to raise the back edge too much, as this might cause the wheel to gouge or grind deeply into the cutting edge. Move the blade slowly back and forth at an angle across the wheel.

Be careful not to overheat the blade. Use only moderate pressure, and dip the blade in water frequently. Watch to see that it is being ground to the desired shape.

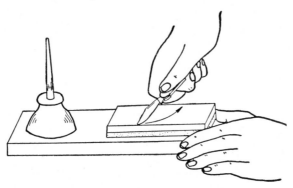

Fig. 179.—After grinding, a knife is whetted on an oilstone. If the blade is not nicked and is only moderately dull, it may be sharpened altogether on an oilstone.

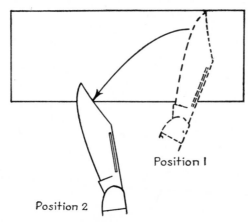

Position 1

Position 2

Fig. 180.—Stroke 1 in whetting a knife. Raise the back of the blade slightly and draw it diagonally across the stone.

After grinding a knife, whet it on an oilstone, using the coarse side first unless a very smooth job of grinding was done.

Whetting a Knife. If a knife is not nicked and is only moderately dull, it may be sharpened on an oilstone without first grinding it. Although it may be sharpened entirely on the fine side of the stone, it is generally faster first to whet on the coarse side and then to finish on the fine side.

To keep a stone from sliding around when in use, it may be mounted on a small board about 1 by 4 by 10 in. long, using small strips tacked to the board to hold the stone in place.

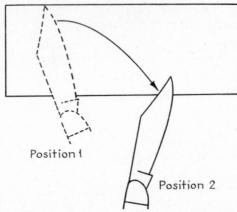

Position 1

Position 2

Fig. 181.—Stroke 2 in whetting a knife. With a little practice, strokes 1 and 2 will blend together into a continuous motion.

Stroke 1. To whet a knife, put a few drops of light oil, such as motor oil mixed with kerosene, on the stone and place the blade flat as shown in position 1, Fig. 180. Raise the back of the blade just enough to make the cutting edge touch the stone. Then draw the knife across the stone diagonally into position 2. Be sure to keep the heel of the blade down against the stone at the beginning of the stroke, and use moderate to heavy pressure.

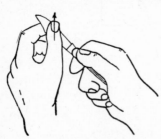

Fig. 182.—Feeling to find out if the knife has a wire edge. Note the motion of the thumb.

Stroke 2. Next, turn the blade over into position 1, Fig. 181. Tilt the blade by raising the back edge slightly, and draw it diagonally over the stone into position 2. Repeat these two strokes several times. With practice, they will blend into one continuous motion.

If the whetting is being done on the coarse side of the oilstone, continue until a fine burr or wire edge has been produced. A wire edge is a thin rough edge that can be felt with the thumb or finger (see Fig. 182), or seen in a good light (see Fig. 186). Also, a wire edge will catch on a cloth while a smooth edge will not.

After a wire edge is produced, turn the stone over and whet on the fine side, using light pressure and the same diagonal drawing strokes.

If the whetting is done properly, the wire edge will quickly disappear and the knife will be really sharp.

Stropping a Knife.   To produce an exceptionally fine keen edge, strop the knife on a piece of smooth leather after it has been whetted

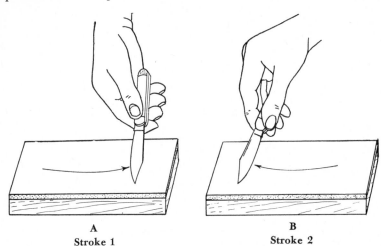

A	B
Stroke 1	Stroke 2

Fig. 183.—To produce a fine, keen edge, strop the knife on leather after whetting on the oilstone.   After stroke 1 as at A, roll the knife over on the back of the blade into position for stroke 2.   With practice, strokes 1 and 2 will blend together into a continuous motion.

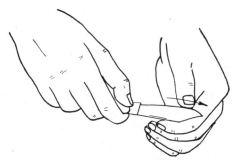

Fig. 184.—An excellent way to test a knife for sharpness.   Draw the ball of the thumb lightly along the edge.   Do not press.   A sharp edge will "take hold" or pull on the tough cuticle.   A dull tool feels smooth and will not "take hold."

on the fine side of the oilstone.   Be sure to use pulling strokes with the cutting edge trailing—not leading (see Fig. 183).   If a strop is not available, the knife may be stropped on a piece of smooth wood, on a shoe sole, or even on the palm of the hand.   A piece of smooth leather glued to a block of wood about 2 in. wide by 6 in. long makes a good strop.

Testing a Knife for Sharpness. Probably the best way to tell whether or not a knife is sharp—the way used by most good mechanics —is to feel it with the thumb. Hold the blade, cutting edge up, in the open hand, and with *very light pressure* move the thumb lengthwise along the edge (see Fig. 184). Do not press against the edge. If the knife "takes hold" or pulls on the calloused skin of the thumb, it is sharp; if it does not take hold, or feels slick and smooth, it is dull.

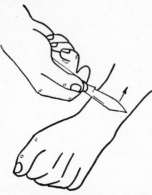

Another test is the shaving test (see Fig. 185). It is not too much to expect a knife to shave. With a little practice, one should be able to put a shaving edge on a knife.

Fig. 185.—**A sharp tool will easily shave.**

Still another way to test the knife for sharpness is to hold it up to the light and rock the blade back and forth (see Fig. 186). If the knife is dull, the edge of the blade can be seen. It will reflect the light and appear as a narrow shiny surface. If the knife is sharp, the edge cannot be seen.

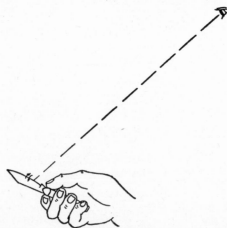

Fig. 186.—**Another test for sharpness. Hold the tool up to the light and move it back and forth. The edge of a sharp tool cannot be seen. A dull edge appears as a narrow, bright, shiny line that reflects light.**

Sharpening Butcher Knives. A butcher knife is sharpened in the manner described in the preceding paragraphs. It is ground if necessary and then whetted on an oilstone. The edge of a butcher knife can be kept in good condition by using a sharpening steel on it occasionally.   The steel does not really sharpen the knife, but simply

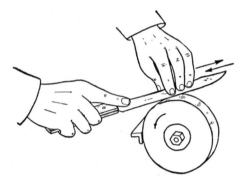

Fig. 187.—A butcher knife may be sharpened in the same way as a pocket knife. Grind it first if necessary and then whet it on an oilstone.

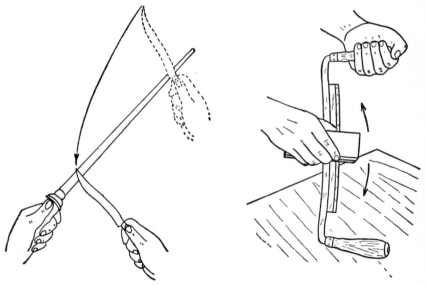

Fig. 188.   Fig. 189.

Fig. 188.—Using the sharpening steel occasionally will help keep the butcher knife in good condition.

Fig. 189.—A drawknife is sharpened in the same manner as a wood chisel or plane bit, except that, in whetting, the stone may be rubbed over the edge of the tool instead of moving the tool back and forth over the stone.

removes bits of fat and tissue and straightens the microscopic teeth that form the edge.

To use the sharpening steel, hold the steel in the left hand with the point up (see Fig. 188).   Tilt the blade slightly so that the cutting edge is in contact with the steel, and beginning at the heel of the blade, draw it down quickly with a diagonal sweeping stroke, cutting edge foremost.   Use only light pressure.   In a similar manner, stroke the other side of the blade on the other side of the steel, and continue, stroking first one side of the blade and then the other.

Sharpening Drawknives.   A drawknife is sharpened in the same manner as other keen-edged knives, except that, in whetting, the stone may be rubbed over the edge, instead of moving the tool back and forth over the stone (see Fig. 189).

3. Sharpening Axes and Hatchets

To sharpen an ax or a hatchet, first grind it on a medium or fine grinding wheel if the edge is blunt or if it is nicked.   After grinding,

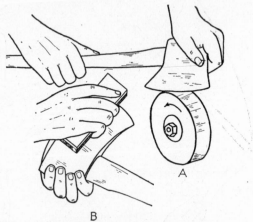

Fig. 190.—Sharpening an ax.   After grinding, a smooth keen edge may be produced by whetting with an oilstone.

whet with an oilstone to produce a keen smooth edge (see Fig. 190). The stone may be rubbed on the tool, or if more convenient, the stone may be placed on a bench and the tool rubbed on the stone.   If the tool is only slightly dull, it may be sharpened altogether with an oilstone.   In grinding an ax for chopping, the blade should be made somewhat thinner than an ax for splitting.   Be careful not to overheat the tool when grinding.

4. Sharpening Plane Bits and Wood Chisels

When a plane bit or wood chisel becomes dull, decide first whether it is necessary to grind it.   If a tool is properly used and cared for, it can be resharpened many times on the oilstone before it will need

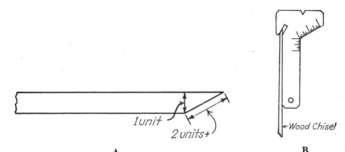

A   B

Fig. 191.—The length of bevel on a plane bit or wood chisel should be a little more than twice the thickness of blade for general-purpose work. B. Checking the angle of bevel with a homemade gage (see also Fig. 217).

regrinding.   The tool should be ground if the edge is unusually dull, nicked, or if a slight bevel has been formed on the flat side by careless whetting.   The tool should be ground, also, if it does not have the desired angle of bevel, or if the cutting edge is not square with the sides.

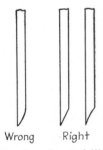

Wrong   Right

Fig. 192.—An edge tool like a plane bit or wood chisel should be ground with a straight or a concave bevel—not with a convex bevel.

Fig. 193.—A plane for general-purpose work should be ground with the corners slightly rounded.

Angle and Shape of Cutting Edge.   For general work, grind plane bits and wood chisels at an angle of about 25 to 30 deg. (see Fig. 191).   For softwoods, the angle may be somewhat less, and for hardwoods, somewhat more.   The smaller the angle, the easier the tool will cut, but the sooner it will become dull.

The bevel should be ground straight or slightly concave—not rounded or convex (see Fig. 192). For general-purpose planes, it is desirable to grind the corners of the blade slightly rounded as shown in Fig. 193. This is to feather the sides of the shaving and avoid scratches or grooves in the surface being planed.

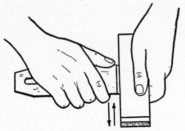

Some mechanics prefer to straighten the edge of a plane bit or wood chisel by rubbing it on the edge of an oilstone before grinding (see Fig. 194).

Fig. 194.—The edge of a plane bit or wood chisel may be straightened or squared on the edge of an oilstone.

Grinding Plane Bits and Wood Chisels.

In grinding a plane bit or wood chisel, the following points are important:

1. Hold the tool against the wheel in a manner that will produce a smooth, even bevel of the desired angle.
2. Adjust the work rest, if possible, so that the tool, when held firmly against the rest, will come in contact with the wheel at the desired angle (see Fig. 195).

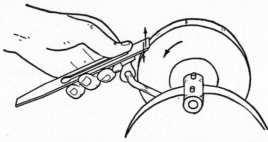

Fig. 195.—A good method of holding the tool while grinding. Adjust the tool rest to give the desired angle of bevel, and hold the tool firmly against the rest. Move the tool back and forth sideways across the face of the wheel.

3. Grasp the tool so that the first finger will come against the work rest, and thus enable replacing the tool in proper position after it has been removed for inspection or dipping into water.
4. Turn the wheel toward the cutting edge—not away from it.
5. Turn at a moderately fast, steady speed. Do not turn so fast that the gears whine or the grinder vibrates.
6. Hold the tool against the wheel with a medium, yet firm, pressure.
7. Move the tool from side to side across the face of the wheel.
8. Dip the tool into water frequently to prevent overheating.
9. Inspect the work frequently to see that the tool is being ground to the proper shape and angle of bevel.

A quick, easy way to check the angle of bevel is to use a simple gage cut from sheet metal (see Fig. 191*B*). (A combination gage for checking the grinding of plane bits and wood chisels, twist drills, and cold chisels is illustrated in Fig. 217.) With a little practice, the angle of bevel can be checked by eye, by remembering that the length of bevel should be a little more than twice the thickness of the blade (see Fig. 191*A*). A rule can be used, of course, for measuring the length of bevel and thickness of the blade.

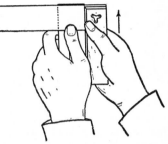

Fig. 196.—Checking an edge tool for squareness of grinding. **Place the square on top of the blade but be careful not to touch the sharp edge against the square.**

To check to see that the cutting edge is square with the edges of the tool, use a try square (see Fig. 196). Place the square on top of the blade, allowing the cutting end to project slightly beyond the square. Be careful not to touch the cutting edge against the square.

It is sometimes easier for beginners to hold the tool against the flat side of the wheel rather than against the usual curved grinding

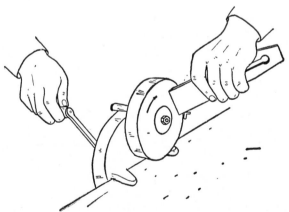

Fig. 197.—**An edge tool may be ground on the flat side of the grinding wheel if desired. This produces a smoother job and is usually more easily done than grinding on the curved surface. More care must be used, however, to prevent overheating.**

surface (see Fig. 197). If the grinding is done on the side of the wheel, a little more care will be required to prevent overheating.

Continue grinding until the dull edge is removed, all nicks are removed, the edge is straight and square, and the bevel is of the desired angle. Remove the burr, or wire edge, which is left from grinding, by whetting on an oilstone.

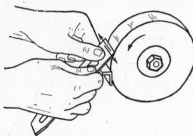

Fig. 198.—Both hands may be used to hold the tool when a power grinder is used. Grasp the tool so that when the first finger of the right hand is against the tool rest the desired angle of bevel will be produced.

Whetting a Plane Bit or Wood Chisel. If the tool has a good bevel on it and is not nicked, but is simply dull, it can be quickly sharpened by whetting on an oilstone without first grinding it. Place a few drops of oil on the coarse side of the stone, and whet the bevel of the tool until a slight burr, or wire edge, is formed.

In whetting, place the tool on the stone with the cutting edge making an oblique angle with the edge of the stone (see Fig. 199*A*).

Keep the bevel of the tool flat on the stone, or with the heel raised

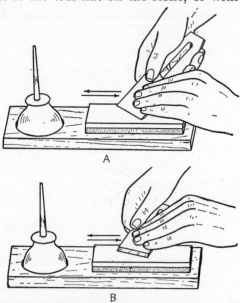

A

B

Fig. 199.—If a plane bit or wood chisel is not nicked and is only moderately dull, it may be sharpened on the oilstone without first grinding. Whet as at A, using the coarse side of the stone. Then whet alternately as at A and B, using the fine side. When whetting in position B, keep the tool perfectly flat against the stone. Note that the cutting edge makes an oblique angle with the edge of the stone in both positions.

only very slightly. Push the tool vigorously forward and backward, bearing down hard on the forward stroke but relieving the pressure on the backstroke.   Be particularly careful to move the hands parallel to the surface of the stone, and not to allow a dipping or scooping motion, as this would round the end of the tool as well as hollow out the stone.   Use full-length strokes.

After a slight wire edge has been produced by whetting on the coarse side of the stone, or by grinding on a grinding wheel, remove the wire edge by whetting on the fine side of the stone.   Place the tool *perfectly flat* on the stone, with the cutting edge making an oblique angle with the edge of the stone, and push it forward (see Fig. 199*B*). A few strokes will remove the wire edge, or turn it from the flat side of the tool to the beveled side.   If the wire edge turns, then turn the tool over to the position shown in *A*, Fig. 199, and whet *lightly* on the beveled edge.   Then reverse the tool, and whet again on the flat side.   In this operation, two points are very important:

1. Keep the tool perfectly flat against the stone when whetting on the flat side.
2. Use very light pressure when whetting the beveled side.

If the whetting is properly done, the wire edge will quickly become smaller and smaller and, after a few reversals, practically disappear. The tool will then be sharp.   Some time may be saved in removing the wire edge by drawing it through a piece of wood.

If the tool is not held perfectly flat when whetting the flat side, a small bevel may be produced on the flat side, and it would then be impossible to put the edge in good condition without regrinding it.   If too much pressure is used while whetting the beveled side, the wire edge may be increased instead of decreased.

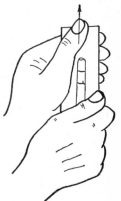

In alternately whetting the flat and the beveled sides, make sure that the wire edge is actually turned back and forth.   For example, if the tool is being whetted on the flat side, be sure the wire edge is actually turned from the flat to the beveled side before reversing the tool for whetting on the beveled side.   A wire edge is easily detected by feeling with the thumb or

Fig. 200.—Feeling to find out if there is a wire edge on a plane bit.   Wire edges are removed by light whetting on an oilstone.

finger.   Place the thumb against the side of the tool, and move it lengthwise of the tool out over the cutting edge (see Fig. 200).

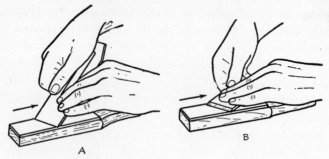

Fig. 201.—A very keen edge may be produced on a tool by finishing on a leather strop. Using draw strokes, strop first on the beveled side and then on the flat side:

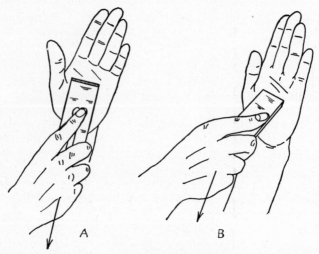

Fig. 202.—If a leather strop is not available, an edge tool may be stropped on the palm of the hand.

Fig. 203.—Testing a plane bit for sharpness. Draw the ball of the thumb lightly along the edge, but do not press against it. A sharp tool will "take hold" and pull on the tough cuticle, while a dull tool feels smooth and will not "take hold."

Stropping a Plane Bit or Wood Chisel. To produce an exceptionally fine, keen edge on a tool, strop it on a piece of smooth leather after it has been whetted on the fine side of the oilstone (see Fig. 201). A few drawing or pulling strokes with the cutting edge trailing, not leading, first on the beveled side and then on the other, is all that is required. If a piece of leather is not available, the tool

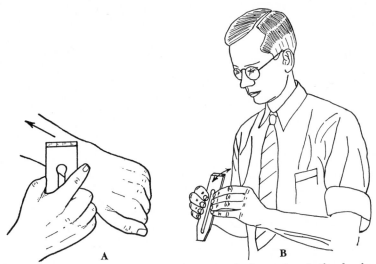

A   **B**

Fig. 204.—Other methods of testing sharpness of edge tools. **A,** the shaving test; **B,** light-reflection test. A sharp edge cannot be seen; a dull edge reflects light and appears as a narrow, shiny surface.

may be stropped on a piece of smooth wood, or even on the palm of the hand (see Fig. 202).

5. Sharpening Wood Scrapers

Filing and Whetting the Edge. The first step in sharpening a hand wood scraper is to drawfile the edges square and straight, rounding the corners slightly (see Fig. 205). Use a rather fine file, such as a smooth mill file.

Next, whet the scraper on an oilstone, first on the edge and then on the flat sides to make the arrises smooth and sharp (see Fig. 206). (An arris is a sharp edge formed by the meeting of two surfaces.)

Forming the Scraping Burr (Burnishing). After filing and whetting, form a small scraping burr by stroking the scraper with a burnisher, or other piece of smooth hard round steel, like a nail set.

Place the burnisher flat against the side of the scraper, and draw it along over the edge (see Fig. 207). Use moderate pressure, and

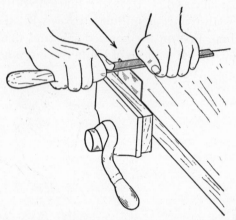

Fig. 205.—The first step in sharpening a very dull scraper is filing it. The edge should be kept square and straight. The corners may be rounded slightly.

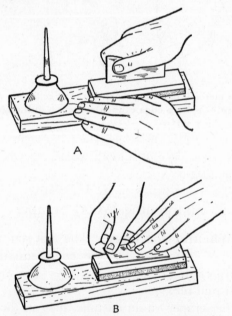

Fig. 206.—After filing, whet the scraper alternately on edge and on the flat side. Use the fine side of the oilstone.

repeat the stroke two or three times. Turn the scraper over, and burnish the other side in the same manner. A drop or two of oil on the burnisher makes it work better.

After burnishing the flat sides of the scraper, burnish the edge. Place the scraper in a vise or hold it firmly in an upright position on the bench.   Hold the burnisher square with the edge, and draw it along with the handle end slightly ahead (see Fig. 207*B* or *C*).   Use

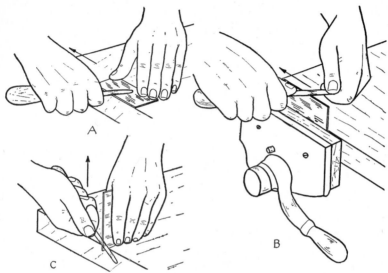

Fig. 207.—Forming the scraping burr on a scraper.   Draw the burnisher over the edge three or four times, use moderate pressure, and keep the burnisher flat against the scraper as at **A**.   Then turn the burr with three or four strokes as at **B** or **C** (see Fig. 208).   A little oil on the burnisher makes it work better.

only moderate pressure.   Heavy pressure is not required.   Be sure to keep the burnisher square with the scraper on the first stroke. Repeat the stroke two or three times, each time tilting the burnisher a little, until the last stroke is made at an angle of about 85 deg.

Fig. 208.—In turning the burrs on a scraper, hold the burnisher at 90 deg. to the scraper on the first stroke, and then gradually tilt it on succeeding strokes until it finally makes an angle of about 85 deg. on the last stroke.

instead of 90 deg.   Then turn the scraper around and burnish from the other side in the same manner.   The scraping burrs are then as shown in *B*, Figs. 208 and 209.

The edge of a scraper may be renewed several times with the burnisher before it will need refiling and whetting.

Sharpening a Beveled-edge Scraper. The blade of a cabinet scraper is usually sharpened to a beveled edge, instead of a square edge. Some mechanics prefer hand wood scrapers sharpened with beveled edges also. When a scraper is sharpened to a beveled edge, the angle of bevel is usually about 45 deg., or about double that of a plane bit.

The first step in sharpening a beveled scraper is to remove the old burr with a file and then to file the edge to the desired bevel. Then whet it on an oilstone, first on the beveled edge and then on the flat side, to remove the wire edge left by filing.

Fig. 209.—A. Shape of scraping burr after burnisher has been used on flat sides of scraper. B. Shape of burr after burnisher has been used on edge of scraper.

Next, burnish the edge in much the same manner as for a square-edged scraper. Burnish first on the flat side. Then, burnish the beveled edge, with the burnisher flat against the bevel on the first stroke, and at a little greater angle on each succeeding stroke, until finally it makes an angle of about 75 deg. with the side of the scraper on the last stroke (see Fig. 210). In case the burr is turned too

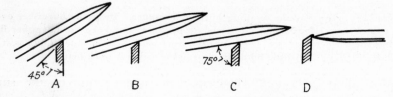

Fig. 210.—Steps in turning a burr on a beveled-edge scraper. In case the burr is turned too much, it may be raised somewhat, as at D.

much, it may be raised somewhat by drawing the point of the burnisher along under the burr as shown in Fig. 210D.

6. Sharpening Auger Bits

A small file, known as an *auger-bit file*, is best for sharpening an auger bit. A small triangular or three-cornered file, or a small flat file, may be used, if the workman is careful.

The following points are important in sharpening an auger bit:

1. File on the inside of the spurs or scoring nibs.
2. File on the top side of the cutting lips (the side next to the shank).
3. Retain the original bevel or suction on the cutting lips, and remove about the same amount of material from each one.

To file the spurs or scoring nibs, hold the bit firmly against the edge of a bench or other support (see Fig. 211).   File on the inside of the nibs.   If they are filed on the outside, they will cut a circle that is too small and the bit will not feed into the wood.   In case the nibs have been bent outward, or burrs have been formed on the outside, they may be filed lightly on the outside, care being taken not to under-cut or bevel the nibs on the outside.

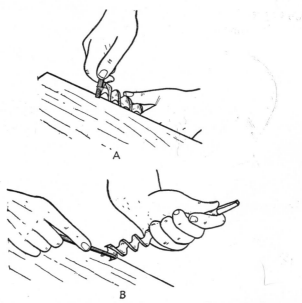

Fig. 211.—Two points are important in sharpening auger bits: A, file the spurs on the inside only. B, file the cutting lips on the top side (next to shank).

To file the cutting lips, rest the bit firmly against the bench top as shown in Fig. 211B.   Be careful to retain the original angle of bevel or suction and to file both lips about the same amount.

7. Sharpening Twist Drills

Sharpening of twist drills is one of the most important jobs in the farm shop.   Drilling equipment enables many repair jobs to be done that otherwise would be impossible.   Yet drilling equipment is practically worthless without sharp drill bits, and if drill bits are used much, they will require frequent resharpening.   Most drilling diffi-culties and most drill breakage can be traced to faulty sharpening.

Proper Shape of Drill Point. Before attempting to grind a twist drill, one should have a clear picture in mind of the shape of a properly sharpened drill. There are two main requirements:

1. The cutting lips must have clearance or be ground off behind the cutting edge to allow the drill to bite into the metal. The proper clearance is about 12 deg.
2. The cutting edges should be exactly the same length and make the same angle with the central axis of the drill, the proper angle being about 59 deg., or practically two-thirds of a right angle. (The axis is an imaginary center line running lengthwise of the drill from one end to the other.)

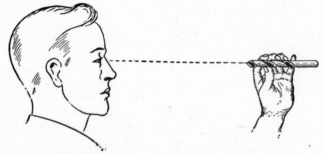

Fig. 212.—To judge lip clearance of a drill, hold it up about level and look straight at the end of it (see Fig. 213).

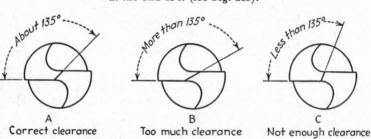

Fig. 213.—Judging lip clearance. With the cutting edges horizontal, the short line across the end of the drill should be about halfway between vertical and horizontal, as at A. If it is much nearer horizontal than vertical, as at B, the lips have too much clearance. If the short line is near vertical, as at C, the lips have insufficient clearance.

If the cutting lips do not have clearance, the drill cannot bite into the metal, and if there is too much clearance, the drill may take too deep a bite or gouge.

If both cutting edges are the same length and make the same angle with the central axis, the point of the drill will be centered. If the point is not centered, the drill will make an oversize hole, and one lip will do more than half of the cutting. /

If the cutting lips are not ground at the proper angle of 59 deg. with the axis, the cutting edges will be slightly curved instead of straight, and the drill will be too pointed or too blunt.

Judging Lip Clearance. To judge lip clearance, hold the drill up about level and look straight at the end of it (see Fig. 212). Rotate

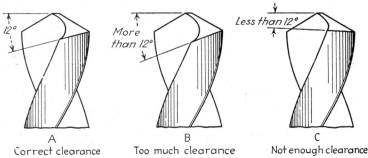

A	B	C
Correct clearance	Too much clearance	Not enough clearance

Fig. 214.—Judging the lip clearance at the outer edge or periphery of the drill.

the drill until the cutting edges are about horizontal or level (see Fig. 213). The short line across the end should then be about half-way between the vertical and horizontal (see Fig. 213*A*). If it is much nearer horizontal than vertical (Fig. 213*B*), the lips have too much clearance. If the short line is much nearer vertical than horizontal (Fig. 213*C*), the lips do not have enough clearance.

The amount of lip clearance, particularly at the outer edge or periphery of the drill, may be judged by simply looking at the outer edge of the cutting end (see Fig. 214), or by standing the drill vertically, point down, against the bench top beside a rule (see Fig. 215) and turning the drill slowly. The heel of the cutting lip should register slightly higher on

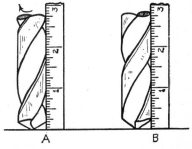

Fig. 215.—Lip clearance at the outer edge may also be judged by standing a drill beside a rule and rotating it. The heel of the cutting end should register higher on the rule than the front cutting edge.

the rule than the front edge. Sometimes drills are ground with proper lip clearance at the outer edge or periphery, yet with improper clearance at the center. Therefore, clearance should not be judged by looking at the outside edge alone.

Checking Length and Angle of Cutting Edges. The length of the cutting edges and the angle between the cutting edges and the central axis of the drill can best be checked by a gage (see Fig. 216). Such a gage can be made easily from sheet metal (see Fig. 217). With some practice, however, one can become reasonably proficient at checking the lengths and angle of the cutting edges by eye. Lengths of cutting edges can be measured, of course, with a rule.

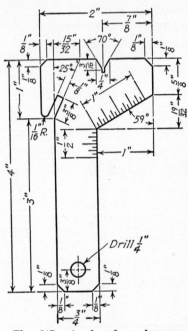

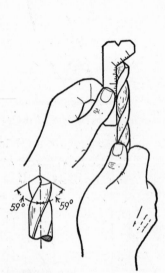

Fig. 216.—Checking the cutting lips for (1) length and (2) angle with the central axis of the drill. Both lips should be the same length and make an angle of about 59 deg. with the central axis. (The notch in the end of the gage is for checking the angle of bevels on a cold chisel.)

Fig. 217.—A plan for a homemade grinding gage for checking twist drills, cold chisels, and plane bits and wood chisels.

Placing the Drill in Position for Grinding. The work rest on the grinder should be at about the same level as the wheel shaft. Adjust it if necessary; then place the drill on the work rest with

1. The drill about level (both ends about the same height).
2. The axis of the drill making an angle of about 59 deg. with the cutting surface of the wheel (see Fig. 218). (It is a good plan to file a line on the work rest at the 59-deg. angle.)
3. One cutting edge horizontal and against the grinding wheel. (Rotate the drill if necessary, but keep the drill about level.)

Grinding the Drill. With the drill in proper position and the wheel turning at normal speed, push the cutting end *slowly*, yet firmly, against the wheel, and then slowly elevate the point by gradually lowering the other end. Keep the point pushed against the wheel. Make slow, deliberate strokes—not fast, quick thrusts. Do not attempt to roll or rotate the drill while it is being ground—at least, not until you have become rather expert—but simply keep the point pushed against the wheel while the other end is slowly lowered.

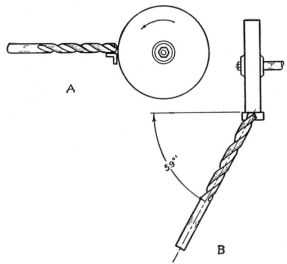

Fig. 218.—Side view (A) and top view (B) of a drill in proper position for grinding. The end of the drill is placed flat on the work rest, with one cutting lip horizontal and against the grinding wheel, and with the axis of the drill making an angle of about 59 deg. with the surface of the wheel.

When one lip is ground, rotate the drill half a round and grind the other lip in identically the same manner. Dip the drill in water frequently to prevent overheating and drawing the temper. Inspect the work every few grinding strokes, to make sure the drill is being ground properly. Check both for clearance and for length and angle of cutting edges.

If more clearance is desired, grind more on the last part of the stroke by using more pressure or slower motion on this last part. The heel of the cutting lip is thus ground away more. If less clearance is needed, use less pressure or a slightly faster motion toward the end of the grinding stroke. If there is a tendency to get too much clearance, possibly the drill is pointed too high up on the wheel at the beginning of the stroke.

If it is found that the two lips are not exactly the same length, then grind a little more on the short one.

Grinding may be done on either a hand-driven or a power-driven grinder. With a hand grinder, hold the drill in the left hand and

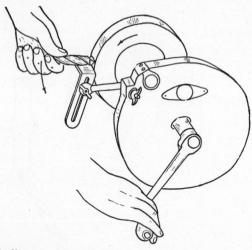

Fig. 219.—Grinding a twist drill with a hand grinder. Force the cutting end of the drill slowly but firmly against the revolving wheel. Gradually raise the cutting end by lowering the other end. Keep point of drill pushed against the wheel.

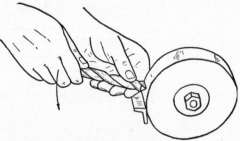

Fig. 220.—With a power-driven wheel the end of the drill may be supported by the first finger of the left hand, which is placed on top of the work rest.

turn with the right (see Fig. 219). Use a moderate, steady speed. Where a power-driven grinder is used and both hands are free to hold the drill, the first finger of the left hand may be placed on the work rest and the drill rested on the finger (see Fig. 220). In this case, be careful not to let the finger come into contact with the grinding wheel.

Instead of grinding on the curved face of the wheel, some mechanics prefer to grind the drill against the flat side. This is all right, but when the side of the wheel becomes grooved or out of shape after considerable use, it will be more difficult to dress than the curved grinding surface.

Grinding a Drill for Soft or Hard Materials.   For drilling brass or other soft metals, or for drilling hard materials where heavy pressure is required, a drill ground with the usual shape has a tendency to gouge.   To prevent this, the cutting lips may be made blunt by grinding narrow flat surfaces on the front edges, the surfaces being

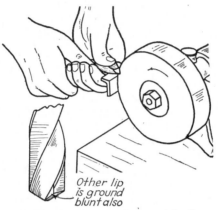

Other lip
is ground
blunt also

Fig. 221.—For drilling extremely hard or soft materials, the cutting edges may be ground blunt to give a scraping action.

parallel to the axis of the drill (see Fig. 221).   The drill then has more of a scraping action, and there is less tendency to gouge.

8. Sharpening Cold Chisels and Punches

Grinding Cold Chisels. For general cutting, a cold chisel should be ground with the bevels on the cutting edge making an angle of about 70 deg. with each other.   For some work, such as cutting thin metal or soft metal, a keener edge may be ground.

To grind a cold chisel, grasp it firmly at such a point that when the first finger comes up against the work rest the chisel will bear against the wheel at the desired angle (see Fig. 222).   A little experimenting may be required to determine the exact place to hold the chisel.   Once it is found, then keep this hold on the chisel whenever it is removed for inspection or dipping into water.   The chisel can then be easily replaced on the grinding wheel at the desired angle.

Hold the end of the chisel firmly against the grinding wheel, and swing it from side to side, pivoting it over the work rest (see Fig. 222). This grinds the corners of the chisel back a little and gives a slightly curved cutting edge. With the corners thus rounded, there is less danger of breaking them off.

When one side of the bevel is ground smooth and even, turn the chisel over and grind the other side in the same manner. As the grinding proceeds, check to see that the desired angle of bevel is being produced. A good way is to use a sheet-metal gage (see Fig. 223 and

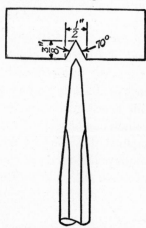

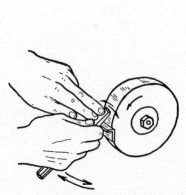

Fig. 222.—Grinding a cold chisel. Hold the chisel firmly against the work rest, with the first finger of the right hand touching the underside of the rest. Press the cutting edge of the chisel against the wheel with the fingers of the left hand and swing the handle of the chisel back and forth with a wrist action of the right hand.

Fig. 223.—Checking the angle of grinding on a cold chisel. For general work, the cutting end should be ground to an angle of about 70 deg. A gage for checking the angle may be made of sheet metal (see also Fig. 217).

Fig. 217). Also, dip the tool into water frequently to prevent over-heating and drawing the temper.

Sometimes, there is a tendency to grind the bevel longer on one edge of the blade than on the other. This is generally due to the fact that when the chisel was made the two opposite flat sides of the stock were not drawn out straight to form the cutting wedge. To avoid or counteract this tendency to grind the bevel longer on one edge, rotate the chisel very slightly about its long axis as it rests against the work rest and the wheel, thus raising one edge of the chisel off the wheel a little. Then hold it firmly in this position as it is swung from side to side in grinding.

Grinding Center Punches. To grind a center punch, place the end flat on the work rest with the axis of the punch making an angle of about 30 deg. with the grinding surface of the wheel. The point will then be ground at an angle of about 60 deg., which is about right. Push the point against the turning wheel, and roll the punch slowly (see Fig. 224). Move the punch back and forth across the work rest to distribute the wear evenly on the grinding wheel.

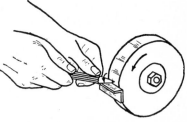

Fig. 224.—Grinding a center punch. Hold the punch against the wheel at the desired angle and roll it slowly. Also, move it from side to side in order to distribute wear on the wheel.

Grinding Pin Punches. If the end of a pin punch has been broken, it may be restored to shape by grinding; or if a small punch is needed and one is not available, possibly a larger punch can be ground to size and made to serve. A punch may be held against the grinding wheel and manipulated in various ways. The main points to observe are not to get the grinding wheel grooved or out of shape and to grind the punch as smoothly and evenly as possible.

9. Fitting Screw Drivers

A screw driver should be ground or filed to a very blunt end. The two flat surfaces should be straight and parallel near the tip, or even

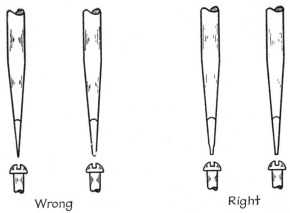

Wrong   Right

Fig. 225.—Right and wrong shapes of screw-driver ends.

slightly concave (see Fig. 225). The end should be square with the broad surfaces, and of a thickness a little less than the width of the screw slot it is to fit.   It should fit the screw slot snugly.   If the

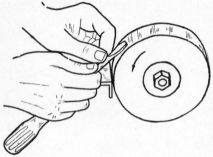

end is rounded or sharpened to an edge like a knife, it will easily slip from the slot and mar the screwhead.

To grind a screw driver, hold it on a grinding wheel as shown in Fig. 226.   Move the blade endwise back and forth a little to grind the face a short distance back from the end.   Turn the screw driver over, and grind the other face in the same manner.   Remove the tool and inspect it frequently to make sure

Fig. 226.—**Grinding a screw driver. Be careful to grind it to the proper shape.**

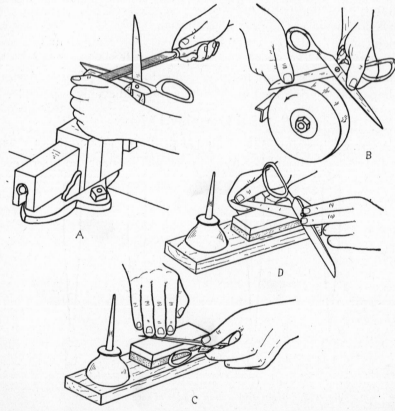

Fig. 227.—**Scissors may be sharpened by filing or grinding the beveled edges and then finishing on the oilstone, alternately whetting the flat side and the beveled edge of each blade.**

that it is being ground to the desired shape.   Also, dip the tool in water frequently to prevent overheating and drawing the temper.

10. Sharpening Scissors and Snips

To sharpen a pair of scissors or snips, grind or file the beveled edges carefully at the original angle, and then finish by whetting on an oilstone.   Some scissors are too hard to file and can be sharpened only by grinding.   In grinding, hold the blade at an angle across the grinding face of the wheel, with the back of the blade tilted just enough to grind at the desired bevel (see Fig. 227).   Move the blade back and forth slowly across the wheel.   If the scissors are not too dull, the beveled edges may be renewed by whetting on the coarse side of an oilstone.   After the beveled edges are renewed (by filing, grinding, or whetting on the coarse side of an oilstone), finish the sharpening by whetting on the fine side of the oilstone.   Be careful to keep the blades perfectly flat when whetting the flat side and at the correct angle when whetting the beveled edge.

11. Sharpening Hoes, Spades, and Shovels

Such tools as hoes, spades, and shovels are easily sharpened by straight filing or by drawfiling (pushing the file sidewise).   Wherever possible, clamp the tool in a vise (see Fig. 228).   If a vise is not at hand, the tool can frequently be held satisfactorily by cramping the handle against a box, a tree, or a fence post.   Be careful to maintain

Fig.   228.—Hoes, spades, and shovels are easily sharpened by plain filing or by draw filing (pushing the file sidewise).

the original angle of bevel, and be careful not to let the hands slip and come in contact with the cutting edge of the tool.

If the end of a spade or shovel has been broken or worn badly out of shape, it can often be restored by cutting the end off straight or to the desired shape with a cold chisel and hammer, and then dressing with a file or grinding wheel.

12. Sharpening Saws

With some study and practice, anyone having a reasonable amount of mechanical ability can learn to sharpen saws quite acceptably.

Before attempting to sharpen a saw, one should have clearly in mind the proper shape of the teeth, and it is desirable also to understand the cutting action of a saw.

Shape of Saw Teeth. There are two principal differences in the shape of ripsaw and hand crosscut saw teeth. One difference is in

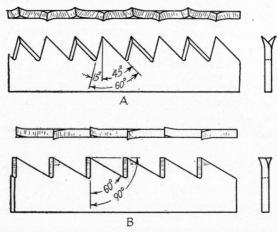

Fig. 229.—A, hand crosscut saw; B, ripsaw. There are two chief differences in the shape of crosscut and ripsaw teeth: (1) crosscut saw teeth are beveled; ripsaw teeth have square edges; (2) the front edges of ripsaw teeth are perpendicular to the tooth line, while the front edges of crosscut saw teeth make an angle of about 15 deg. with the perpendicular.

the hook or pitch of the teeth. The front edge of a ripsaw tooth is perpendicular to the tooth line of the saw, while the front edge of a crosscut saw tooth makes an angle of 15 deg. with the perpendicular (see Fig. 229).

The second chief difference is that the crosscut saw tooth is beveled, while the ripsaw tooth is not. In filing ripsaw teeth, therefore, the file is pushed straight across the blade; and in filing crosscut saw teeth, the file must make an angle with the saw blade to form the bevel. The usual angle is between 45 and 60 deg. A 45-deg. angle gives a wider bevel and a keener edged tooth, which is desirable for sawing softwood; while a 60-deg. angle gives a blunter tooth that will stay sharp longer in sawing hardwood. The narrower bevel, produced by filing at about 60 deg. to the blade, is usually preferred for saws in the farm shop.

The front edge and the back edge of a handsaw tooth make an angle of 60 deg. with each other, regardless of whether it is a ripsaw or a crosscut saw tooth.

Set of Teeth. It will be noted upon examining a saw that the points of the teeth are bent outward, one tooth in one direction and the next tooth in the opposite direction. This alternate bending of the teeth gives a saw what is called *set* and causes it to cut a kerf (groove) that is slightly wider than the thickness of the saw blade, thus preventing binding or pinching.

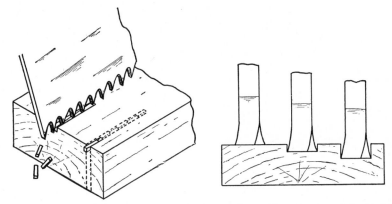

Fig. 230.—Cutting action of a ripsaw. The teeth act like a series of small chisels following each other and cut off the ends of the fibers.

Cutting Action of Handsaws. The ripsaw is used for cutting wood lengthwise of the grain or fibers. The teeth act like a series of small wood chisels following each other and cut off the ends of the fibers (see Fig. 230).

The crosscut saw, of course, is used for cutting across the grain or fibers. It first makes two parallel incisions with the points of the teeth, thus severing the fibers, and then forces out the wood between the incisions in the form of sawdust (see Fig. 231).

Sharpening a Handsaw. There are three chief operations, or steps, in fitting a handsaw, namely, (1) jointing, (2) setting, and (3) filing. A fourth operation, that of side dressing or side jointing may be performed, but is generally omitted.

Jointing a Handsaw. Jointing has a twofold purpose: (1) to make the teeth all the same length and (2) to serve later as a guide to indicate when the teeth are filed enough. A good job of saw fitting is often spoiled by filing some teeth too much. Jointing leaves a flat shiny surface on the point of each tooth. When this surface *just* disappears in filing, the tooth is sharp and should not be filed further.

To joint a saw, run a mill (flat-type) file over the ends of the teeth, lengthwise of the saw. Be very careful to keep the file square with

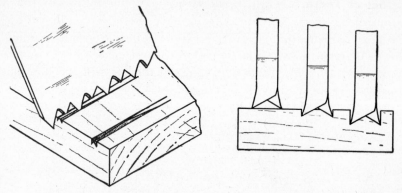

Fig. 231.—Cutting action of a hand crosscut saw. The points of the teeth make two parallel cuts, severing the fibers.

Fig. 232.—A saw may be jointed with a file held in the hands if care is used to keep it square with the saw blade.

the saw blade. This can be done by grasping the file in both hands, holding it by the edges, thumbs on top and first fingers underneath (see Fig. 232). A wooden square-edged block or a special tool may also be used to hold the file in the proper position (see Fig. 233).

File the tooth line straight, or curve it out slightly in the middle of the saw to give it a "breast" effect. File until there is a small shiny point on the end of each tooth, except possibly an occasional very short tooth.

Many experienced mechanics omit jointing when the saw teeth are uniform in length and when they require only light filing. By filing every tooth the same number of strokes, they are able to keep the teeth uniform in length without first jointing them.

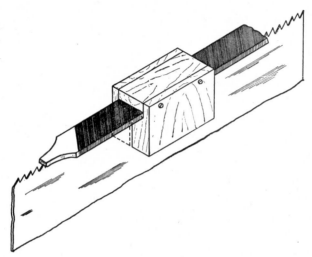

Fig. 233.—A homemade holder for holding the file while jointing a saw.

Setting a Handsaw. After jointing, the saw teeth are next set, unless they are very uneven in shape and size. In this case, the teeth should be filed to approximately the correct shape and size before setting and then filed again after setting.

A saw need not be set every time it is filed, particularly if only a light filing is required. A saw can sometimes be filed two or three times before it needs to be reset.

Setting is commonly done with a small tool known as a *spring saw set*. To use it, simply place the set over a tooth and squeeze the grips or handles. Use only moderate pressure, and do not raise up or yank on the handles. Too much pressure may mash the end of the tooth out of shape, and raising up on the handles may give the tooth too much set. When the handles are squeezed, a small plunger is forced against the end of the tooth, bending it over against a small anvil.

To set a saw, place it in a saw vise or clamp with the teeth projecting an inch or two above the jaws of the clamp. Begin at one end of the saw, and set every other tooth, using the spring saw set (see Fig. 234). Be sure to bend the teeth in the same direction they were originally set. When half the teeth are set, reverse the saw in the clamp and set the remaining teeth.

Most saw sets are adjustable, and when using one with which you are not familiar, it is best to set a few teeth and then examine them closely before setting the whole saw. If the teeth are not set enough, or too much, adjust the tool accordingly.

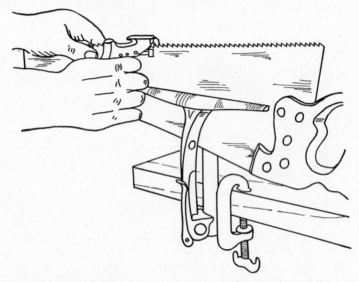

Fig. 234.—Setting a saw with a spring saw set. Set half the teeth from one side; then reverse the saw in the clamp and set the other half.

Only about one-third to one-half the length of a tooth should be bent in setting. If the depth of set is more than this, some teeth may be broken out, or the blade may be kinked or cracked.

The amount of set that a saw should have depends upon the kind of wood to be sawed and upon the thickness of the saw blade above the teeth. The better saws are ground thinner above the teeth and therefore require very little set. Green or wet wood will require more set than dry well-seasoned wood, and softwoods more than hardwoods. For average work, bending the teeth out about $\frac{5}{1000}$ in. should be ample. Too much set causes the saw to cut too wide a kerf, resulting in poorer control of the saw and extra work to push it. Too little set causes pinching or binding in the kerf.

Filing a Hand Crosscut Saw. Clamp the saw securely in a saw vise or clamp with the teeth projecting between ⅛ and ¼ in. above

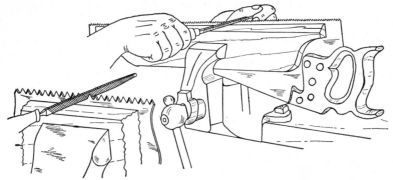

Fig. 235.—Filing a crosscut saw held in a homemade clamp and machinist's vise.

Fig. 236.—For best results, clamp the saw at about the height of the armpits and work in a good light.

the jaws—just enough for the file to clear the jaws easily. If the teeth project too far above the clamp, the saw will chatter and the file will screetch.   A clamp made of two 1 by 4's and used in an ordinary vise is satisfactory (see Fig. 235).

For best results, the top of the saw clamp should be at about the height of the armpits or possibly an inch or two lower.   One must be constantly on guard while filing to maintain the desired shape of teeth, and this can best be done when the sides of the teeth are easily seen; hence, the rather high position of the saw while filing it.

In order to avoid eyestrain and to ensure a good job of filing, *good light is absolutely essential*.   It is usually best to work in front of a window where the light will shine on the teeth and it will be easy to see the reflections from the small shiny surfaces left by jointing.   If a lamp is needed, it should be placed above and a little in front of the workman.   The light rays should shine down on the saw teeth (see Fig. 236).

The kind and size of file to use depend upon the size of the saw teeth and the preference of the mechanic.   In general, 6-in. slim-taper saw files are recommended for saws with seven to nine points per inch, and 7-in. slim-taper files for saws with five to six points per inch.   Some mechanics prefer blunt files instead of tapered ones.

The following points are important in manipulating the file:

1. Hold the file handle in the right hand (assuming the workman is right-handed) and only moderately tight.
2. Hold the tip of the file lightly between the thumb and first finger of the left hand.
3. Exert pressure on the forward strokes only.
4. Lift the file slightly on the return strokes, or at least release the pressure.
5. Make long, slow cutting strokes—not short, fast ones.
6. Keep the file level or pointed up very slightly if necessary to prevent screetching.
7. Use enough pressure to make the file cut, but no more.

If the file slides along without cutting, try a little more pressure and a slower stroke.   Be sure to release the pressure on the back-stroke.   If this does not make the file cut, it is probably dull and should be discarded.

First Position.   Study Fig. 237 very carefully to get the proper starting position for filing a saw.   The following points are important:

1. Place the tip end of the saw in the saw clamp and with the handle to the right.
2. Find the first tooth in the end of the saw that is bent out toward you. Place the file in the first gullet (V notch between two teeth) to the left.

3. Point the file across the saw blade at an angle of about 60 deg. and with the point toward the saw handle.
4. If the teeth are of the proper shape (see Fig. 229), press the file firmly down into the gullet and let it find its own bearing against the two teeth. If the front edges of the teeth are not about 15 deg. from the vertical, however, rotate the handle of the file so that the front edges will be filed at this angle.
5. Keep the file level, or pointed upward a little if necessary to prevent screetching.
6. Push the file forward, cutting the front edge of one tooth and the back edge of the adjacent one.
7. Lift the file slightly on the backstroke, or at least release the pressure.

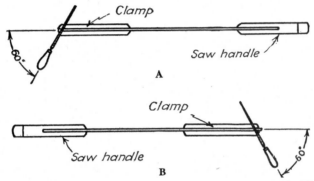

Fig. 237.—The two positions for filing a hand crosscut saw: A. First position. Begin at the tip of the saw and work toward the handle. Hold the file at an angle of about 60 deg. to the saw blade and be sure that the file cuts the front edge of each tooth at an angle of about 15 deg. from the vertical. B. Second position. After half the teeth are filed, turn the saw around in the clamp and file the other half from the other side.

8. File until about half the flat shiny surfaces made by jointing are filed away.
9. Then move the file two gullets to the right (toward the saw handle), and file in a similar manner. Continue filing every other gullet until you reach the handle.
10. Inspect your work frequently to be sure you are getting the teeth properly shaped. Remember that the front edges should be about 15 deg. from the vertical. This angle is changed by rotating the file handle slightly in the hand.

Second Position. When half the teeth have been filed from the first position, turn the saw around in the clamp with the handle to the left (*B*, Fig. 237). Find the first tooth in the end of the saw that is set out toward you, and place the file in the first gullet to the right.

Point the file at an angle of about 60 deg. across the saw blade and toward the saw handle. File in a manner similar to that used in the first position, filing in every other gullet until the handle of the saw is reached.

It is better to make about the same number of strokes with the file in each gullet, even though not quite all the flat shiny surfaces (left from jointing) are removed. Some of these shiny surfaces may well be left until the final touching up.

Touching Up. After going over all the teeth once, examine the saw carefully to see if any blunt shiny points still remain from jointing. If they do, as they most likely will, then place the saw in the first position again (*A*, Fig. 237), and go over the saw, filing *only those teeth which have shiny points*. Then reverse the saw, placing it in the second position again (*B*, Fig. 237), and go over the saw, filing only those teeth which have shiny points. In a similar manner, go over the saw three or four times, if necessary, filing only the teeth that have shiny points, until all teeth are sharp.

Keeping Proper Shape and Angle of Teeth. It is very important that the front edges of the teeth be filed to a pitch of about 15 deg. from the vertical. It is important also that the file be held at a uniform angle of about 60 deg. with the saw blade in order to produce the proper bevel. By placing the file in a gullet between two properly filed teeth and letting it seek its own position, the desired angles for filing can be determined. One must be constantly on guard, however, to keep this position after it is established.

A common trouble among beginners is to file the teeth in pairs with a broad short tooth next to a long slim one. This trouble is usually caused by failure to maintain the proper slope of about 15 deg. on the front of each tooth. To correct the trouble, therefore, first be sure that the file is held so as to give the proper angle, and then press it firmly against the broad tooth and lightly, if at all, against the narrow one, as it is pushed through the gullet between them.

Filing a Ripsaw. The same general procedure is used for filing ripsaws as for hand crosscut saws. There are two points of difference, however, that need to be kept in mind:

1. The front edges of the teeth are perpendicular to the tooth line instead of at an angle of 15 deg. from the perpendicular.
2. The edges of the teeth are not beveled but are square with the saw blade (see Fig. 229).

Except for the difference in angles of filing required by these two differences, a ripsaw is filed exactly the same as a crosscut saw.

Side Dressing or Jointing a Handsaw. To side dress or side joint a saw, lay it on a flat board or bench, and rub the sides of the saw teeth with the edge of an oilstone or with a fine file. The objects of side dressing are to smooth any rough edges left from filing and to even up the set in the teeth. Side dressing is usually not necessary for general sawing.

If after trying a saw, it is found to be set unevenly and tends to run to one side of the line, it may be side dressed on the side that leads away from the line. Side dressing may also be used to reduce the amounting of set in case the saw cuts too wide a kerf.

Points on Sharpening Handsaws

1. Always work in a good light so that the points of the teeth may be easily seen.
2. Set only the points of the teeth—not more than one-third to one-half the length of the teeth.
3. Work with the saw at about the height of the armpits.
4. Clamp the saw with ⅛ to ¼ in. projecting above the jaws of the clamp— just enough to allow the file to clear the jaws of the clamp.
5. File on the forward stroke only. Lift the file slightly, or at least release the pressure, on the backstroke.
6. Use long, slow, rhythmic strokes. If the teeth are reasonably uniform in shape and size, file each gullet about the same number of strokes.
7. Just barely file away the flat shiny surfaces on the points of the teeth left by jointing.
8. Have no slope on the front edges of ripsaw teeth. Make them perpendicular to the line of teeth.
9. Slope the front edges of crosscut saw teeth about 15 deg. from a perpendicular to the line of teeth.
10. In filing a crosscut saw, point the file toward the handle of the saw. keeping the angle between the file and the saw blade at about 60 deg.
11. In filing a ripsaw, file straight across, keeping the file at right angles to the saw blade.
12. Do not point the file upward or downward, but keep it level, unless the file screetches and elevating the point slightly will stop the screetching.
13. If the teeth tend to become uneven in size, first be sure the file is held to give the proper slope on the front edges of the teeth, and then press harder against the big ones and lightly or not at all against the small ones.

Sharpening Bucksaws and Pruning Saws. The same general principles used in setting and filing crosscut handsaws apply in setting and filing bucksaws and pruning saws. The shape and angles of the teeth of bucksaws and pruning saws are different from handsaws, however, and therefore the file must be held differently when filing them. The proper shape of teeth can be determined by looking at some of the teeth near the ends of the saw that have not been used a great deal. Holding the file firmly between two of these teeth near the end of the blade will give the correct angles for filing.

Sharpening Crosscut Timber or Log Saws. To sharpen timber or log saws, one should have a combination jointer and raker gage and a saw set for such saws. It is possible, however, to set a log saw with a hammer and setting block. A mill (flat-type) file about 8 in. long is used to file the teeth.

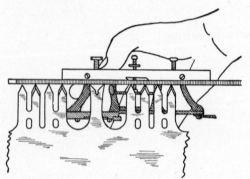

Fig. 238.—Jointing a crosscut log saw.

The operations or steps in fitting a log saw are: (1) jointing, (2) filing down the raker teeth, (3) setting the cutting teeth, and (4) filing the teeth.

Jointing. To joint a log saw, place a file in the jointing tool (see Fig. 238) and file down the teeth until the shortest of the cutting teeth is reached. If the saw is in fair shape, very light jointing is all that is required. It is very difficult to hold a file in the hands and joint the teeth of a log saw because the teeth are too long.

Filing Down the Rakers. The rakers, or those teeth which shave off and carry out the wood between the two parallel incisions made by the cutting teeth, should be a little shorter than the cutting teeth, the exact amount varying with the kind of wood to be sawed. In general, they should be about $\frac{1}{64}$ in. shorter for hardwoods, and about $\frac{1}{32}$ in. shorter for softwoods. To file down the rakers, place the raker gage on the saw and file down the points of the rakers even with the gage (see Fig. 239).

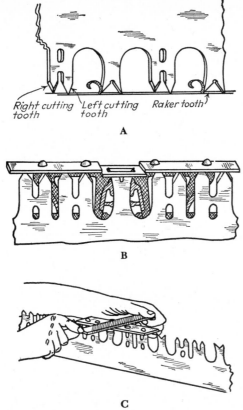

Right cutting tooth   Left cutting tooth   Raker tooth

A

B

C

Fig. 239.—A. How a crosscut log saw works. The right and left cutting teeth make incisions and the raker teeth cut out the wood between the incisions. B. A raker gage in place ready to file down the rakers. C. Filing down the rakers. File them $\frac{1}{64}$ to $\frac{1}{32}$ in. shorter than the other teeth.

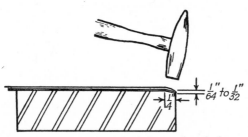

$\frac{1}{4}''$ $\frac{1}{64}''$ to $\frac{1}{32}''$

Fig. 240.—A homemade setting block for setting timber or log saws.

Setting the Teeth. The teeth may be set with a hammer and setting block or with a spring saw set. Many prefer the spring set. Not more than $\frac{1}{4}$ in. on the end of a tooth should be bent. A homemade setting block can be made easily by filing a corner of a piece of iron as shown in Fig. 240.

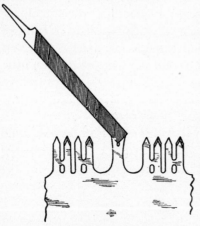

Fig. 241.—File the gullets of the raker teeth so that the square end of the file will fit. Be careful not to file the rakers too short.

Filing the Teeth. File the rakers on the inside of the end notch with a flat file. The angle at the center of the notch should be about a right angle and may be checked with the square end of the file (see Fig. 241). Be careful not to file the teeth shorter than the length to which they were jointed. It is a good practice to file half the rakers from one side of the saw and the other half from the other side.

File the cutting teeth in much the same manner as the teeth of a crosscut handsaw (see Fig. 242). File half the teeth from one side of the saw and half from the other. The proper angle of the points and the width of the bevel on the cutting teeth depend upon the kind of wood to be sawed. For softwoods, a long point with a wide bevel

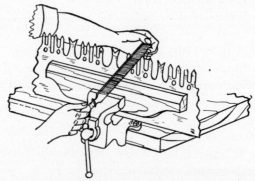

Fig. 242.—Filing the cutting teeth of a crosscut log saw.

is recommended, and for hardwoods or knotty or frozen wood, a blunter point with a narrower bevel.

Gumming a Log Saw. After several filings, the gullets between the teeth of a log saw become so shallow that they cannot well hold all the sawdust made by the teeth. Consequently, the sawdust binds against the sides of the saw kerf and makes the saw pull hard. The saw should then be gummed, that is, the gullets should be filed or ground deeper. It is a very slow, tedious job to gum a saw with a file. The best method is to use a special thin grinding wheel made for this purpose. A special stand or work rest can be made in front of the grinder to support the saw while it is being gummed. It is best not to do all the gumming in a gullet at one time, but to grind a little in one gullet and then proceed to the next, going over the saw three or four times to complete the job. This avoids overheating the saw and drawing the temper.

Sharpening Circular Saws. In sharpening a circular saw, the same general operations are performed as in fitting a handsaw or log saw. The operations or steps are: (1) jointing or truing up, (2) gumming, if needed, (3) setting, and (4) filing. Not all these operations need be done every time a saw is fitted.

Jointing. A circular saw is jointed to make it truly circular, or to make the tooth points all the same distance from the center of the saw. Jointing is usually best done by leaving the saw mounted on its own shaft and turning it slowly backward by hand while holding a file firmly yet lightly against the ends of the teeth. An easy method of holding the file in the proper position is to adjust the saw so that the teeth barely project through the slot in the saw table and then hold the file flat on the table over the slot.

Gumming. A circular saw may be gummed in the same manner as a log saw. It may be done with a round file or with a flat file with a round edge; but a much easier way is to use a special saw-gumming wheel. Circular saws with certain kinds of teeth, such as fine-toothed crosscut saws, do not require gumming.

Setting. The teeth of a circular saw may be set with a hammer and setting block or with a large spring saw set. The amount and depth of set required will depend upon the kind of a saw, whether it is a ripsaw, a cutoff saw, a cordwood saw, etc., and upon the kind of wood to be sawed. If, after fitting a saw, it is found to bind in the saw kerf, it is a simple matter to give it a little more set.

Filing. The kind of file to use in filing a circular saw will depend on the shape and size of the teeth. For a crosscut saw, usually a three-cornered file about 8 in. long is used. For a ripsaw or a cordwood saw, a mill (flat-type) file about 8 or 10 in. long is used.

In filing a circular saw, be careful to preserve the original angle of bevel and the original pitch of the teeth. The proper position for the file can be determined by pressing the file into a gullet, or against the side of a properly filed tooth, and allowing it to "seat" against the tooth.

13. Replacing Handles in Tools

Tools with broken handles are sometimes discarded when they could be restored to practically perfect condition by simply replacing the handles. Although new handles can be made by a careful workman, it is usually much more practical to buy the handles ready to be put on the tools. Fortunately, it is usually possible to buy handles that can be fitted to the tools with a minimum of time and work and with the assurance that they will fit satisfactorily.

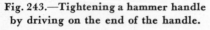

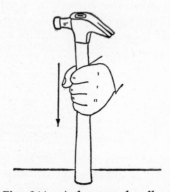

Fig. 243.—Tightening a hammer handle by driving on the end of the handle.

Fig. 244.—A hammer handle may also be tightened by ramming the handle down against the bench top.

Tightening Loose Handles. Handles of tools like hammers, axes, and hatchets often become loose and simply need to be retightened. This is usually an easy job and can be done by first driving the handle tightly into the head of the tool (see Fig. 243), and then driving the wedges tighter into the end of the handle. Another practical way of tightening the handle into the head is to ram the handle endwise down against a bench or some other solid object (see Fig.

244). Steel wedges for driving into the end of the handle and expanding it in the eye of the tool are usually available at hardware stores. Steel wedges with roughened sides are easily made in the shop, as are wooden wedges. Either type is satisfactory, provided it is of suitable size and is carefully made and driven into place.

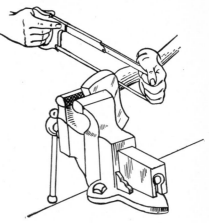

Replacing Ax Handles. To replace a broken ax handle, first remove the old pieces of the handle from the ax. A good method is to saw close to the head of the ax with a hacksaw, then remove the wood from the eye by drilling first with a metal-drilling twist drill, and then by punching (see Figs. 245 and 246).

Fig. 245.—To remove a broken handle from an ax, first saw close to the head with a hack saw.

Next, carefully work the end of the new handle down to size, using a rasp (see Fig. 247) and trying it in the head of the ax frequently as the work progresses. A rasp is

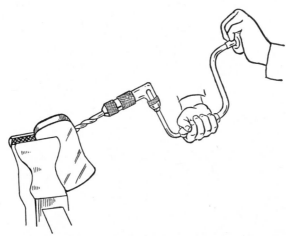

Fig. 246.—The pieces of the old handle that remain in the eye of the ax head may be removed by drilling with a twist drill and then punching.

preferred to a plane or drawknife, as such tools tend to follow the grain and are more difficult to control. The head of the ax should fit back to within about $\frac{1}{2}$ to $\frac{3}{4}$ in. of the biggest part of the handle.

It is also important that the handle be so fitted that when the ax is held vertically on a bench with the end of the handle touching the bench the cutting edge will touch at a point about two-thirds its length from the outer end (see Fig. 248).

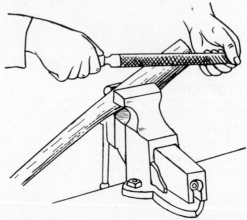

Fig. 247.—Work the new handle down to size with a rasp.   Frequently try the handle in the head of the ax for fit as the work progresses.

After the end of the handle is worked down to size, rip the end (see Fig. 249), if a wooden wedge is to be used.   The ripped part should extend a little more than halfway through the ax when the handle is put in place.

Next, drive the handle firmly into place, using a mallet (see Fig. 250).   If the grip on the end of the handle is not round or knob-

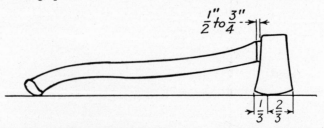

Fig. 248.—The handle should be fitted so that a straight line from the end of the handle tangent to the cutting edge of the ax head, will touch the edge at a point about two-thirds of its length from the outer end.

shaped, but is of the style that is cut off at an angle, there may be some danger of splintering or marring it with a mallet.   This danger may be largely avoided by chamfering or rounding the end somewhat with a file before driving with the mallet.   After the handle is driven into place, make a thin wooden wedge and drive it tightly into the

end of the handle, and saw off the protruding end of the handle and
wedge with a hacksaw (see Figs. 251 and 252). In case steel wedges

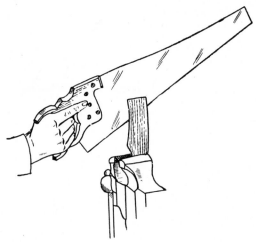

Fig. 249.—After fitting the end of the handle, rip a kerf for the wedge if a wooden
wedge is to be used.

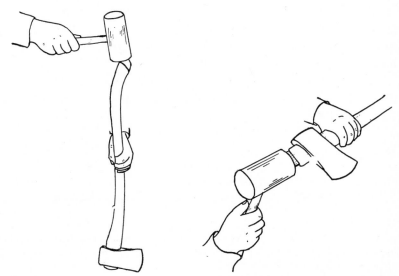

Fig. 250.—Driving the fitted handle
tightly into place.

Fig. 251.—Driving a wooden wedge into
the kerf in the end of the handle.

are used, the end of the handle need not be ripped as they can be
easily driven into place after the end of the handle has been cut off
even with the head of the ax.

Replacing Shovel Handles. The procedure for replacing a shovel handle will depend somewhat upon the type of handle used. Most handles are held in place by rivets. To remove the old broken

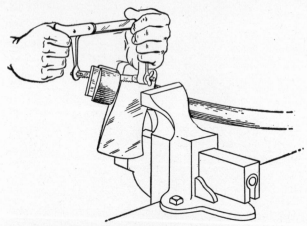

Fig. 252.—After the handle is fitted and tightly wedged in place, saw off the protruding part of the handle and wedge with a hack saw.

handle, cut off the rivet heads by grinding, or by cutting with a sharp cold chisel or sawing with a hacksaw, and then drive out the rivets (see Fig. 253). Next, fit the new handle, drill holes for the rivets,

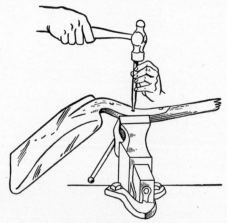

Fig. 253.—To remove a broken shovel handle, grind or cut off the heads of the rivets and remove them by punching.

insert them, and hammer the ends down with a ball-peen hammer. Strike one or two heavy blows first with the flat face of the hammer, and finish by light peening with the ball peen (see Fig. 254). It is

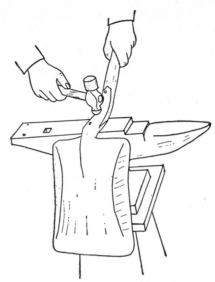

Fig. 254.—After fitting a new shovel handle, drill holes, insert rivets, and hammer the rivet heads down.

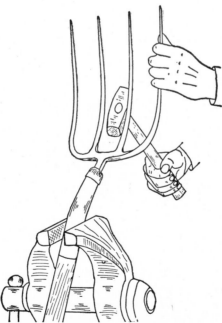

Fig. 255.—Most fork handles can be replaced by driving the fork out of the end of the old handle and driving it into the end of the new handle.

usually possible to buy new handles that require little or no fitting, other than the drilling of the holes.

Replacing Fork Handles. It is usually possible to buy fork handles complete with metal ferrules and with the ends bored, so that all that is required in replacing a broken handle is simple driving the fork from the old handle and driving it into the new one (see Fig. 255). In case such fitted and bored handles are not available, the ferrule or straps can be removed from the old handle and mounted on the new one after the end is trimmed to fit. The end of the handle is then carefully bored or drilled to a suitable size. If in doubt as to the proper size of bit, use one that is too small rather than too large, and if necessary, enlarge the hole later with a larger bit. The fork should fit very tightly, yet it should not split the handle.

Replacing Hoe and Rake Handles. The style of hoe or rake will suggest the exact method of replacing a broken handle. The general procedure is the same as for forks and shovels. Sometimes the handles of garden tools, like hoes and rakes, break off where they fit into the socket or ferrule. In such a case, it is frequently practical to remove the pieces of wood from the socket or ferrule and to refit and use the old handle. Care should be taken to make the end of the old handle fit snugly into the socket or ferrule.

In the case of a tool attached to the handle by means of a slightly tapered square tang driven into the end of a handle, it is sometimes advisable to drill a small hole straight through the end of the handle—ferrule, tang, and all—and insert a small rivet. This keeps the tool from pulling off the handle.

Replacing a Handsaw Handle. The best way to repair a saw with a broken handle is to remove the handle and install a new one. To install a new one, carefully mark and drill the holes through it, and fasten it in place with the special screws used on the original handle. A new handle can be made in the shop if necessary. To do this, lay the old handle on a piece of suitable wood, and mark out the shape with a pencil. Then saw out the handle approximately to shape with a coping saw, compass saw, or band saw, and finish with rasp, files, scraper, and sandpaper. Carefully saw the kerf to receive the end of the saw blade, and attach the handle as outlined above for a new handle.

14. Cleaning Tools

If grime, dirt, and grease accumulate in use, the tools should be cleaned by wiping with a cloth moistened in gasoline or kerosene.

Tools that have become rusty may usually be cleaned by rubbing with a rubbing compound used in automobile repair shops for finishing

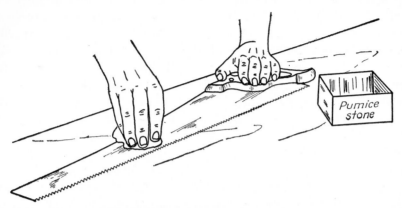

Fig. 256.—A rusty saw may be cleaned by rubbing with pumice stone and water; after which it should be wiped dry and oiled.

painted surfaces. This is a rather mild abrasive and will not scratch like emery cloth or sandpaper. For long-neglected and badly rusted tools, it may be necessary to resort to faster cutting abrasives like

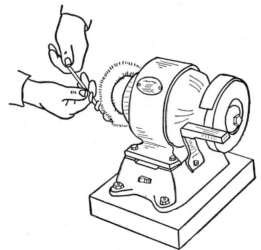

Fig. 257.—A rotary wire brush is excellent for removing rust and cleaning many tools.

emery or sandpaper. In such cases, it is well to finish with a fine rubbing compound.

Another abrasive that may be used is pumice stone. It may be applied to a moistened cloth and then rubbed on the tool. This is a good way to clean a rusty saw (see Fig. 256). After such treatment, the tool should be dried thoroughly and then given a light coat of oil.

A rotary wire brush or a buffing wheel on a small motor-driven grinder is excellent for removing rust and polishing tools (see Fig. 257).

To prevent rusting, tools should be wiped occasionally with an oily cloth. In case tools are to be placed in storage, they should be coated with a heavy oil or light grease.

Jobs and Projects

1. Learn how to sharpen a pocketknife quickly and easily. (A blade in fair condition usually can be made really sharp in three or four minutes.) If you carry a pocketknife, make it a practice to keep at least one blade very sharp, possibly even sharp enough to shave with.

2. Make a study of the grinders in the shop. Note the methods of adjusting the work rests, wheel guards (if any), and bearings (if adjustable). Are the bearings easily and conveniently lubricated? Are safety-glass eye shields used? If not, could they be installed easily?

3. Of about what grain size are the grinding wheels? Do the wheels appear to be of about the proper grade for the grinding that has been done on them? What are indications of too hard a wheel, and of too soft a wheel?

4. Do any of the grinding wheels need dressing? How may one tell when a wheel needs dressing? Dress a wheel if you find one that needs it.

5. Adjust the work rest on a grinder. How close should the rest be to the wheel? Why should a rest be kept carefully adjusted?

6. Inspect the axes and hatchets about your home or farm. If any need new handles or need sharpening, take them to the shop and put them in first-class condition.

7. Inspect other keen-edged tools, like planes, wood chisels, and drawknives, which you may have at home. If any need grinding, sharpening, or repairing, take them to the shop and put them in first-class condition.

8. Make it a point to keep all the tools you use in the shop sharp and in good working condition. If in doubt as to the method of sharpening, consult the text or your instructor.

9. Inspect the twist drills in the shop. Make a list of any defects in sharpening (grinding) that you find.

10. Grind some twist drills, making it a point to grind them properly and to avoid the common mistakes in grinding. Practice until you can grind a drill properly in two or three minutes. Whenever you start to use a

drill, first inspect it, and if it needs grinding, stop and grind it.   It takes just a minute or two and really saves time.

11. Compare rip and crosscut handsaws.   Note the difference in the bevel on the edges of the teeth.   Note also the difference in the hook of the teeth, or the angle that the front edges of the teeth make with the tooth line.   These differences are very important and must be kept in mind in filing saws.

12. Examine the saws in the shop, and determine, simply by inspection if you can, which ones are sharp and which ones are dull.   Try the saws, and verify your conclusions.

13. Practice saw sharpening on short practice blades if such are available in the shop.   If you do not get a good job the first time, joint the teeth and try again.

    Remember that it is important to watch the angle on the front edges of the teeth in order to produce the proper "hook."   The front edge of a ripsaw tooth should be straight up and down; and the front edge of a crosscut handsaw tooth should be about 15 deg. from vertical.

    Remember also to watch the small shiny surfaces left on the tooth points by jointing and to stop filing a tooth just as the shiny surface disappears.

14. After you have become reasonably proficient in filing a practice blade, sharpen a handsaw, which you may bring from home or which may be assigned you by the instructor.

15. Sharpen any other saws, such as bucksaws, pruning saws, log saws, or circular saws, about your home farm or about the shop and that may need sharpening.

16. Inspect the forks, hoes, shovels, spades, rakes, etc., about your home or farm.   If any need new handles, minor repairs, or sharpening, take them to the shop and put them in good condition.

6. Rope Work

Every farmer has occasion to use rope, and it will be well worth his time to master some of the common knots, hitches, and splices, so that he can make them easily and quickly. It is better to learn a few of the more useful ones thoroughly than to acquire only a superficial knowledge of many and be unable to use them when needed.

Rope work is not difficult, although it may at first appear to be complicated and involved. There is usually a system underlying most knots, hitches, and splices, and once the system is understood, the processes are simple and easy. It is, therefore, important in studying a particular unit of ropework to learn the system or principles involved.

MAJOR ACTIVITIES

1. Finishing the Ends of a Rope

2. Tying the Ends of Rope Together

3. Tying Loop Knots

4. Making Hitches

5. Shortening Rope

6. Splicing Rope

7. Making Rope Halters

8. Making Livestock Tackles

9. Using Blocks and Tackle

10. Taking Care of Rope

1. Finishing the Ends of a Rope

Relaying Strands. The ends of a rope frequently become untwisted, and the strands need relaying before finishing the end or making an end knot or splice. The process of relaying strands is also frequently used in splicing rope.

To relay the strands of a rope, hold the rope in the left hand, and with the right hand twist one strand tightly to the right and wrap it part way around the rope to the left (see Fig. 258). Then move the left thumb up the rope and hold the strand in place. In the same

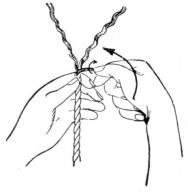

manner, twist and wrap the other two strands in turn, continuing the process until the strands are relaid. It is best not to rotate the rope in the hand, but simply to move the hand straight up the rope to hold the various strands in place as they are relaid. With a little practice, strands can be relaid in practically as good condition as they were originally in the rope.

Whipping the Ends of a Rope. Whipping is a neat and effective method of preventing the ends of a rope from untwisting and is recommended when the rope must be passed

Fig. 258.—Relaying strands. Each strand in turn is twisted to the right and then wrapped to the left and part way around the rope.

through small holes. When done with strong durable cord, like fish line, whipping is quite permanent. There are various ways of whipping a rope end. One of the best methods is as follows:

1. Unlay one strand of the rope back an inch or two from the end (see Fig. 259).

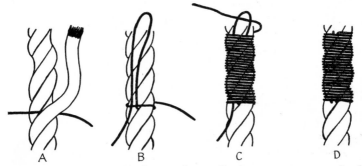

| A | B | C | D |

Fig. 259.—Whipping an end of a rope is an effective way to prevent fraying.

2. Place one end of a strong cord, $2\frac{1}{2}$ to 3 ft. long, under the raised strand, leaving the short end of the cord 6 or 8 in. long; and then relay the strand.
3. Hold the end of the rope up, letting the short end of the cord hang down.
4. Wrap the long end of the cord once around the rope, just above the short end.

5. Pull the short end of the cord toward the end of the rope and turn it back, forming a U, or a bight. It is best to lay the sides of this U in a groove in the rope.
6. Wind the long end of the cord around the rope and the U turn in the cord, keeping the turns tight and close together.
7. When the wrapping has progressed as far as desired, pass the long end of the cord through the end of the U loop in the cord, keeping it tight.
8. Pull on the short end of the cord, drawing the U loop back under the wrapping to about the center. Cut off the loose ends.

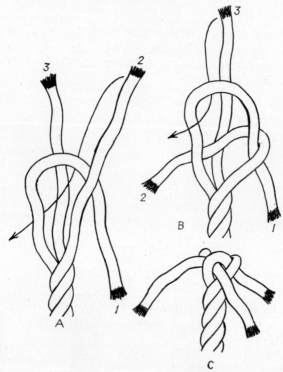

Fig. 260.—The crown knot.

Making the Crown Knot. This knot is used principally as the first step in making the crown splice. The steps are as follows:

1. Unlay the ends of the rope about five turns.
2. Bring strand 1 down between strands 2 and 3, forming a U loop (see Fig. 260).
3. Place strand 2 around behind the U loop and in front of strand 3.
4. Pass strand 3 through the loop, and draw the strands down even and tight.

Making the Crown or End Splice.   This knot is one of the most useful and permanent end treatments for a rope.   It is made by first making a crown knot and then proceeding as follows:

1. Place strand 1 over the first adjacent strand in the main part of the rope, and under the next strand (see Fig. 261).
2. In a similar manner pass strand 2 over the strand in main part that is adjacent to strand 2 and under the next.
3. Then pass strand 3 over the strand in the main rope that is adjacent to it and under the next.   *Strand 3 should come out at the same place strand 1 went in,* and when properly done, the three strands will come out of the main part equally spaced and no two in the same place.
4. Draw each of the strands up tight.   It is a good plan to pull the end of each strand back up toward the end of the rope slightly.   The strands,

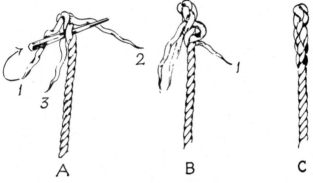

Fig. 261.—The crown splice.

as they are tucked or woven into place, should keep amost at right angles to the strands of the main part and work diagonally around the rope, not straight down it.

5. Continue the process outlined in steps 1, 2, 3, and 4 until each strand is tucked under strands of the main rope three or four times.
6. When the strands have been woven far enough, the loose ends are cut off about ¼ in. long and the splice is smoothed by rolling it under the foot on the floor.

If it is desired to make a tapered splice, part of each strand may be cut out before taking the last one or two tucks.

Untwist the Rope Slightly to Facilitate Tucking.   In order to facilitate tucking in making the end splice, and other splices as well, it is a good plan to untwist the rope slightly.   Hold the end of the rope between the palm and last three fingers of the left hand, leaving the thumb and first finger free (see Fig. 262). Untwist the rope slightly with a twist of the right hand. Then open up a place for the strand to be tucked, using the first finger of the left hand. Place the end of the strand against the end of the first finger, and with the left thumb, push it through the opening made for it. As the finger recedes back out of the hole, the strand is thus made to follow it through.

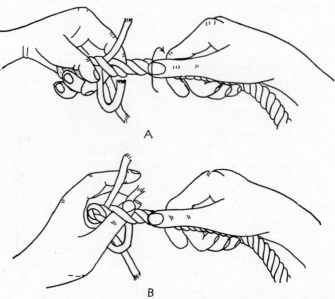

Fig. 262.—An easy method of tucking. A. The rope is untwisted slightly and then the first finger of the left hand is placed under a strand, opening up a hole. B. As the finger is withdrawn, the thumb pushes the strand through the hole.

Making the Wall Knot. This is generally used as the first part of a Matthew Walker knot, although it is sometimes used alone as an end knot for a rope. To make the wall knot, proceed as follows:

1. Unlay the strands about 3 or 4 in., and hold the rope in the left hand, loose ends up.
2. Bring strand 1 halfway around and across in front of the rope, holding it in place with the left thumb (see Fig. 263).
3. Pull strand 2 down and around the end of strand 1, releasing strand 1 from under the left thumb and placing the thumb on strand 2 to hold it in place temporarily.
4. In a similar manner, pull the end of strand 3 down and around the end of strand 2 and pass the end of strand 3 up through the loop of strand 1.

5. Draw the strands up even and tight.

Making the Matthew Walker Knot. This is one of the most useful and permanent end knots. It is made as follows:

1. First make a wall knot, but leave it loosely constructed (see Fig. 264).
2. Then pass end 1 up through loop *A* ahead of it, end 2 up through loop *B*, and end 3 up through loop *C*.
3. Draw the strands up even and tight.

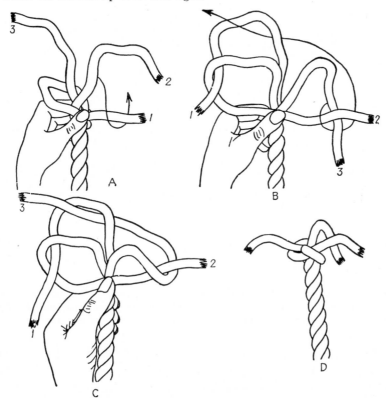

Fig. 263.—The wall knot.

The figure-8 knot (Fig. 265) may be used where a large bulky knot is needed on the end of a rope to keep it from drawing through a hole or through a pulley. It is easy to untie after pressure is released from it.

The overhand knot (Fig. 266) is sometimes used as an end knot in a rope, but more frequently as a step or a part of other knots or hitches. It is quickly and easily tied. As an end knot it is somewhat bulky, and after pressure has been applied, it may be difficult to untie.

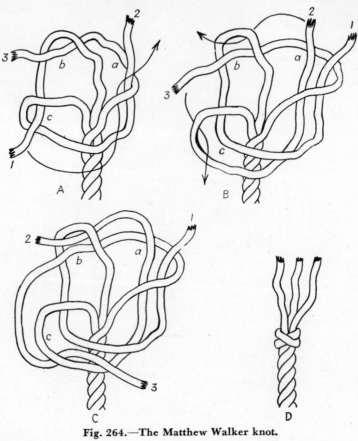

Fig. 264.—The Matthew Walker knot.

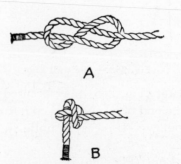

Fig. 265.—The figure-8 knot.

2. Tying the Ends of Rope Together

The square knot (Fig. 267) is one of the most useful knots for joining the ends of twine, string, or rope. The following very simple rule may help in learning to tie the square knot:

Start the knot by crossing the ropes, left over right. Then cross right over left. (Left over right, then right over left.)

Fig. 266.—The overhand knot.

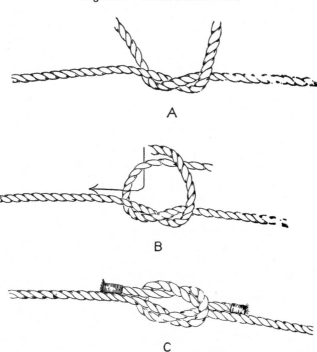

A

B

C

Fig. 267.—The square knot is a very useful knot. Do not confuse it with the granny knot.

The granny knot (Fig. 268) has much the same appearance as the square knot and is often mistakenly tied for the square knot. The granny knot will slip under strain and should normally not be used.

Fig. 268.—The granny knot slips under load. It is often mistakenly tied instead of the square knot.

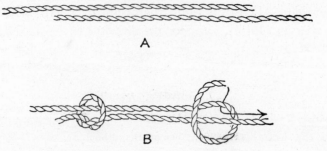

Fig. 269.—The fisherman's knot.

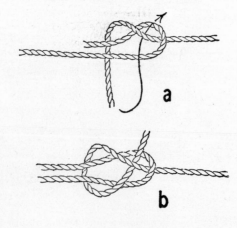

Fig. 270.—The weaver's knot.

It should be carefully studied so that it will not be tied when the square knot is desired.

The fisherman's knot (Fig. 269) is used for joining silk lines or guts on fishing tackle. It may be used also for tying two ropes together. To make the fisherman's knot, place the two ends to be joined side by side. Then tie each end around the other, using an overhand knot.

The weaver's knot, also called the *sheet bend* or the *becket bend*, (Fig. 270) is good for joining two ropes, particularly ropes of different sizes. It remains secure without drawing up tight and is easy to untie.

3. Tying Loop Knots

The slip knot (Fig. 271) is used for tying a loop around an object. To tie a slip knot, make a U loop or bight, and then tie the end around the main part of the rope, using an overhand knot.

Fig. 271.—The slip knot.

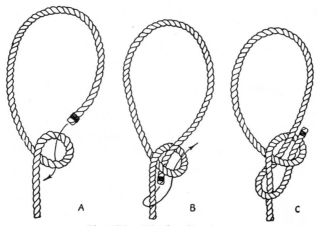

Fig. 272.—The bowline knot.

The bowline knot (Fig. 272) is one of the most useful of knots. It holds securely, yet will not slip or draw up tight. To make it, form a loop near the end of the rope, pass the end through the loop, around the main part of the rope, and back through the loop again.

The tomfool's, or double-bow, knot (Fig. 273) is sometimes called a trick knot although it is very useful.   It is often used for holding hogs

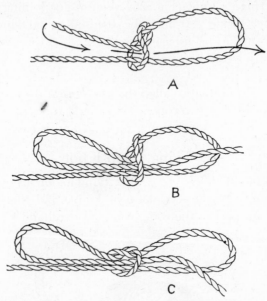

Fig. 273.—The tomfool's, or double-bow, knot.

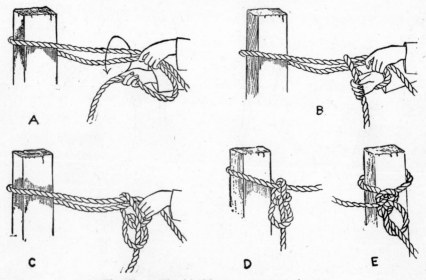

Fig. 274.—The hitching, or manager, knot.

while ringing them.   The first loop is slipped over the hog's upper jaw, and the main part of the rope fastened to a post.   The knot is untied, and the hog released, by simply pulling on the end that goes to the second bow or loop.

The hitching or manger, knot (Fig. 274) is commonly used for tying an animal to a hitching post or a manger. To make it, start as in typing a slip knot. Instead of drawing the rope entirely through

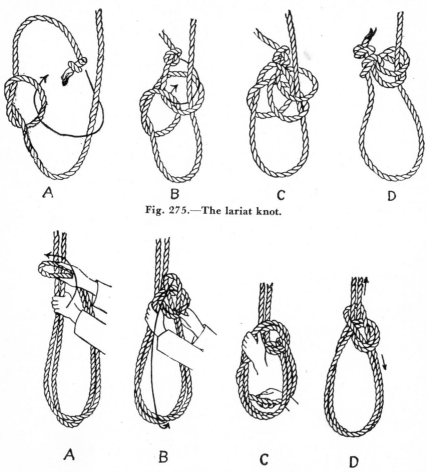

Fig. 275.—The lariat knot.

Fig. 276.—The bowline on the bight.

and completing the slip knot, however, leave a loop, and then pass the end of the rope through this loop.

The lariat knot (Fig. 275) is made as follows: Tie two overhand knots, one at the end of the rope and drawn up tight, and the other back about a foot and drawn up loosely. Pass the end of the rope around the main part and through the loop as at *B*, and then again around the main part and through the loop as at *C*.   Finish by drawing the parts up tight.

     The bowline on the bight (Fig. 276) is used for forming a loop in the middle of a rope or near the end where it has been doubled.   It is tied as follows: Form a loop in the doubled rope and pass the doubled end of the rope through the loop, as at *B*.   Then bring the doubled end down, and slip it over and back up around the loop.

4. Making Hitches

     Hitches are, in the main, quickly made temporary fastenings that depend upon the pull on the rope to keep them tight.

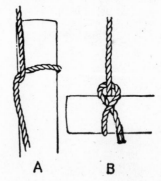

A   B

Fig. 277.—Two forms of the half hitch.

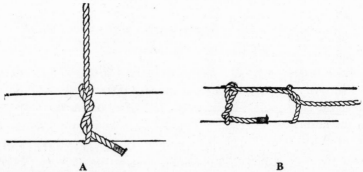

A   B

Fig. 278.—A, the timber hitch.   B, the timber hitch and half hitch combined.

     The half hitch is shown in two forms in Fig. 277.   The half hitch is more often used in combination with other hitches, or as steps in making other hitches, than alone.

     The timber hitch (Fig. 278*A*) is similar to the half hitch but is made more secure by wrapping the loose end once or twice through the loop. The timber hitch and half hitch are often used in combination for holding or moving logs (see Fig. 278B).

The clove hitch is one of the most useful ways of fastening a rope to a post or stake. Two methods of making the hitch are illustrated in Fig. 279.

The miller's, or grain-sack, knot (Fig. 280) is by far the best knot for tying grain sacks. It is a very useful knot, particularly for farmers. To tie the knot, place the string across the front of the neck of the bag, under the last three fingers, but over the first finger of the right hand, as at *A*. Take the string around the neck of the sack, under the heel of the right hand and under all fingers, but over the

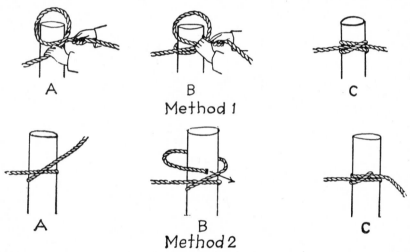

A   B   C
Method 1

A   B   C
Method 2

Fig. 279.—The clove hitch.

end of the string, as at *B*. Wrap the string around the neck of the sack a second time in the same manner. Then with the first finger of the right hand, pull the end of the string up under the loop formed when the knot was started (see Fig. 280C). Leave the loop as shown at *D*, or draw it on through as shown at *E*, depending upon how securely the sack is to be tied. If the loop is left as at *D*, the knot is ordinarily secure enough and yet it can be untied easily by pulling on the end of the string. For long shipment or rough handling, the knot should be completely formed as at *E*.

The Blackwall hitch (Fig. 281) is used for fastening a rope temporarily to a hook.

The cat's-paw (Fig. 282) is also used for fastening a hook to a rope. It may be used in the middle of a rope as well as near a free end.

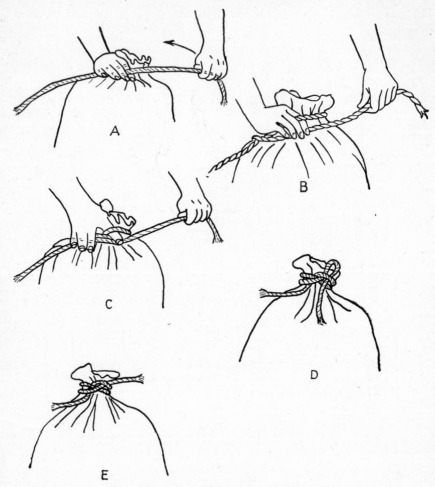

Fig. 280.—The miller's, or grain-sack, knot.

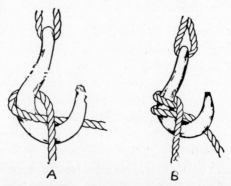

Fig. 281.—Two forms of the Blackwall hitch.

Pull may be applied to either or both ends. It is quickly made or detached.

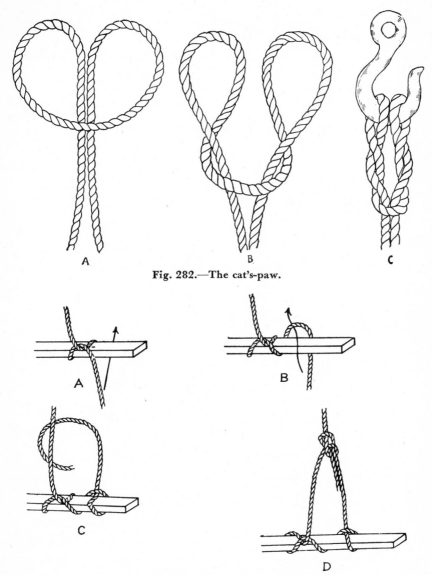

Fig. 282.—The cat's-paw.

Fig. 283.—The scaffold hitch.

The scaffold hitch may be made by wrapping the rope around a board as shown at *A*, *B*, and *C*, Fig. 283, and then fastening the end to the main part of the rope with a bowline as shown at *D*.

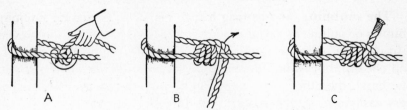

Fig. 284.—The snubbing, or running, hitch.

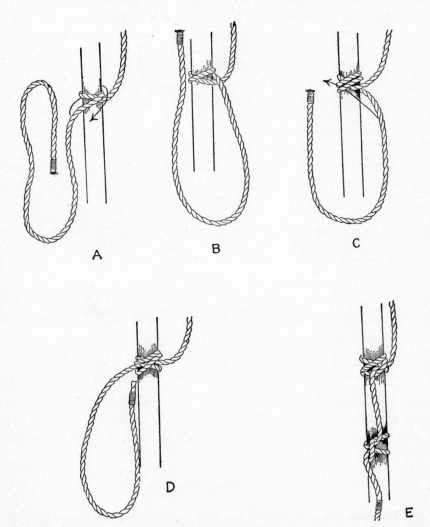

Fig. 285.—The well-pipe hitch.

The snubbing, or running hitch (Fig. 284), is used for snubbing animals or pulling in a rope against a force and fastening it to prevent it from being pulled out again. If a pull comes on the rope, the hitch will not slip; but by holding the rope between the hitch and post with one hand, the other hand can slip the hitch out away from the post very easily.

The well-pipe hitch (Fig. 285) is useful for securely fastening a rope to a pipe or other cylindrical object. It may be made as follows: Wrap the rope around the pipe, making a half hitch, as at *A*. Make a second wrap, wrapping downward and over the main part of the rope as at *B*. Then make a third wrap, this time going under the main part of the rope and up between the last two turns, as at *C* and *D*. Complete the hitch by making a clove hitch in the end of the rope. Pull on the rope should be parallel with the pipe.

5. Shortening Rope

The sheepshank (Fig. 286) is useful for shortening a rope temporarily. To make the sheepshank, gather up the rope, forming two

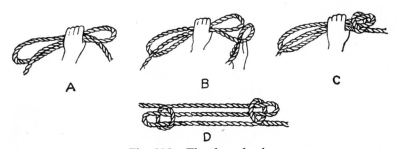

Fig. 286.—The sheepshank.

loops of the desired length as at *A*. Then make a half hitch around the end of each loop as at *B* and *C*. If tension is kept on the rope, the sheepshank will remain secure. In case the tension is not steady, and yet it is desired to keep the sheepshank, then the ends of the rope may be threaded through the ends of the loops.

6. Splicing Rope

The short splice is made where a considerable enlargement in the rope would not be objectionable or where only a short length of rope can be spared for making the splice. It may be made as follows:

1. Unlay the strands for six or seven turns on the ends to be joined.
2. Place the two ends together so that the strands from one end alternate with the strands from the other. Be sure that every strand branches outward from the main rope directly and without crossing over the center of the rope (see Fig. 287).
3. Tie each strand from one rope with the corresponding strand from the other rope, using a simple overhand knot. When all three strands are thus tied, draw them all up even and tight (see Fig. 288, *B* and *C*).
4. Tuck the strands from each rope under the strands of the other, using the method outlined for the crown or end splice (Fig. 261, page 203). Tuck the strands alternately, making a single tuck on one strand, then a

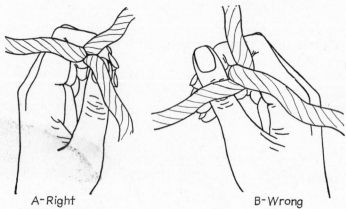

A-Right   B-Wrong

Fig. 287.—Strands spread out preparatory to placing two ropes together for splicing.

single tuck on the next strand, etc. Each strand should be tucked ultimately three or four times.

If a tapered splice is desired, part of each strand may be cut out before taking the last few tucks. To facilitate tucking, use the method of partly untwisting the rope as illustrated in Fig. 262. The strands, as they are tucked or woven into place, should keep almost at right angles to the strands of the main rope, and work diagonally around the rope, not straight down it. To make a firm, tight splice, it is a good plan to keep the strands drawn up tight as they are tucked, and to pull the ends of the strands back toward the middle of the splice occasionally as the work proceeds.

The long splice (Fig. 289) should be made if the rope is to pass through pulleys or if a neat splice is desired that will not appreciably increase the size of the rope. The following directions are for making the long splice in a three-strand rope. Four-strand ropes are spliced in a similar manner.

1. Unlay the strands of each rope about fifteen turns.
2. Place the two ends tightly together in exactly the same manner as for the short splice (see Fig. 289A). Be sure that the strands of one end alternate with the strands from the other end.

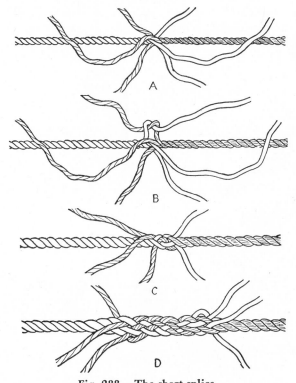

Fig. 288.—The short splice.

3. Tie overhand knots in two pairs of the strands (Fig. 289B). Draw the knots up tight. Be careful that the strands are properly paired.
4. Unlay one strand of the pair not tied, and lay the other strand in its place, twisting it tightly as it is laid (see Fig. 289C). When all but about 6 in. of the strand is laid, tie the two strands, using a simple overhand knot (see Fig. 266).
5. Turn the rope end for end, and untie either pair of strands that were tied at the beginning of the splice. Unlay the strand that comes from the right and lay the other strand in its place. When all but about 6 in. of the strand is laid, tie the two strands, using a simple overhand knot. The splice will then appear as in Fig. 289D.

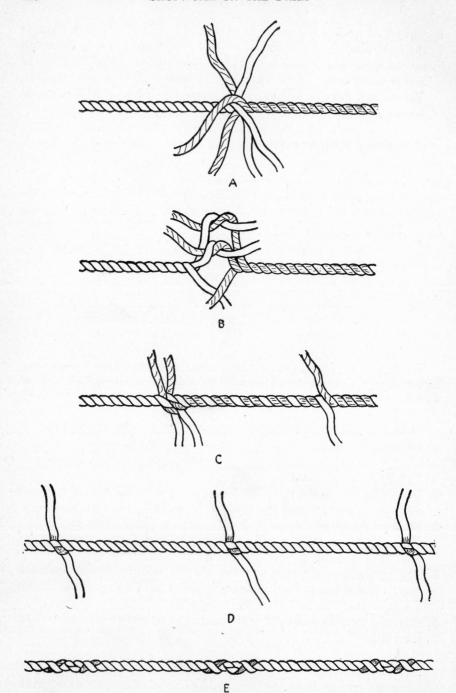

Fig. 289.—The long splice.

6. Cut off all strands to about 6 in. long, and weave or tuck each one into the rope, using the same general method of tucking as used in the crown or end splice and in the short splice. Each strand should go over the first adjacent strand of the main rope and under the next (see Fig. 290). Take at least three such tucks with each strand. Then cut off the strands, leaving the ends about ¼ in. long.

Repairing a Broken Strand. To repair a rope with one broken strand, unlay the broken strand five or six turns each way from the break, and then lay in its place a good strand from a rope of the same size, or from the end of the rope being repaired. Join the ends of the new strand to the ends of the broken strand and tuck them in the same manner as in the long splice.

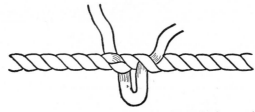

Fig. 290.—Tying and tucking the ends in finishing the long splice. After tying, each strand is placed over the first strand of the main rope and under the next. In a similar manner, each strand is tucked at least twice more, making three tucks in all.

The eye splice is used in many places where a permanent loop is to be made in the end of a rope. It is made as follows:

1. Unlay the strands about five turns.
2. Double the end of the rope back, forming a loop of the desired size, and place the unlaid strands as shown in Fig. 291A.
3. Place strand 1 under a strand of the main rope (see Fig. 291B).
4. Place strand 2 over the strand under which 1 was placed, and under the next (see Fig. 291C).
5. Place strand 3 over the strand under which 2 was placed, and under the next. Strand 3 must come out of the rope at the same point that strand 1 went in (see Fig. 291D). (*This is important.*)
6. Continue weaving and tucking the strands as in the crown splice and the short splice.

A modification of the eye splice, called the *side splice*, is used in splicing the end of one rope into another rope at a point other than its end.

Fig. 291.—The eye splice.

The loop splice is used for making a loop or eye in a rope at a point other than the end. It is made as follows:

1. Raise two strands of part 2, and pass part 1 through (see Fig. 292).
2. Raise two strands in part 1, *at a point that will give the desired size of finished loop*, and pass part 2 through.
3. Draw up tight.

Serving a Splice. Splices are sometimes served or wrapped tightly with a strong cord. This keeps the ends of tucked strands

Fig. 292.—The loop splice.

from pulling out, gives the splice a neater and more finished appearance, and makes it more durable. In cases where the length to be served is not great, or where the serving is near the end of the rope, it may be done in exactly the same manner as whipping, as explained on page 201. Where serving must be placed at some distance from the end of the rope, however, or where several inches are to be served, then a similar method, described below and illustrated in Fig. 293, may be used.

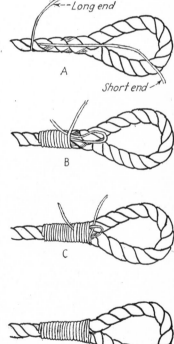

Fig. 293.—Serving an eye splice.

1. Raise one strand of the rope, and pass the cord under it.
2. Lay the short end of the cord along the rope in the direction of the serving.
3. Wrap the long end of the cord tightly and snugly around the part to be served and over the short end of the cord.
4. When the serving is almost completed, lay the short end of the cord back, forming a loop (see Fig. 293B).
5. Wrap six or eight turns more with the long end of the cord, and place the end through the loop (see Fig. 293C).
6. Draw the short end of the cord up tight, and cut off the exposed ends.

7. Making Rope Halters

A nonadjustable halter is illustrated in Fig. 294. It is quickly and easily made, only a loop splice, a side splice, and an end splice

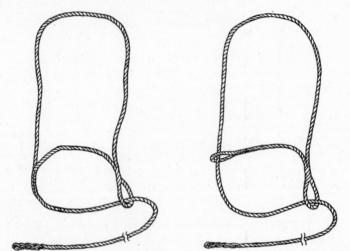

Fig. 294.—A nonadjustable halter.　　Fig. 295.—An adjustable halter.

being required. The lengths of the various parts of the halter are given in Table 4.

TABLE 4.　Approximate Dimensions for Rope Halters

Kind of halter	Diameter of rope, in.	Total length of rope, ft.	Length of parts, in.	
			Nosepiece	Headpiece
Horse..........................	½ or ⅝	14	8	16
Cow............................	½	12	7	14

An adjustable halter, illustrated in Fig. 295, is likewise simple and easily made. A loop splice, an eye splice, and an end splice are the only splices required. It is a good plan to make both the loop splice and the eye splice quite small, so that the halter will not be easily worked out of adjustment as the animal moves his head about.

A nonadjustable halter with guard loop, illustrated in Fig. 296, is a good type of permanent halter. The guard loop prevents the halter from loosening or tightening beyond a certain amount. After the first side splice is made (*B*, Fig. 296), relay the strands for a distance of 5 or 6 in., and then make the final side splice.

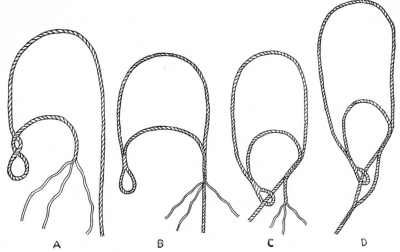

A   B   C   D

Fig. 296.—A nonadjustable halter with guard loop.

Fig. 297.—A calf halter.

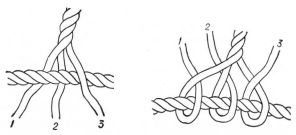

Fig. 298.—Detail of halter splice used on calf halter.

A calf halter, popular with calf-club members, is illustrated in Fig. 297. To make it, first make a loop splice, and then fasten the end of the headpiece to the main rope with a special, tight-fitting, adjustable loop. To make this special loop, unlay the strands back about 7 in. from the end. Then wrap them around the main rope as shown in Fig. 298, and weave them back into the head piece in the same manner as in an end splice or a side splice.

An emergency or temporary halter is easily and quickly made as illustrated in Fig. 299. To make it, turn the end of the rope back upon itself, forming a loop. Then tie an overhand knot in the doubled part, as at 1, Fig. 299. Next tie the end of the nose piece to the main part with a bowline, as at 2.

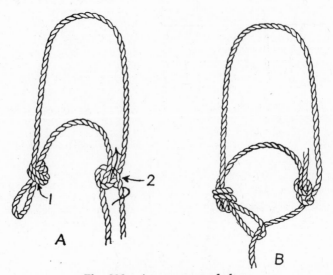

Fig. 299.—An emergency halter.

8. Making Livestock Tackles

Making the Leading or Tying Tackle. The tackle illustrated in Fig. 300 is effective in breaking a colt to lead. It is also good for breaking horses from pulling back when tied. Make the loop around the body so that it will loosen promptly when the tie rope is slackened. In case the manger or hitching rack is low, run the tie rope through a loop or strap in the halter ring, as at *A*, Fig. 300, instead of through the halter ring itself.

Making Casting Tackle for Horses. The tackle illustrated in Fig. 301 is recommended for throwing horses with safety both to the animal and the workman. To make the tackle, tie a bowline on the bight (see page 211) and place it over the animal's head, fitting it much like a horse collar. Then run the ends of the rope between the

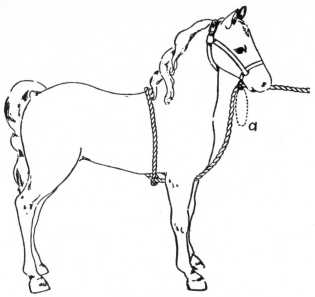

Fig. 300.—A leading or tying tackle.

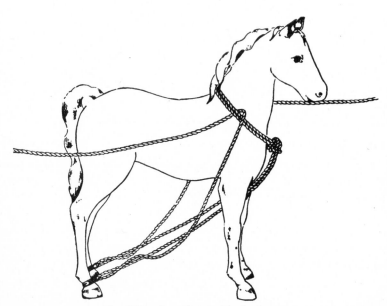

Fig. 301.—A casting tackle for horses.

front legs to rings in ankle bands on the hind feet.   After leaving the
rings, wrap each rope once around itself, and pass it under the doubled
rope around the neck, extending the rope on one side to the rear, and

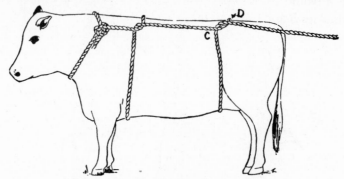

Fig. 302.—A casting tackle for cattle.

the rope on the other side to the front.   To throw the animal, back
him and tighten the ropes at the same time.

Making Casting Tackle for Cattle.   A simple casting tackle
for cattle is illustrated in Fig. 302.   To
make the tackle, tie the rope around the
animal's neck by means of a bowline.
Then make two half hitches, one at the
front of the body and the other at the
rear.   The rope at *C*, Fig. 302, should
come in front of the hipbone, and behind
the other hipbone at *D*.   To throw the
animal, pull backward and toward the
side on which it is to be thrown.

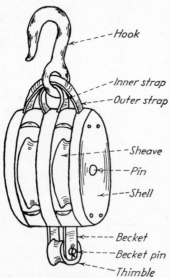

Fig. 303.—Parts of a block.

9. Using Blocks and Tackle

A *block* is a case or shell containing a
sheave or grooved pulley, or a set of them,
over which rope is run (see Fig. 303).
A block with one sheave is called a single
block, one with two sheaves a double
block, one with three sheaves, a triple
block, etc.   A *tackle* is a set of blocks and rope for raising, lowering, or
moving heavy loads (see Fig. 304).

The mechanical advantage of a tackle is equal to the number of
plies of rope passing to and from the moving, or load, block.   Thus

the mechanical advantage of tackle A (Fig. 304) is 2, that of tackle B is 3, and that of tackle C is 4.   In other words, neglecting friction losses, a pull of 50 lb. on the fall rope, that is, the free end of the rope, of tackle A would lift a load of 100 lb., while the same pull on the fall ropes of tackles B and C would lift loads of 150 and 200 lb., respectively.

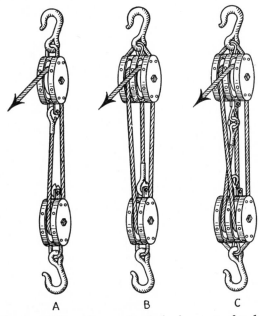

A   B   C

Fig. 304.—Simple tackles.   The mechanical advantage of tackle A is 2, that of tackle B is 3, and that of tackle C is 4.   Note the number of plies of rope supporting the movable block in each case.

In moving a load with a tackle, the end of the fall rope always moves farther than the load itself, the movement of the two being expressed as a formula as follows:

Travel of fall rope = travel of load × mechanical advantage

It is evident, therefore, that in using blocks and tackle, increase in force or pull is gained at the expense of movement.   In other words, a tackle enables a small force, by moving a greater distance, to lift or move a heavier load a short distance.

Using Blocks and Tackle with Safety.   Care should always be used in moving heavy objects.   Some of the more important precautions to be observed in using blocks and tackle are as follows:

1. Always fasten tackle securely.
2. Never stand where you could be hurt in the event of accidental loosening or breaking of some support or fastening.
3. Never straddle or stand near a rope that is being paid out or drawn in under tension.
4. Wear gloves to prevent burns or abrasion of the skin by sharp ends of rope fibers.
5. Keep hands and clothing away from moving ropes or pulleys.
6. Avoid the use of faulty equipment, such as a block with a hook that has started to straighten or a block with a bent, broken, or cracked shell.

Reeving a Set of Blocks. By reeving a set of blocks is meant passing the rope through the blocks to form a tackle. The procedure for reeving a set of blocks is usually obvious. A good way is to place the two blocks reasonably close to each other, and then pass the end of the rope backward through the blocks, opposite to the direction the rope will move when the tackle is used in lifting a load (see Fig. 305). Finish by attaching the end of the rope to the becket (the ring or fastening on the block opposite from the hook), preferably with an eye splice, or the block becket bend, as illustrated in Fig. 306.

Fig. 305.—To reeve a set of blocks, pass the rope backward through the blocks opposite to the direction it will move when lifting a load.

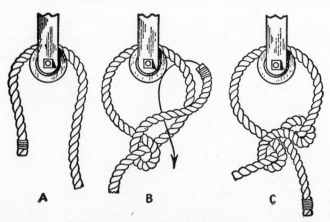

A   B   C

Fig. 306.—The block becket bend.

The block becket bend (Fig. 306) is used for temporarily fastening the end of a rope to a block.   It is in reality a form of clove hitch. To make the block becket bend, pass the rope through the becket (see Fig. 306A), and then take a turn around the main rope outside the loop (Fig. 306B).   Next make a turn around the main rope and through the loop, thus completing a clove hitch.

10. Taking Care of Rope

Coiling and Uncoiling a Rope.   Rope should be coiled in a clockwise direction so as to untwist the strands and prevent kinking.

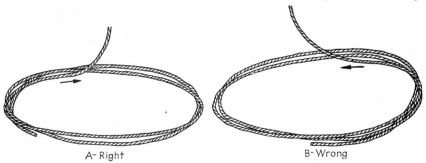

A- Right   B-Wrong

Fig. 307.—Right and wrong methods of coiling a rope.   To prevent kinking and tangling, the rope should be coiled in a clockwise direction and should be uncoiled in a counterclockwise direction.

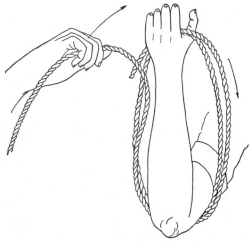

Fig. 308.—A good way to coil a small rope is to wrap it loosely around the flexed left forearm in a right-hand direction.

It may be coiled on the ground or floor, winding clockwise (see Fig. 307), or in a loose coil around the flexed left forearm, winding between the thumb and fingers of the open hand and around the elbow (see Fig. 308).

To take a rope out of a coil, unwind it in a counterclockwise direction. If it starts to uncoil in a clockwise direction, turn the coil over and pull the end of the rope up through the center from the other side. Kinking as a rope is taken from a coil is an indication that it is being pulled from the wrong side of the coil.

The following method is a good one for keeping rope that is reeved in blocks from becoming tangled during moving or storage: Pull the two blocks apart until the free end of the rope extends only a little beyond a movable block. Bring the blocks side by side, forming a loop, in the ropes between the blocks. With the free end of the rope, make one or two half hitches around the loop, and then pass the free end through the loop (see Fig. 309).

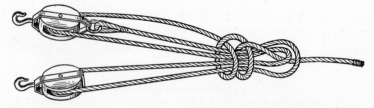

Fig. 309.—A good way to keep a tackle from tangling when not in use.

Relieving Kinks. A load should not be placed on a rope that is kinked, as this would put severe strain on certain fibers and weaken them, or possibly break them, and thus shorten the life of the rope. A new rope frequently tends to kink or twist. This is because the right- and the left-hand twists of the fibers, yarns, and strands have not become equalized. The trouble can be remedied, in the case of a short rope, by fastening one end to an overhead beam or support and tying a weight to the free end. If care is used, a long rope may be dragged *slowly for a short distance* over a *smooth* surface, thus allowing the free end to revolve and equalize the twists in the parts of the rope. It is extremely important that the rope be dragged *slowly* and *not too far*. Otherwise, the rope may be damaged by abrasion. Rope should not be dragged over sandy or gritty surfaces, as small particles of grit may become embedded in the rope and later cause serious internal wear between the fibers.

Avoiding Mechanical Injury. A rope should not be drawn over rough or sharp objects as this might break the outer fibers. Nor should ropes be allowed to rub against each other when in use as in a tackle. Rope should not be bent too sharply or used over pulleys that are too small. The diameter of the pulley should be at least eight times the diameter of the rope.

Storing Rope. Ropes should be stored in dry, well-ventilated places. They should not be exposed to dampness or moisture more than necessary when in use, and they should not be left damp under poor conditions for drying, as dampness causes rotting of the fibers. It is best to hang rope in loose coils on wooden pegs or arms, rather than leave it piled on a floor.

Chemicals or fumes from chemicals should be carefully avoided, as they cause very rapid deterioration of rope fibers. Ropes should be kept out of reach of animals and away from fertilizer, manure, decayed vegetable matter, etc.

Inspecting Rope. To determine the general condition of a rope, grasp it with the two hands a few inches apart and untwist the strands slightly. Internal wear is indicated by rope dust, exposed ends of broken fibers, and distinct edges on the inside of the strands. Extreme softness or the presence of mildew or mold suggests a weakened condition of the rope.

Jobs and Projects

1. Make a general study of the various methods that may be used for keeping the ends of a rope from untwisting and becoming frayed. Compare the different methods as to advantages, disadvantages, ease of making, permanence, etc. Learn and be able to demonstrate quickly two or three or four of the ones you consider to be best for farm use.

2. Make a list of six or eight knots and hitches (exclusive of end knots) that you consider to be most useful on the farm. Practice making them until you can make them quickly and easily.

3. Under what conditions would you recommend the short splice, and under what conditions the long splice? What is the difference between the eye splice and the loop splice? How could you splice a rope into a second one at a point some distance from its ends? How could a broken strand be repaired?
   Gather up ropes you can find that need splicing or repairing, and put them in first-class condition. Practice making the long splice, the short splice, and the eye splice until you can make them quickly and easily.

4. What knots, hitches, or splices should one know in order to make good rope halters? Compare the various halters described in the foregoing pages, and list those features you consider to be good, and those you consider not so good. Make a rope halter you consider to be particularly good, possibly one of your own design.

5. What is the best way to take care of a rope used in a barn hay fork and carrier system when it is not in use? Make a list of all the rope you keep about your home or farm. Is it stored so as to best preserve it when not in use? If not, make provisions for its proper storage.

   Outline methods for removing kinks from ropes. How may ropes be coiled and uncoiled so as to avoid kinking? Why does a new rope tend to kink?

6. Outline methods you would use for casting and holding a steer. Also outline a method for casting a horse.

7. Outline safety precautions to be observed in using blocks and tackle. How may a set of blocks and tackle be arranged for storage or moving so as to avoid tangling?

   What is meant by the mechanical advantage of a set of blocks and tackle? How may the mechanical advantage of a set be determined at a glance?

8. Take two blocks and a rope, and demonstrate exactly how to reeve the blocks.

7. Harness and Belt Work

REPAIRING of harness is not difficult for anyone with a moderate amount of mechanical ability or one who commonly does shop-work in wood or metal. There is available on the market a wide range of harness hardware, such as buckles, loops, hame staples, and repair links, which may be installed on a harness with the use of only such tools as a hammer, vise, file, cold chisel, and punches. With such pieces of equipment, and with a knowledge of harness riveting and leather sewing, a farmer can easily keep harness in good repair.

MAJOR ACTIVITIES

1. Making a Waxed Thread

2. Making a Sewed Splice

3. Making a Riveted Splice

4. Attaching Snaps and Buckles

5. Replacing Worn and Broken Harness Parts

6. Cleaning and Oiling Harness

7. Selecting Belts

8. Taking Care of Belts

9. Determining Pulley Sizes and Speeds

10. Lacing a Belt with Rawhide Thong

11. Lacing a Belt with Metal Hooks

12. Making a Belt Endless

13. Lagging a Pulley

1. Making a Waxed Thread

To make a sewed splice, it is first necessary to make a waxed sewing thread. Three to five strands of No. 10 linen thread are needed.

Tearing the Thread.   Draw the desired length of thread, usually about 5 ft., from the ball and tear it off.   Do not cut it off.   To tear the thread, untwist it for a distance of 6 to 8 in. at the point to be torn, and then jerk it apart.   It will tear at about the middle of the untwisted part.   A good way to untwist the thread for tearing is to roll it between the palm and the right thigh (see Fig. 311*A*), while holding with the left hand at a point about 8 in. back from the thigh. A few short quick jerks (see Fig. 311*B*) is all that is required to tear the untwisted thread.

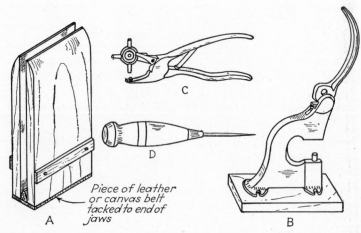

Fig. 310.—Harness repair tools: A, homemade harness sewing clamp for use in a vise; B, hand machine for using tubular rivets; C, leather punch; D, harness sewing awl.

Thread should be torn instead of cut, in order to give a long tapered end that can be threaded easily into the eye of a sewing needle.

To keep the supply of thread from becoming tangled, the ball may be kept in a small can with a hole punched in the lid for the thread to be drawn through.

Assembling the Threads.   After the desired number of strands, usually three or four, have been withdrawn from the ball and torn off to approximately the same length, assemble them with the ends offset or staggered (see Fig. 312), extending the first strand about 1½ in. beyond the second, and the second about 1½ in. beyond the third, and so on.   Staggering the ends gives a long, finely tapered point to the sewing thread.

Waxing the Ends of the Threads.   After the strands have been assembled, wax the tapered ends by rubbing them with a small piece of *harness-maker's wax* (see Fig. 313).   A good way to keep and use

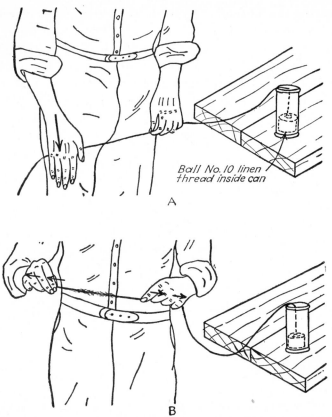

Ball No. 10 linen
thread inside can

A

B

Fig. 311.—Harness thread should be torn, not cut. First draw the desired length from the ball and untwist it by rolling over the thigh, as at A. Then tear it apart with a few short, quick jerks, as at B.

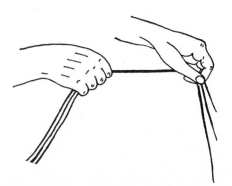

Fig. 312.—To give a fine tapering point to the finished thread, the individual strands are assembled with ends offset or staggered.

the wax is to put it on a pad of scrap leather.   In cold weather, it may be desirable to warm the wax a little.

Twisting and Waxing the Threads.   After the ends are waxed, stretch the thread over a nail or hook and twist it, one end at a time, by rolling it over the right thigh with the palm of the right hand (see

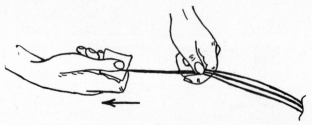

Fig. 313.—After assembling the strands, wax the ends.

Fig. 314).   Hold one end of the thread tight in the left hand, and catch and hold the twist as it is made in the other end by rolling. After both ends are twisted, pull the thread back and forth over the hook in order to even the twist in the two ends.   A little practice will indicate how much to twist the thread.   Too much twist will cause serious kinking and tangling, and too little twist will cause the thread

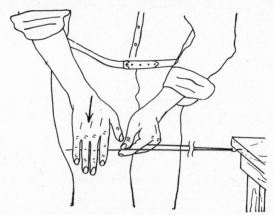

Fig. 314.—Twist the thread by running it around a nail or hook and then rolling it over the thigh, one end at a time.

to be flat instead of round.   The thread should kink slightly when slack, however.   Otherwise, it will be flat and not round and firm when sewed into leather.   It is a common mistake among beginners not to twist enough.

After the thread is twisted, wax it by rubbing the wax back and forth along the thread several times (see Fig. 315).   Then rub the thread with a piece of leather or with the thumb and fingers to work the wax in and distribute it evenly. Apply enough wax to make the thread black all over after it is well distributed. In case the thread is sticky, rubbing it with a piece of *beeswax* will help, but keep beeswax off of the ends that go through the needles.

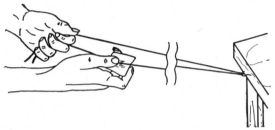

Fig. 315.—After twisting, wax the thread. Apply only a moderate amount of wax and work it in well.

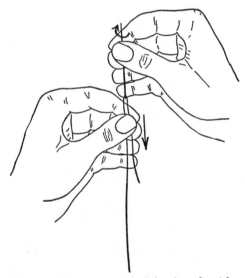

Fig. 316.—Fastening a needle to the end of the thread. After threading, roll the needle in a clockwise direction and slowly move the left thumb and finger down the thread.

Attaching the Needles. Fasten a blunt-pointed harness-sewing needle on each end of the waxed thread in the following manner:

1. Put the end of the thread through the eye of the needle, and push the needle back on the thread about as far as it will go without ruffling up the thread, usually about 2 or 3 in.

2. Fold the pointed end of the thread back alongside the main thread, and hold the doubled thread close to the needle with the thumb and first finger of the left hand (see Fig. 316).

3. Twist the needle clockwise by rolling it between the right thumb and finger while the left thumb and finger move slowly down the thread from the needle.

The end of the thread is thus twisted around and worked into the main thread, and the wax holds it in place. When carefully done the thread just back of the needle is about the same size as the needle.

2. Making a Sewed Splice

Preparing the Ends to Be Spliced. Cut the ends of the pieces of leather square and skive or bevel them back about 2 in. (see Fig. 317). Use a sharp knife or carpenter's plane, and do all beveling on

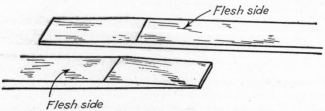

Fig. 317.—Ends of straps beveled or skived ready for sewing.

the rough, or flesh, side of the leather. The hair, or smooth, side is tougher and stronger and should not be cut away in beveling. Do not make the tip ends of the pieces too thin, but leave them about half the normal thickness of the leather.

Stitching the Splice. After the ends are beveled, lap them together about 3½ or 4 in., and place them in a harness-stitching vise or clamp. Be sure to have the hair side of both pieces to the right, and the end nearest you to the right. Place the ends in the clamp so that their top edges project above the jaws of the clamp about ¼ in. or a little less.

A simple homemade clamp (Fig. 310A) used in a woodworking or metalworking vise is quite satisfactory for harness sewing, although a regular harness-maker's clamp or harness-stitching horse might be justified where more than an average amount of harness repairing is to be done.

Punch an awl hole in the single strap beyond the splice, place one needle through the hole and draw the thread about halfway through (see Fig. 318). Punch the second hole through both pieces and about

$\frac{3}{16}$ in. from the first. Insert the left needle, and pull the thread through a little way. Then insert the right needle and pull both threads up tight, keeping the awl in the right hand all the time. In this manner, continue stitching to the end of the splice.

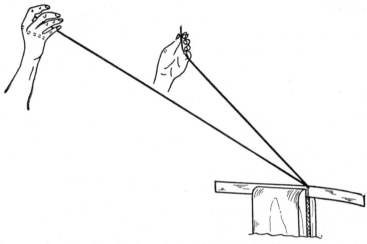

Fig. 318.—Starting the stitching. Place the hair side of both pieces to the right and the end nearest the worker to the right. Punch an awl hole in the single strap just beyond the splice, insert one needle, and draw the thread halfway through.

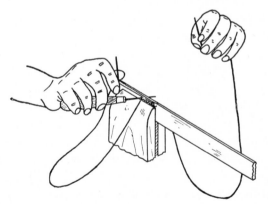

Fig. 319.—Keep the awl in the right hand all the time while sewing.

Punching the Awl Holes. If desired, a line may be marked or creased in the leather about $\frac{3}{16}$ in. from the edge to serve as a guide for the awl holes. This line may be marked with a creasing tool, or a pencil, or one leg of a pair of dividers drawn along a straight edge.

The spacing of the holes may be marked off with a stitching wheel or with a pair of dividers, although this is hardly necessary. With a little practice, the awl holes may be spaced evenly by eye and without first measuring and marking them.

The awl makes a diamond-shaped hole. Hold the awl so as to place the long axis of the diamond about halfway between vertical and horizontal (see Fig. 320). Make only one hole at a time, follow-

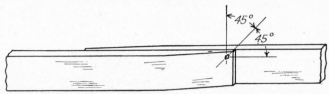

Fig. 320.—The awl makes a diamond-shaped hole. Hold the awl so as to place the long axis of the diamond about halfway between vertical and horizontal.

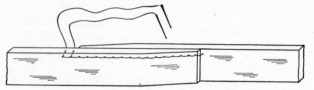

Fig. 321.—Step one in making the crossover.

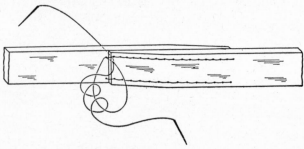

Fig. 322.—Step two in making the crossover. Turn the splice upside down and end for end in the clamp.

ing up immediately with the sewing thread. If several holes are punched before sewing, there will be difficulty in keeping the holes aligned in the two pieces of leather.

Making the Crossover. When the stitching has proceeded across one side of the splice, punch an awl hole through the single strap just beyond the lap. Put both needles through it in the usual manner. Punch a second hole through the single strap, and put the right needle through it. This brings both threads out on the left, or flesh, side of the leather (see Fig. 321).

Turn the splice upside down and end for end in the clamp. Punch two holes in the single strap just beyond the lap. Put one needle through each of these holes, bringing both threads to the right, or hair, side of the leather (see Fig. 322). Take the needle coming

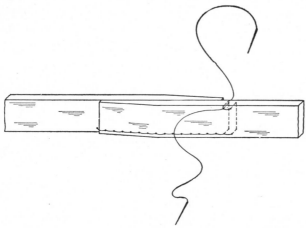

Fig. 323.—The crossover completed and the splice ready to be stitched along the second edge.

through the farthest hole and put it back through the hole nearest the lap, leaving a thread on each side of the splice (see Fig. 323). Then stitch the second edge of the splice in exactly the same manner as the first.

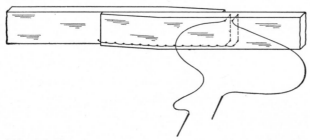

Fig. 324.—Making the lock stitch to prevent the sewing from loosening.

Anchoring the Threads. To prevent the stitching from loosening, the threads may be anchored as follows: Take one regular stitch through the single strap just beyond the splice. Then make an anchor stitch back in the double part of the splice, in line with the last stitch in the double part, but about $\frac{3}{16}$ in. in toward the center of the straps (see Fig. 324). To make this anchor stitch, place both threads through the hole in the usual manner, but draw only one thread, say the one coming through from the left side, up tight. Then wrap this left thread twice in a counterclockwise direction through the loop made by the other thread. Draw both threads up tight. This wraps the threads around each other in the hole. Then cut the threads off.

Some prefer not to make the crossover at all, but to stitch each side of the splice separately and anchor the threads at the end of each side.

After the sewing is done, the splice may be hammered lightly with a smooth-faced hammer or rubbed with a piece of leather to work the stitches in and smooth the wax.

3. Making a Riveted Splice

To make a riveted splice, first bevel the ends of the pieces and lap them in the same manner as for stitching. If solid rivets are to be used, punch holes with a hollow-bit punch (Fig. 310C) or with a punch-type knife blade that will cut a round hole. To complete the job, insert the rivets, place burrs (washers) on the ends, and rivet

Fig. 325.—Styles of harness rivets: A, solid rivet with burr; B, tubular rivet; C, split rivet.

them in place. This kind of riveting is recommended where permanent repairs are being made.

Riveted splices may be quickly and easily made with a hand riveting machine (Fig. 310B) which uses hollow or tubular rivets (Fig. 325B). Such rivets cut or punch their own holes. They are commonly used where repairs must be made quickly, and they are excellent for this purpose.

To use a hand riveting machine, simply place a rivet in the machine, put the pieces of leather to be joined in the machine, and press down on the handle. Use at least two rivets in each splice.

Another type of rivet sometimes used for riveting leather is the split rivet, which has two prongs (Fig. 325C). To use such a rivet, simply drive it through the leather with a hammer, and then bend the prongs over to clinch them and hold the rivet in place. By careful manipulation, or by placing the leather against a piece of iron while the rivet is being driven, the ends of the prongs may be turned back and made to clinch themselves in the leather.

4. Attaching Snaps and Buckles

A buckle or a snap may be attached to a part of a harness by (1) sewing, (2) riveting, (3) both sewing and riveting (see Fig. 326), or (4) using the conway loop (Fig. 327*F* and *G*), or other special harness

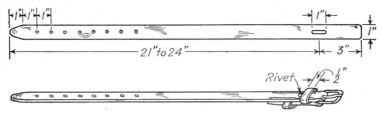

Fig. 326.—A homemade hame strap. The buckle is attached by sewing and riveting.

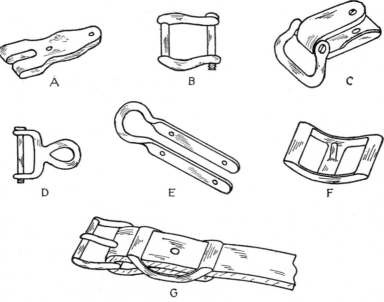

Fig. 327.—Harness repair hardware: A, buckle repair clip; B, screw hame loop; C, bottom hame clip; D, screw cockeye; E, hame clip; F, conway loop; G, buckle attached with conway loop.

hardware, such as the buckle-repair clip (Fig. 327*A*). Sewing and riveting is probably best, although simply riveting or the use of a conway loop is faster and easier and for many purposes practically as good.

To attach a buckle to a leather strap, first cut the end of the strap off square, and round the corners slightly. Then bevel the strap on the rough, or flesh, side back about an inch from the end. Make the slot for the tongue of the buckle by punching two holes in the leather about an inch apart and then cutting out between the holes with a sharp knife. If desired, a leather loop or keeper may be sewed into place by careful work (see Fig. 326).

5. Replacing Worn and Broken Harness Parts

It is usually a simple matter to replace harness parts like cockeyes on the ends of the traces, hame staples, hame buckles, and hame clips.

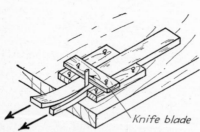

Fig. 328.—A convenient method of splitting leather. A knife blade stuck into the bench top does the cutting, and a few strips of thin wood tacked to the bench guide the strap as it is pulled through.

Such items of harness hardware (see Fig. 327) are usually available at hardware stores and harness-repair shops. The exact methods of removing the old parts and installing the new are easily determined. Heads of old rivets may be cut off with a file, hacksaw, cold chisel, or grinding wheel. Old rivets are then easily punched out, and new parts attached, new rivets inserted and riveted into place. A vise of some kind is valuable for this kind of work. Although such repairs are simple and easy to make, careful and thorough work will pay dividends in improved appearance and durability.

6. Cleaning and Oiling a Harness

Harness may be cleaned by washing with warm water and mild soap or warm water and sal soda. A good method is to take the harness apart and soak it in a tub of warm water into which a handful of sal soda has been dissolved, and then to scrub with a stiff brush. A sloping drainboard or a shallow wide trough with the upper end supported on a sawhorse and the lower end on the tub makes an excellent place to wash and scrub the parts of a harness.

After washing, hang the harness in a warm, dry place. While still slightly damp, apply a good grade of harness oil by rubbing with a sponge or cloth. As the moisture dries out, the oil penetrates into the leather.

Only a good harness oil or a compound of animal oils, such as neat's-foot oil and tallow, should be used in oiling leather. Motor oil or machine oil should never be used because of its detrimental effect on leather.

7. Selecting Belts

Leather belts are usually considered superior to other kinds under conditions favorable to their use.   They are more expensive, however, and they should be used only where they will be protected against moisture, steam, and oil.   Their use is therefore somewhat limited on farms.   The best grades of leather belts are made from the backs of hides.

Rubber belts are made of alternate layers of canvas and rubber vulcanized together.   They are cheaper than leather belts, are not injured by moisture or heat, and are therefore more widely used on farms.

Canvas belts are made of layers of canvas folded and stitched together.   They are treated with materials to make them waterproof. They will stand considerable abuse and are widely used as drive belts on machines like threshers and corn shellers.   They are not recommended for use over pulleys a fixed distance apart, because of their tendency to stretch.   As a drive belt on portable machines, however, they are not objectionable, since the tractor can be "backed into the belt" tighter to compensate for stretching.

8. Taking Care of Belts

Keep belts clean and free from machine oil and grease.   If they become dry and hard after a period of use, clean them and treat them with a suitable dressing.   Use a dressing recommended by the maker of the belt if possible.   Neat's-foot oil makes a good dressing for leather belts.   Another good dressing for leather belts is one made of two parts of edible beef tallow and one part of cod-liver oil.   Melt the tallow, and allow it to cool a little; then add the cod-liver oil, and stir until the mixture is cold.

Rubber belts usually need no dressing.   Washing with soap and water will generally keep them in good condition.   In case the surface of a rubber belt does become hard and dry, however, a light dressing of castor oil may be applied after the belt has first been cleaned.

Treat canvas belts occasionally with a light application of castor oil or raw linseed oil.

The object of applying dressing to a belt is to keep it soft and pliable.   In this condition, it can better conform to the shape of the pulleys and transmit power with the least slippage. The use of rosin, tar, or similar sticky materials is not recommended for belt dressings.

Run leather belts with the hair, or smooth, side next to the pulleys, and rubber belts with the seam on the outside away from the pulleys. The larger the area of contact between a pulley and a belt, the less the chance for slippage. For this reason, large pulleys are better than small ones and crossed-belt drives are generally preferred over straight or open-belt drives. Wherever possible, avoid the use of vertical drives with one pulley directly above the other. Such an arrangement makes it difficult to keep the belt tight on the lower pulley.

9. Determining Pulley Sizes and Speeds

Pulley speeds are designated in revolutions per minute (abbreviated r.p.m.). For most efficient service, a driven machine usually needs to be operated at a definite speed; likewise, the tractor or the motor used to drive it should run at a definite speed. Therefore, a problem frequently arises as to what size of pulley to use on a driven machine to make it run at the desired speed, or just what the speed of the machine will be if a certain size of pulley is used on it. The following simple rule is used for both purposes:

The r.p.m. of the driving pulley $\times$ its diameter = the r.p.m. of the driven pulley $\times$ its diameter

When any three of the factors are known, the fourth is easily found. For example, suppose a tractor pulley is 15 in. in diameter and runs 650 r.p.m. and it is desired to know the speed it will drive an ensilage cutter if the cutter pulley is 18 in. in diameter. Substituting in the formula,

$$600 \times 15 = 18 \times \text{r.p.m. of cutter}$$

Solving by simple algebra,

$$\text{R.p.m. of cutter} = \frac{600 \times 15}{18} = 500$$

The rule may also be stated in other forms, two of which are as follows

(1) R.p.m. of driven pulley = (r.p.m. of driving pulley $\times$ its diameter) $\div$ diameter of driven pulley

(2) Diameter of driven pulley = (r.p.m. of driving pulley $\times$ its diameter) $\div$ r.p.m. of driven pulley

10. Lacing a Belt with Rawhide Thong

Lacing a belt with rawhide thong, although requiring careful work, is not at all difficult. Various types of laces are used. For

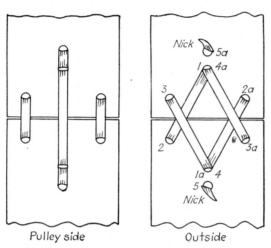

Fig. 329.—The single straight lace for thin narrow belts. The numbers at the holes indicate the order of placing the thong through the holes.

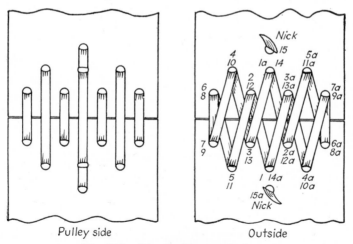

Fig. 330.—The double straight lace.

light belts, the single straight lace (Fig. 329) is good. For moderate duty, the double straight lace (Fig. 330) is probably the most generally used. For belts that work around small pulleys, the double hinge lace (Fig. 331) is recommended; and for heavy duty, the double lock lace (Fig. 332).

Punching the Holes for Laces. In preparing a belt for lacing, first cut the ends off square, and mark the locations for the holes very

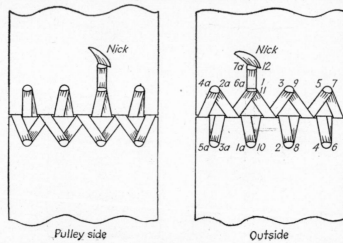

Fig. 331.—The double hinge lace. This lace is recommended where belts work over small pulleys.

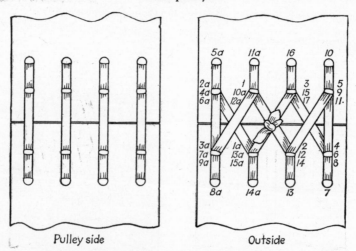

Fig. 332.—The double lock lace is good for heavy duty.

carefully. Figure 333 gives suggested locations and spacings of holes for some common widths of belts. Holes should be no larger than necessary on account of the weakening effect on the belt. Holes $\frac{3}{16}$ in. in diameter are large enough for laces $\frac{1}{4}$ in. wide. Oval holes with the long axis parallel with the edges of the belt are better than round holes.

Threading the Laces. With most styles of lacing, the thong is started in one of the holes in the middle of the belt. One end is then threaded back and forth from one end of the belt, across the joint to the other, working to one edge of the belt and then back to the center. The other end of the lace is threaded to the other edge of the belt and then back to the center. The order of lacing through the various holes is indicated by the numbers in the Figs. 329 to 332. One end of the lace goes from hole 1, to hole 2, to 3, to 4, etc. The other end goes from hole 1a, to hole 2a, to 3a, to 4a, etc.

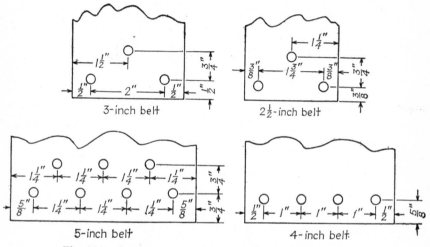

Fig. 333.—**Suggested spacing of holes for common belt laces.**

In making the hinge lace, the thong is threaded through the holes in exactly the same manner as for straight laces, except that it is always passed through the joint instead of directly across.

Fastening the Ends of the Laces. The ends of the laces must be anchored or fastened to prevent them from pulling back out of the holes. Where no idler pulleys are used and the outside of the belt does not run against a pulley, the ends may be tied together with a square knot (see Fig. 332). A good method that may be used regardless of idlers is to draw the end of the lace through a very small hole and then cut a small diagonal nick along the edge of the lace to form a small fish-hook type of barb (see Figs. 329 to 331). As the lace tends to pull back through the hole, this barb flares out and prevents it from going farther,

11. Lacing a Belt with Metal Hooks

Where many belts are used, it will probably pay to buy and use commercial metal belt laces or hooks rather than to depend altogether on leather-thong laces. Some metal belt laces may be applied with

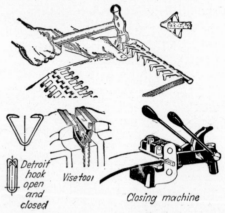

Detroit hook open and closed   Vise tool   Closing machine

Fig. 334.—Metal belt laces and method of applying them: alligator above, clipper below.

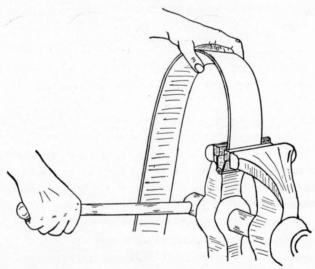

Fig. 335.—Applying metal belt hooks to the end of a belt by means of a special tool used in a vise.

a hammer, while others require a small machine or a tool that may be used in an ordinary vise. Metal belt laces or hooks are usually more easily and quickly applied than leather-thong laces and, in many cases, last longer and are more generally satisfactory.

To apply a lacing of metal hooks, first cut the ends of the belt off square. Then select a suitable size of hooks, put them in place on the end of the belt, and drive them in, with a hammer, a vise, or other tool as may be required. Simple directions accompanying packages of metal laces or machines for applying them indicate any particular precautions that should be observed.

12. Making a Belt Endless

Leather, canvas, and rubber belts may be bought endless if so ordered. Leather belts may be cemented or glued together and made endless in the shop. The lap or splice should be about as lcng as the width of the belt. The splice may be made as follows: Cut the ends off square, being careful, when allowance for the lap is made, that the belt will be of the desired length. Then bevel the ends to a feather edge with a sharp plane. Bevel the flesh side of one end and the hair side of the other. Scrape the bevels with a sharp wood scraper or the edge of a freshly cut piece of glass. Apply a suitable belt cement or glue, and clamp the pieces together. A clamp may be made of two boards held together with screw clamps.

13. Lagging a Pulley

By lagging a pulley is meant covering its face with leather or other belting material. The purpose of lagging is to increase the gripping ability of the pulley and to decrease belt slippage.

A good way to lag a pulley is as follows: Drill rows of holes at intervals across the face of the pulley to receive rivets for fastening the covering to the pulley (see Fig. 336). Make the holes of such a size that the rivets will fit tightly. Cut the end of the pulley covering off square, and punch holes to fit the first row of holes drilled in the pulley. Rivet the end of the covering to the pulley. Punch another row of holes across the covering to fit the next row of holes in the pulley. Locate the holes so that the covering will have to be stretched tightly for the holes to align with the holes in the pulley. With the pulley locked so it cannot turn, stretch the covering tightly and insert the second row of rivets. In a similar manner, continue stretching and riveting the covering at intervals around the pulley.

A good way of stretching the pulley covering while it is being riveted is as follows: Use a piece of lagging material somewhat longer than required to go around the pulley. Punch holes in the end of the material, and secure it to some sort of pry bar or lever by means of leather-thong belt lacing. A pressure or a pull applied to the lever or pry bar will then stretch the material. Another method of temporarily securing the end of the lagging to the pry bar is tying with rope.

Be careful to drive the rivets in straight and to make their heads lie flush or slightly below the surface of the pulley covering. It is important that the rivets fit the holes tightly. If leather is to be used

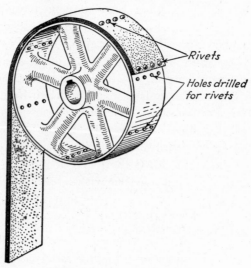

Rivets

Holes drilled for rivets

Fig. 336.—Lagging a pulley.

for lagging, it is a good plan to soak it in water for two or three hours before it is applied. Rubber or fabric belting should be applied dry.

Jobs and Projects

1. Make a waxed harness-sewing thread, fasten on the needles, and make one or two practice splices in short pieces of leather.
2. Examine your work carefully for such points as uniform stitches, well-waxed and twisted thread which does not become flat, line of stitching uniform distance from edge of leather, and secure anchoring of ends of thread.
3. Clean, repair, and oil a harness. Take it apart and wash it, and restitch or repair with rivets wherever needed before oiling.
4. Make a hame strap, a dog collar, or some other piece of leather work, possibly even a halter, for which you have need.

5. Using two pieces of cardboard to represent the ends of a belt, lay out and punch holes for a suitable type of leather-thong lacing. Lace the cardboard with shoe string.

6. Inspect any canvas, rubber, or leather belting that may be used on machines about your home or farm. Are they clean, soft, and pliable, and are the lacings or joints in good repair? If not, clean them, apply such dressings as you deem advisable, and make such repairs to lacings as may be needed.

7. Study various types of belt laces that are used in your community, comparing them as to cost, ease of applying, and general suitability.

8. A tractor has a belt pulley $9\frac{1}{4}$ in. in diameter, and it runs at a speed of 1,078 r.p.m. What size pulley should be used on an ensilage cutter to be driven by the tractor in order to give a cutter speed of 550 r.p.m.?

9. An electric motor runs at a speed of 1,725 r.p.m. It is to drive a hammer mill at a speed of 2,200 to 2,600 r.p.m. Suggest sizes of pulleys that would be suitable for the motor and the mill.

10. A tractor with a belt pulley $12\frac{3}{4}$ in. in diameter and running at a speed of 975 r.p.m. is to be used to drive a wood saw at a speed of 2,000 r.p.m. What size pulley should be used on the saw?

11. An electric motor with a speed of 1,725 r.p.m. and a pulley 3 in. in diameter is to be used to drive a grinder that has a pulley $2\frac{1}{2}$ in. in diameter. How fast will the grinder run?

8. Concrete Work[1]

CONCRETE is simply a mass of sand and gravel held together by a cement paste that has hardened.   The cement paste is made of Portland cement and water.   If the cement paste is rich and strong and good concrete practices are followed, the concrete will be strong. On the other hand, if the paste is weak and watery or if the work is carelessly done, the concrete will be weak and poor.

Concrete is an ideal material for foundation walls, floors, walks, steps, water tanks, and countless other jobs on the farm.   It is economical, strong, durable, sanitary, and attractive in appearance. Although passably good results may be achieved even with poor workmanship, good workmanship really pays and results in a definitely superior product.   It is just as easy to make good concrete as it is to make poor concrete.   Following a few simple principles makes the difference.

MAJOR ACTIVITIES

1. Selecting Good Materials

2. Determining Proportions of Materials

3. Estimating Quantities of Materials Needed

4. Building and Preparing Forms

5. Reinforcing Concrete

6. Measuring Materials

7. Mixing Concrete

8. Placing Concrete

9. Finishing Concrete

10. Protecting Fresh Concrete While Curing

11. Removing Forms

12. Making Watertight Concrete

---

[1] The illustrations in this chapter are furnished through the courtesy of Portland Cement Association.

1. Selecting Good Materials

Good-quality concrete depends upon the use of good materials. The materials for making concrete are (1) fine aggregate (usually sand), (2) coarse aggregate (pebbles, gravel, crushed stone, etc.), (3) Portland cement, and (4) water.

Fine aggregate includes all particles from very fine (exclusive of dust) up to and including those which will just pass through a screen having meshes ¼ in. square. Coarse aggregate includes all pebbles or broken stone ranging from ¼ in. up to 1½ or 2 in. The maximum size of coarse aggregate to be used is governed by the nature of the work. Usually, the largest dimension of any piece of aggregate

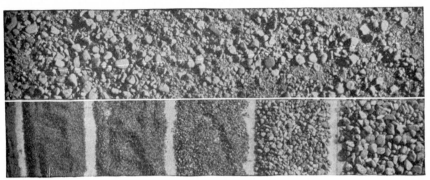

Fig. 337.—A sample of well-graded sand before and after it has been separated into various sizes. Particles vary from fine up to pebbles that will just pass through a ¼-in. mesh screen.

should not exceed one-third the thickness of the slab or wall in which it is used.

Sand. Sand should be hard and clean, that is, free of dust, loam, clay, and vegetable matter. These foreign materials are objectionable because they prevent a good bond between the cement and the particles of sand, and thus reduce the strength of the concrete and increase its porosity. Concrete made with dirty sand or pebbles hardens slowly at best and may never harden satisfactorily.

Sand should be well graded, that is, the particles should not all be fine or all coarse, but should vary from fine to the size that just pass a screen having meshes ¼ in. square (see Fig. 337). If the sand is well graded, the finer particles help to fill the spaces between the larger particles. A given amount of cement will then bind together a greater amount of aggregate, and thus increase the amount of concrete that can be made from a sack of cement.

Coarse Aggregate. Pebbles or crushed stone to be used in a concrete mixture should be tough, fairly hard, and free from any of the impurities that would be objectionable in sand. It is desirable also that coarse aggregate be well graded for the same reasons that it is desirable that sand be well graded (see Fig. 338).

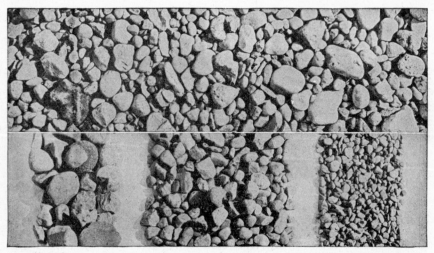

Fig. 338.—A sample of well-graded coarse aggregate before and after it has been separated into various sizes. Note the variety of sizes, the smaller stones filling in the spaces between the larger ones.

Bank-run Gravel. The natural mixture of sand and pebbles as taken from a gravel bank is usually referred to as *bank-run material*. In bank-run material, fine and coarse aggregates are seldom present in the right proportion to produce a good, workable, economical mixture. Most gravel banks contain too much sand. Money can usually be saved by screening out the sand and then recombining the materials in the correct proportions.

Cement. Many brands of Portland cement are on the market. Practically all are made to meet standards adopted by the United States Government and the American Society for Testing Materials. Therefore, cement that is bought from reputable dealers and stored properly may be expected to give satisfactory results.

Since cement readily absorbs moisture from the atmosphere, it should be stored in a dry place. Any cement containing lumps so hard that they do not readily pulverize when struck lightly with a shovel should not be used.

Water. Water used to mix concrete should be clean and free from oil, alkali and acid. In general, water that is fit to drink is good for concrete.

Testing Sand and Gravel for Silt. Sand and gravel may be easily tested to determine if they contain injurious amounts of fine clay or silt as follows (see Fig. 339):

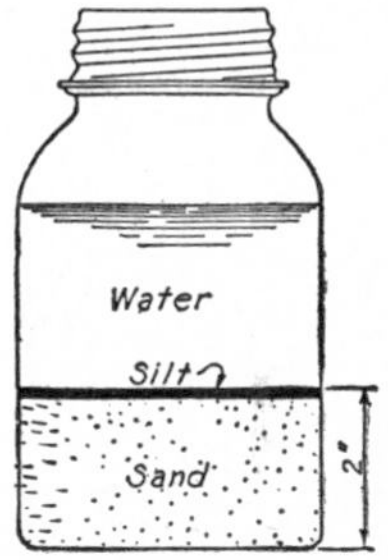

1. Place 2 in. of a representative sample of the sand or gravel in a pint fruit jar.
2. Add water until the jar is almost full, fasten the cover, shake vigorously, and then set the jar aside until the water over the material becomes clear.
3. Measure the layer of silt on top of the sand or gravel. If the layer is more than ⅛ in. thick, the material is not clean enough for concrete.

Fig. 339.—If a 2-in. sample of sand contains more than ⅛ in. of silt, the sand should be washed or rejected.

Washing Sand and Bank-run Gravel. Sand and gravel containing too much silt or clay may be washed on a wide, shallow,

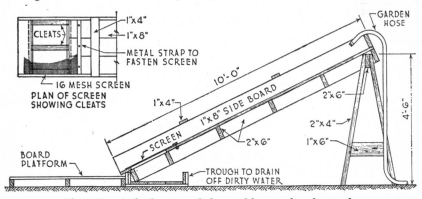

Fig. 340.—A sloping trough for washing sand and gravel.

sloping trough (see Fig. 340). To wash the material, shovel it on to the high end, and drench it with water by means of a hose, pipe, or pail, washing out the objectionable silt or clay. Retest the material after washing to make sure that it is clean.

Testing Sand and Gravel for Vegetable Matter. A test to see whether sand or gravel contains too much decomposing vegetable matter for use in concrete may be made as follows:

1. Place ½ pint of water in a colorless 1-pint fruit jar, and dissolve a heaping teaspoonful of household lye in it.
2. Pour ½ pint of a representative sample of the sand or gravel into the jar containing the lye water.
3. Cover the jar, and shake vigorously for 1 to 2 min.
4. Set the jar aside for 24 hr., and then inspect it in a good light.
5. If the water is clear or colored not darker than cider vinegar, the material is suitable for use in concrete. If the water is darker than this, however, the material should not be used unless it is washed first to remove the objectionable vegetable matter.

2. Determining Proportions of Materials

Table 5 gives suggested mixes for various kinds of concrete work on the farm. The 1-2¼-3 mix (1 sack of cement to 2¼ cu. ft. of sand

TABLE 5. SUGGESTED CONCRETE MIXES*

Use of concrete	U.S. gal. of water per sack cement with average moist sand	Sand and gravel per sack cement		Largest size of gravel, in.
		Sand, cu. ft.	Gravel, cu. ft.	
Most farm construction such as floors, steps, basement walls, walks, yard pavements, silos, grain bins, and water tanks..............	5	2¼	3	1½
Concrete in thick sections and not subject to freezing. Thick footings, thick foundations, retaining walls, engine bases.............	5½	2¾	4	1½
Thin reinforced concrete such as milk-cooling tanks, fence posts, thin floors, most uses where concrete is 2 to 4 in. thick...............	5	2¼	2½	¾
Very thin concrete such as top course of 2-course floors and pavements, concrete lawn furniture, most uses where concrete is 1 to 2 in. thick............................	4	1¾	2¼	⅜

* These are trial mixes for average conditions. *It is particularly important to use not more water per sack of cement than shown in the table.* If sand is very wet decrease amount of water used 1 gal. per sack of cement. If sand is dust dry increase amount of water ½ gal. per sack of cement. Change proportions of sand and gravel slightly if necessary to get a workable mix.

and 3 cu. ft. of gravel) is used for most jobs. The mixes suggested in the table are for average conditions and may be used as trial mixes. The proportions of sand and gravel may be changed slightly if necessary to give workable mixes. Do not use more water per sack of cement than suggested, however, as this will produce a weaker concrete. By a workable mix is meant one that is smooth and plastic and that will place and finish well. It should not be so thin that it runs, nor so stiff that it crumbles (see Fig. 341). It should be rather sticky when worked with a shovel or trowel. For most jobs, a workable mix is one that is "mushy" but not "soupy."

3. Estimating Quantities of Materials Needed

The amounts of cement, sand, and coarse aggregate needed for a given job may be estimated closely by first figuring the volume of concrete to be made, and then using tables such as Table 6, 7, or 8. The quantities given in these tables are for average conditions, and, in any particular case, the quantities needed may vary 5 or 10 per cent from those given.

TABLE 6.   APPROXIMATE AMOUNTS OF MATERIALS REQUIRED
PER CUBIC YARD OF CONCRETE

Use of concrete	Sacks of cement	Sand, cu. yd.	Gravel, cu. yd.	Largest size of gravel, in.
Most farm construction such as floors, steps, walks, tanks, and silos. 1-2¼-3 mix	6¼	⅔	¾	1½
Concrete in thick sections and not subject to freezing. Thick footings and foundations, etc. 1-2¾-4 mix	5	⅔	¾	1½
Thin reinforced concrete such as milk cooling tanks, fence posts, slabs 2 to 4 in. thick. 1-2¼-2½ mix	6½	⅔	¾	¾
Very thin concrete as for lawn furniture top course of 2-course floors, concrete 1 to 2 in. thick. 1-1¾-2¼ mix	8	⅔	¾	⅜

Suppose, for example, that a concrete tank is to be built and the volume of the forms is figured to be 46 cu. ft., or approximately 1¾ cu. yd. As shown in Table 6, 1 cu. yd. of 1-2¼-3 concrete requires approximately 6¼ sacks of cement, ⅔ cu. yd. of sand, and ¾ cu. yd. of gravel. Since 1¾ cu. yd. of concrete is needed, the *approximate* amounts of materials may be found as follows:

$$1\tfrac{3}{4} \times 6\tfrac{1}{4} = 11 \text{ sacks of cement}$$
$$1\tfrac{3}{4} \times \tfrac{2}{3} = 1\tfrac{1}{4} \text{ cu. yd. of sand}$$
$$1\tfrac{3}{4} \times \tfrac{3}{4} = 1\tfrac{1}{3} \text{ cu. yd. of gravel}$$

It is usually good practice to allow a little extra, say 5 to 10 per cent, for waste and variations in the work.

Table 7 may be used for figuring the amounts of materials needed for floors, walls, or other plain flat slabs of concrete of the most com-

TABLE 7. APPROXIMATE AMOUNTS OF MATERIALS REQUIRED PER 100 SQ. FT. OF 1-2¼-3 MIX CONCRETE

Thickness of concrete, in.	Concrete, cu. yd.	Sacks of cement	Sand, cu. yd.	Gravel, cu. yd.
4	1⅓	7¾	¾	1
6	2	11⅔	1	1⅓
8	2½	15½	1⅓	1¾
10	3	19⅓	1¾	2¼
12	3¾	23	2	2⅔

TABLE 8. APPROXIMATE AMOUNTS OF MATERIALS REQUIRED PER 100 SQ. FT. OF PORTLAND CEMENT MORTAR OR CONCRETE

Thickness of mortar or concrete, in.	Amount of mortar or concrete, cu. yd.	Mix proportions				
		1-3		1-1¾-2¼		
		Sacks of cement	Sand, cu. ft.	Sacks of cement	Sand, cu. ft.	Gravel (⅜ in.), cu. ft.
⅜	⅛	1	3			
¾	¼	2	6			
1	⅓	2¾	8	2⅔	6	7
1½	½	...	..	4	8	10
3	1	...	..	8	16	19

monly used mix, 1-2¼-3. For example, the materials needed for a basement floor 20 by 30 ft. and 4 in. thick may be computed as follows: From Table 7, 100 sq. ft. of concrete of this mix and 4 in. thick requires 7¾ sacks of cement, ¾ cu. yd. of sand, and 1 cu. yd. of gravel. For 600 sq. ft. (20 × 30 = 600), the approximate quantities of materials needed would be

$$7¾ \times 6 = 46½ \text{ sacks of cement}$$
$$¾ \times 6 = 4½ \text{ cu. yd. of sand}$$
$$1 \times 6 = 6 \text{ cu. yd. of gravel}$$

It will be noted from Table 6 that approximately ¾ cu. yd. of gravel and ⅔ cu. yd. of sand are required for 1 cu. yd. of concrete.

This applies to all the various grades or kinds of concrete listed in the table. Therefore, if these figures can be remembered, approximate amounts of gravel and sand may be easily estimated without a table.

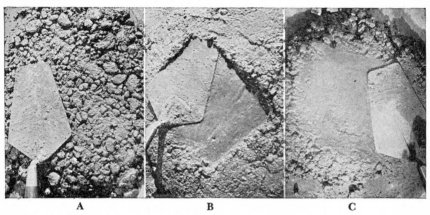

A   B   C

Fig. 341.—Mixes with different degrees of workability. A does not contain enough cement-sand mortar to fill the spaces between pebbles. This mix is hard to work, rough, and porous. B has an excess of mortar. It is plastic and workable but is expensive owing to the small yield of concrete per bag of cement. It is also likely to be porous. C contains the correct amount of mortar. With light troweling the spaces between pebbles are filled. Note how it hangs together on the edges of the pile. This mix gives maximum yield of concrete and produces smooth, dense surfaces.

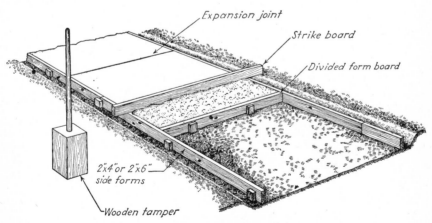

Fig. 342.—Forms for a concrete sidewalk. Concrete walks are often poured in alternate sections. After these harden, the intermediate sections are then poured.

Simply compute the volume of the forms or the volume of concrete required, and multiply by $\frac{3}{4}$ to determine the amount of gravel and by $\frac{2}{3}$ to determine the amount of sand needed. It will be noted also from the table that about six sacks of cement are required for 1 cu. yd. of concrete of the mixes recommended for most farm jobs.

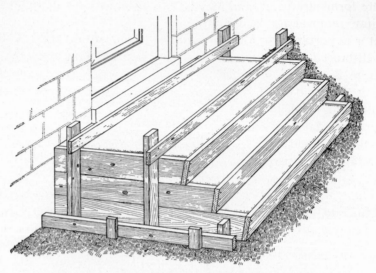

Fig. 343.—Simple forms for building concrete steps.

4. Building and Preparing Forms

Forms, generally made of wood, hold the concrete in place until it hardens. Rough lumber may be used where appearance is not

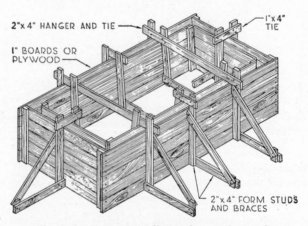

2"x 4" HANGER AND TIE

1"x 4" TIE

1" BOARDS OR PLYWOOD

2"x 4" FORM STUDS AND BRACES

Fig. 344.—Forms for a livestock watering tank.

important, but where a smooth finish is desired, the forms should be carefully built of good smooth lumber. Tongue-and-groove lumber or shiplap is commonly used to give tight joints. Construct forms so that they can be easily removed without damage to the fresh concrete and with the least possible damage to the form lumber. Where forms are to be used again, they may be built in sections to facilitate removal and reassembly.

It is important that forms be built tight and strong and that they be well-braced in position to prevent bulging or leaning when the wet concrete is tamped into place. Wood forms are commonly oiled with used engine oil to prevent warping and to prevent the concrete from sticking to them.

For foundation work below ground, forms are not necessary if the soil will stand without caving or crumbling and the excavation is carefully dug to size (see Figs. 157 and 158, page 120).

5. Reinforcing Concrete

Concrete, like stone, is strong in compression and can support very heavy loads that tend to mash or crush it. Steel rods or other

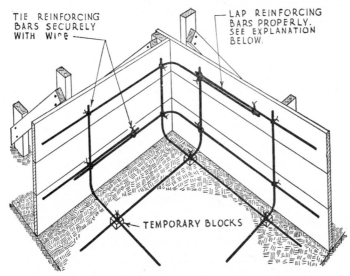

TIE REINFORCING BARS SECURELY WITH WIRE

LAP REINFORCING BARS PROPERLY. SEE EXPLANATION BELOW.

TEMPORARY BLOCKS

Fig. 345.—Reinforcing bars should be carefully lapped at the joints and securely tied and supported in place to ensure proper location after the concrete is poured.

forms of reinforcement should be used, however, where the loads or forces tend to pull it apart or bend it. It is important that the reinforcement be placed where it will do the most good in helping to resist the pulling or bending forces. For example, in a concrete beam, as a lintel over a window or a door, place the reinforcement near the lower side, for this is the side that tends to stretch or pull apart when the beam is loaded. Important or elaborate structures, such as floors aboveground, beams, columns, and retaining walls, should be designed and supervised by a competent engineer or contractor.

Steel reinforcing bars and wire mesh are generally used for reinforcement. Do not use scrap iron and rusty fence wire. Reinforcement should be free from rust and other coatings. Place reinforcement carefully as may be specified in plans, and then securely wire or anchor it to prevent dislodging when the concrete is poured (see Fig. 345). Lap the ends of steel bars a distance of 12 in. for each ¼ in. of diameter. For example, ¼ in. bars should be lapped 12 in.; ⅜ in. bars, 18 in.; ½ in. bars, 24 in.; etc.

6. Measuring Materials

All materials, particularly water, should be accurately measured. A pail marked in gallons and half gallons is convenient for measuring water. Sand and gravel can be easily measured by using a bottomless box or frame made to hold exactly 1 cu. ft. or any other desired volume (see Fig. 346). To measure materials, place the bottomless box on the mixing platform and fill. When the required amount of the material has been placed in it, lift the box and the material remains on the platform.

Cement is usually bought in sacks, each sack holding 1 cu. ft. Quantities less than 1 cu. ft. are easily measured as portions of a cubic foot in a bottomless 1-cu. ft. box or in pails.

Pails may also be used for proportioning small batches of materials. For example, a 1-2¼-3 batch could be made by taking 1 pail of cement, 2¼ of sand, and 3 of pebbles or stone.

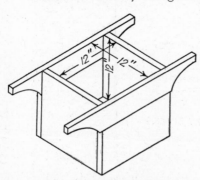

Fig. 346.—A bottomless box holding a definite amount, such as 1 cu. ft., makes a convenient measure for proportioning sand and gravel.

7. Mixing Concrete

Although machine mixing is preferred, first-class concrete can be mixed by hand. Whichever way is used, continue the mixing until every pebble is completely coated with a thoroughly mixed mortar of cement and sand. In machine mixing, run the mixer 1 to 2 min. after all materials, including water, have been placed in it.

Fig. 347.—Steps in mixing concrete by hand. For best results, concrete must be thoroughly mixed.

A tight barn floor or some paved area makes a good surface on which to mix concrete by hand. If such a place is not available, a watertight mixing platform should be made. If much concrete is to be mixed, have the platform large enough for two men, using shovels, to work upon at one time. A good size is 7 ft. wide and 12 ft. long. The platform is preferably made of matched lumber so that the joints will be tight. Nail strips along three sides to prevent materials from being pushed off in mixing.

The usual procedure for mixing by hand is as follows (see Fig. 347):

1. Spread the measured amount of sand on the platform, and then spread the required amount of cement evenly over the sand.
2. Turn the material with square-pointed shovels until the color becomes uniform and the sand and cement are thoroughly mixed. There should be no streaks of brown and gray, as such streaks indicate incomplete mixing.
3. Next, measure out the required amount of coarse aggregate, and spread it in a layer on top of the cement-sand mixture.
4. Mix until the pebbles have been uniformly distributed throughout the mass.
5. Then make a depression or hollow in the middle of the pile, and add the correct amount of water while the materials are turned. Continue mixing until all materials are thoroughly and uniformly mixed.

8. Placing Concrete

Concrete should be placed in the forms within 30 min. after mixing. Deposit it in layers of uniform depth, usually not exceeding 6 in. Tamp well or spade as it goes into the forms (see Fig. 348). By "spading" is meant working a spade or chisel-edged board in the concrete, particularly near the forms, working the spade up and down and to and fro. This works the larger pieces of aggregate away from the forms into the mass of concrete and ensures a smooth, dense surface when the forms are removed. Care should be used not to dislodge any reinforcement that may have been placed in the forms.

If work must be stopped before the forms are filled, roughen the surface with a stiff broom before it hardens. Then before resuming the placing of concrete, wet the surface and cover it with a layer of cement mortar about ½ in. thick. This helps to make a tight joint between the old and new work. The mortar may be made of 1 part of cement, 2½ parts of sand, and enough water to give a "mushy" (not "soupy") consistency.

9. Finishing Concrete

After concrete is placed, level it off in the forms with a straight-edged board or wood float, and then smooth it with a wood float to make an even surface. Further finishing should be delayed until the concrete is quite stiff.

Fig. 348.—Thorough spading of concrete as it is placed in the forms helps assure smooth, watertight walls.

If a gritty nonskid surface is desired, as on walks or barn floors, use a wood float for final finishing. Using a steel trowel would result in an oversmooth surface that would be slippery when wet.

Where a smooth surface is desired, as in a poultry-house floor, do the final finishing with a steel trowel. Overtroweling should be avoided, however, as this produces a surface that tends to check and dust off after hardening.

Where more than normal traction is desired, as on cattle walks or some pavements, the surface may be finished with a stiff broom.

10. Protecting Fresh Concrete While Curing

Concrete needs moisture to cure or harden properly. New concrete, therefore, should be protected from drying out for at least

Fig. 349.—After concrete becomes quite stiff but is still workable, use the wood float to make a compact smooth surface. No further finishing is required to produce a gritty nonslip surface.

Fig. 350.—To produce a smooth, dense surface, finish with a steel trowel after the concrete has become quite stiff.

5 days. Cover floors and other horizontal surfaces with burlap, earth, straw, etc., and keep this material wet for the required time. Cover walls with burlap or canvas, and keep such covering wet. Some protection may be afforded by leaving the forms in place.

Concreting in Cold Weather. New concrete should be protected from freezing for at least 3 days. If concrete work is done in freezing temperatures, heat the mixing water and the aggregate and protect the work after pouring by covering with straw, manure, paper, or canvas, or by building enclosures around the new work and heating with stoves. Aggregate may be heated by building a fire inside an old smokestack, metal culvert, or steel barrel laid on its side and piling the aggregate over it. Cement should not be heated, nor should water hotter than 150°F. be mixed with cement.

Fig. 351.—A broomed finish makes a good nonslip surface where more than normal traction is desired, as on cattle walks.

11. Removing Forms

Forms should not be removed until the concrete has hardened sufficiently to be self-supporting and until there is no danger of damage to the concrete in removing the forms. The time will vary from 1 day to 2 weeks or more, depending upon the weather, the nature of the work, etc. In summer, wall forms may generally be removed after 1 or 2 days, and in colder weather in 4 to 7 days. Forms for roofs and floors over basements should not be removed in less than 7 days in summer and 14 days in cold weather.

If rough or honeycombed surfaces are found when the forms are removed, patch them by working a stiff cement mortar into the holes or crevices with a wood float. One part of cement and $2\frac{1}{2}$ parts of sand makes a good mortar. It is better to avoid rough or honeycombed surfaces, however, by using a good workable mix and spading it well into the forms.

12. Making Watertight Concrete

Watertight concrete can be made by using a fairly rich, properly proportioned mix and doing a good job of mixing, placing, and curing. No special ingredients are required. Any of the mixes listed in Table 5, except the 1-2¾-4 mix, will make watertight concrete, if the work is carefully and properly done. It is particularly important that no more water be used in the mix than that specified in the table and that the concrete be kept moist for at least 7 days.

Jobs and Projects

1. Test two or three samples of sand for the presence of silt, using the fruit-jar or washing test.
2. Test samples also for the presence of harmful amounts of vegetable matter, using the lye-water test.
3. Using a ¼-in. screen, separate a sample of bank-run material into fine and coarse aggregate. Take a definite amount of material, such as a gallon, and measure the coarse and fine material in suitable measures, as pints. Does this sample of bank-run material have about the right proportions of coarse and fine material for average good farm concrete? (See Table 5 for trial mixtures.)
4. Examine samples of sand and of coarse aggregate for gradation or variation in size of particles. If suitable size screens are available, separate the samples into piles by screening, and then weigh or measure the amounts of material in each pile. If screens are not at hand, simply spread the material out thin on a paper and note by inspection the amounts of fine, medium, and coarse particles. Refer to Figs. 337 and 338 for appearance of good samples.
5. Make a small trial mix of concrete that would be suitable for average farm work such as for floors or steps. See Table 5 for suggested proportions. Decide upon just how much water should be used, and be careful to use no more than just this amount.
   Trowel the material thoroughly, and see if it makes a good workable mix that is smooth and plastic. Is it somewhat sticky when worked with the trowel? Is it somewhat "mushy" (not "soupy")? Does the material hang together reasonably well on the edges of the pile? (See Fig. 341 for the proper appearance of a workable mix.)
   If your trial mix is not satisfactory, just what should be done to make it better? Add the materials to change the batch, but do not change the proportions of water and cement.

6. How much cement, sand, and coarse aggregate will be needed for each of the following jobs?

A drive 7 ft. wide, 6 in. thick, and 150 ft. long.

A basement floor 4 in. thick, 30 ft. wide, and 36 ft. long.

A foundation wall 8 in. thick and 30 in. high for a poultry house 20 ft. wide and 30 ft. long.

7. How many square feet of sidewalk 4 in. thick can be built with one sack of cement?

8. Make a list of a few improvements made of concrete that you would like to have about your home or farm, such as steps, walks, well curbs and platforms, and poultry-house or garage floors.

Select one or two of these jobs that you think you could do, and make a list of cement, sand, and gravel, and also a list of form lumber that would be required. Estimate the costs.

If arrangements can be made to do one or two of these jobs, plan the work carefully and do them.

If it should not be feasible to do some concrete job as mentioned above as an individual project, it may be possible to do it as a class project. If so, each member of the class should help with the planning as well as the doing of the job, and when it is done, make a short, well-organized report, outlining all steps of the job, and including estimates of amounts of materials required, amounts actually used, costs, etc.

9. Soldering and Sheet-metal Work

SOLDERING is one of the most useful processes commonly done in the farm shop. Many valuable and timesaving appliances, as well as repairs, can be made with a very small outlay for materials and equipment. Soldering is really easy to do, but it is often not done well, because of failure to understand the process or careless methods of doing it.

When two pieces of metal are joined by soldering, the molten solder runs between them, fills up the spaces, and fuses with and penetrates into the surface of the pieces. Upon cooling, the solder solidifies and binds the pieces together. Soldering is essentially an alloying process, or a process in which metals are fused and mixed together at the point of connection. It is therefore important that conditions be kept favorable for alloys to form while soldering. The two most important conditions are as follows:

1. The metals to be joined must be thoroughly cleaned of all grease, dirt, and oxide or tarnish and kept clean (usually with the aid of fluxes).
2. The pieces themselves must be heated and kept somewhat above the melting temperature of solder for a short time.

If the surfaces to be joined are dirty or coated with oxide, solder cannot alloy or mix with them. Likewise, if insufficient heat is applied to the pieces, the solder will not intermix well with the surfaces, and inferior work will result.

The solder most commonly used is composed of equal parts of lead and tin and has a melting temperature lower than that of either lead or tin. It is available in the form of bars, solid wire, hollow wire filled with a flux core, or ribbon. For large jobs requiring considerable solder, it is usually more economical to buy it in bars. For the occasional job, acid-core or paste-core solder is convenient and satisfactory. Flux-core wire solders are more expensive than plain bar, wire, or ribbon solder, but where only a small amount of soldering is to be done, the added convenience of having the flux in the solder is well worth the extra cost.

MAJOR ACTIVITIES

1. Operating a Gasoline Blowtorch

2. Cleaning Surfaces to Be Soldered

3. Applying Fluxes

4. Cleaning, Tinning, and Using Soldering Irons

5. Soldering Different Metals

6. Repairing Small Holes

7. Patching Large Holes

8. Soldering a Seam or Joint

9. Repairing Tubing

10. Laying Out Sheet-metal Work

11. Cutting Sheet Metal

12. Folding and Forming Joints

13. Riveting Sheet Metal

14. Fastening Sheet Metal with Self-tapping Screws

1. Operating a Gasoline Blowtorch

The gasoline blowtorch is useful not only for heating soldering irons and pieces being soldered, but also for many other jobs about the farm, such as heating a nut that is stuck, thawing a frozen pipe, or warming the intake pipe of an engine on a cold morning. The details of a blowtorch are shown in Fig. 352. Ordinary untreated motor fuel may be used in a blowtorch, although stove or lamp gasoline is much superior and should be used whenever available. Ordinary motor fuel has a tendency to form gum and clog the small passages in the torch. Gasoline treated with lead should not be used because of its poisonous nature.

The torch is operated in the following manner:

1. *Fill the fuel chamber* about three-fourths full with clean gasoline through the filler plug in the bottom. Use only moderate pressure in tightening the plug. If gasoline leaks around it, rub a little laundry soap on the threads.

2. *Pump air into the chamber.* Ten or twelve strokes of the pump will usually be enough if the pump is in good condition.

3. *Fill the priming cup* by opening the control valve a little and placing the thumb over the end of the burner to deflect the gasoline down into the cup. Be careful not to let gasoline overflow onto the torch or bench. If it does, wipe it up thoroughly before lighting.

4. *Light the gasoline* in the priming cup, and protect the flame from any strong winds or drafts. It is essential that the burner be well heated to vaporize the gasoline.

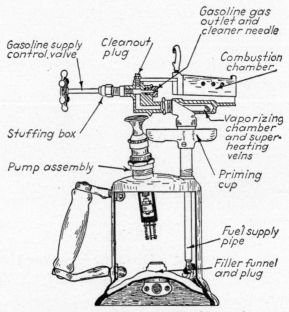

Fig. 352.—Details of a gasoline blowtorch.

5. *Light the torch* by opening the control valve just before the gasoline in the priming cup burns out. Do not open the valve too soon. Give the burner time to heat. If necessary use a match, applying it to the air holes in the side of the burner and not at the end of the burner.

6. *To turn the torch out, turn the control valve just enough to stop the flow of gas,* and no tighter. Screwing the valve too tight is likely to damage the seat. The sheath around the needle valve will contract when the torch cools, forcing the valve very tight against its seat and possibly damaging the valve or the seat.

Remedying Blowtorch Troubles. *Pump Troubles.* If the pump fails to pump air, it is usually because of the drying out of the pump

leather. In such a case, remove the pump plunger and oil the leather. It is a good plan to put a drop or two of oil on the leather occasionally to keep it soft and pliable. After the leather becomes worn, it should be replaced.

Another common cause of pump trouble is improper action of the check valve on the bottom of the pump. The valve is usually made of cork and is held against its seat by small springs. If the valve becomes cracked or if dirt prevents it from seating, the air in the torch will leak back through the pump and the plunger will not stay down. Sometimes a check valve will stick shut and not open to admit air from the pump to the fuel chamber.

In case of trouble with the valve, remove the pump and examine the valve. If it is only dirty, simply clean it; if it is cracked, install a new piece of cork.

Gasoline Leaks. If gasoline leaks around threaded joints, unscrew them and apply common laundry soap to the threads. If leaks occur around the stuffing box of the control-valve stem, tighten the stuffing box nut *slightly* with a wrench.

Torch Burns with Pulsating Red and Blue Flame. If there is a strong pulsating flame, first red and then blue, the burner is not hot enough. The flame should be turned out, the priming cup refilled, and the torch regenerated. The cup probably cannot be filled from the torch itself, and gasoline will have to be supplied from some outside source, as a squirt can. Sometimes a torch can be made to warm up by pointing the flame straight down against the ground. The burner is thus heated so that it can better vaporize the gasoline.

Torch Burns with Weak Flame. If the flame is weak and cannot be increased by opening the control valve further, then there is either (1) not enough air pressure in the chamber, or (2) the gasoline passages are partly clogged. Pump air into the chamber. If this does not remedy the trouble, then it can be assumed that the control-valve opening or some of the other gasoline passages are partly clogged. If a small particle of carbon or dirt is lodged in the control-valve opening, closing the valve and then opening it two or three times will usually dislodge the particle.

If this fails to remedy the trouble, it is likely that the fuel-supply tube or some of the passages in the vaporizing chamber are clogged with dirt, gum, or carbon, especially if the torch is old or has been used a long time. Many torches have a cotton wick in the fuel-supply pipe to strain the gasoline and to prevent pulsation of the flame.

After long use, the wicking may disintegrate or become clogged with dirt and have to be replaced.

If the passages are clogged, take the torch apart very carefully and clean it. Soaking the parts in kerosene or in alcohol will help to clean them. The passages may be blown out with compressed air or with a tire pump. Running small wires through the passages will sometimes help. Care must be used not to damage the small parts, particularly the control-valve orifice and the threaded plugs and openings. Coat all threads with laundry soap before reassembling.

Manufacturers will ordinarily repair torches at reasonable cost if they are sent back to the factory. This may be more satisfactory than attempting to clean and repair them at home.

2. Cleaning Surfaces to Be Soldered

Solder will not stick to metal that is dirty or coated with oxide or tarnish. The first step in soldering, therefore, is to clean the parts

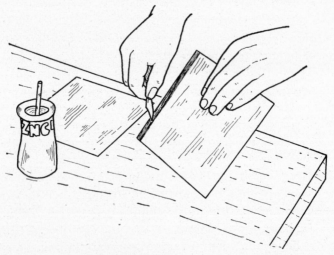

Fig. 353.—One of the first steps in soldering is to clean the work thoroughly. This may be done by scraping with a dull knife.

thoroughly and to remove all oxide or tarnish. This is commonly done by mechanical means, as scraping with a dull knife (see Fig. 353), filing, or rubbing with steel wool or emery cloth, or by the use of fluxes, or by the use of both mechanical means and fluxes.

Metals oxidize or tarnish to some extent when exposed to air even for short periods. When metal is heated as it must be in soldering, the oxidation takes place much more rapidly. A flux is therefore needed to remove the last trace of oxides just at the instant of solder-ing. A flux applied to a surface after cleaning and before heating helps exclude air and thus keeps the formation of oxides to a minimum.

A flux also fills the space between the soldering iron and the piece being soldered and thus better enables the heat to flow from the iron to the work. A flux, therefore, may be considered as aiding the soldering process in the following ways: (1) removing oxides, (2) preventing or minimizing oxidation while the work is being heated, and (3) aiding the flow of heat from the soldering iron to the work.

Kinds of Fluxes. Various materials in the form of pastes, liquids, or powders are used as soldering fluxes on different metals.

Soldering pastes under different trade names are available at hardware stores. They are compounded from various materials, and most of them make excellent fluxes for most common metals. They are easily applied and are generally less messy and less corrosive than liquid fluxes.

Muriatic acid (commercial hydrochloric acid), when diluted with equal parts of water, is a very effective flux for soldering galvanized iron and zinc. It may be bought at drugstores. Because of its corrosive nature, muriatic acid must be used sparingly and with care. Muriatic acid may be used also as a cleaner for some metals like iron and steel, after which a suitable flux, such as zinc chloride, is applied. Muriatic acid may also be used to etch stainless steel before applying a flux.

Zinc chloride, or *cut acid*, as it is frequently called, is a common flux that can be used on most metals. It may be prepared as follows:

1. Drop small pieces of zinc into a bottle about half full of muriatic acid, adding more pieces from time to time until no more zinc will dissolve and there is a slight excess of zinc left in the bottle. The resulting liquid is zinc chloride. Zinc may be obtained from an old fruit-jar lid or the shell of an old dry-cell battery. Zinc from such sources should be carefully cleaned before using.
2. After all chemical action has stopped, strain the zinc chloride through a cloth, or allow the dirt to settle and pour off the clear liquid.
3. Dilute the zinc chloride with one-fourth to one-half its volume of water.

Care should be taken not to get the acid on the hands or clothing. *Neither acid nor zinc chloride should be kept around tools;* nor should zinc chloride be made around tools, as the vapors or fumes will cause severe corrosion. If acid or other flux should be spilled on tools, wipe it off at once, wash with strong soap, rinse, and apply a coating of oil.

Rosin is sometimes used for soldering bright tin. A small quantity of powdered rosin is sprinkled on the part to be soldered, and when the hot soldering iron is applied, it melts and spreads over the surface. Another method of using rosin is to dissolve it in alcohol or gasoline and apply as a liquid. Rosin is a very mild flux. It is used where extreme caution must be taken against corrosion.

Tallow is a good flux for soldering lead. After the lead is thoroughly scrapped, it should be heated slightly, after which the tallow is applied to the warm surface.

Sal ammoniac is a good flux for cleaning and tinning soldering irons. It may be obtained in cakes, in lumps, or in powdered form. A teaspoon of powdered sal ammoniac, or the equivalent in lump form, dissolved in a pint of water makes a good cleaning solution into which a hot soldering iron may be dipped quickly and only for an instant and thus cleaned.

Small cakes of sal ammoniac, especially prepared for cleaning and tinning irons, are available at hardware stores. These are quite satisfactory, and their use is generally recommended.

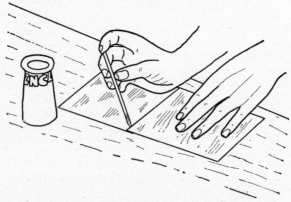

Fig. 354.—**After cleaning the work, apply flux. Liquid flux can be applied easily with a medicine dropper or a hollow glass tube.**

3. Applying Fluxes

Liquid fluxes may be easily applied with a medicine dropper or a hollow glass tube. To use a hollow glass tube, lower it down into the bottle of the flux, and then place a finger tightly over the upper end. A small amount of flux is thereby trapped in the lower end of the tube and may be easily transferred to the work. To release the flux, simply remove the finger from the upper end of the tube (see Fig. 354). Small brushes may be used for applying fluxes that are not too corrosive.

Spread paste fluxes on the work with a small piece of wood, such as a matchstick, preferably after the work has been heated slightly.

Do not use more flux than necessary, and be careful not to get flux on parts that are not to be soldered, because many fluxes are corrosive, and all of them are somewhat messy.

4. Cleaning, Tinning, and Using Soldering Irons

Soldering irons are really made of copper, because of its resistance to oxidation and corrosion and because of its ability to absorb and

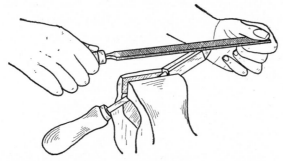

Fig. 355.—A clean well-tinned iron is essential. If the surface of the iron is pitted or rough from overheating, it may be smoothed and cleaned by filing while either hot or cold.

give up heat readily. The best size of iron for average farm shopwork is the one that weighs about 1 lb. In general, the larger the iron that can be conveniently handled, the better. Large irons require heating less often. An iron that is too large, however, is clumsy and cannot be handled with ease.

An electric soldering iron is internally heated by current passing through a heating element built into the iron. Electric irons are very convenient, particularly where small soldering jobs occur frequently. They are much more expensive than common soldering irons, however, and are somewhat more fragile and subject to damage. They are therefore not used extensively in farm shops.

By tinning an iron is meant simply coating the faces of the pointed end with solder. Good work cannot be done unless the iron is kept well tinned. If the tinning becomes burned, then the iron cannot transfer heat readily from itself to the work being soldered. Also, an untinned iron is usually a dirty iron and is a hindrance to good work.

If the surface of an iron is pitted and rough from overheating, smooth it and clean it by filing, either while hot or cold (see Fig. 355). In extremely bad cases, hammer the end of the iron, either while hot or cold, to smooth and reshape the point. Be careful not to get the point too long or too short, but to retain the original shape.

After cleaning, tin the iron by heating and applying flux and solder. Probably the best way to do this is to rub the hot iron in 2 or 3 drops of molten solder on a cake of sal ammoniac (see Fig. 356). Another way is to dip the hot iron quickly into and out of a cleaning

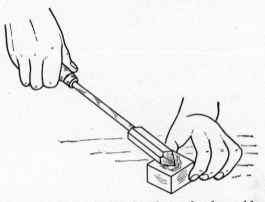

Fig. 356.—Rubbing a clean hot iron in a few drops of molten solder on a cake of sal ammoniac is a good way to tin it.

fluid (which can be made by disolving a teaspoon of sal ammoniac in a pint of water) and then rub it in molten solder.

Heating the Soldering Iron. The gasoline blowtorch is quite satisfactory for heating soldering irons. It furnishes a clean, intense flame, and it can be readily taken to wherever it is needed. Any kind of heat that is reasonably clean, however, may be used for heating soldering irons. Where gas is available, a small gas-heated bench furnace is ideal. Such a furnace may be easily made by mounting a burner from an old gas stove in a sheet-metal enclosure on a metal stand.

To heat a soldering iron with a blowtorch, place the whole end of the iron—not just the tip—in the flame (see Fig. 357B). Once the iron is up to operating temperature, it may be pulled back out of the flame, as at A, Fig. 357, to prevent overheating and yet keep it hot.

Keeping the Iron at Proper Working Temperature. A good workman is always careful to keep his iron *clean, well-tinned, and at a good working temperature.* Only poor work can be done with an iron that is too cold. The solder will melt and spread slowly and unevenly, and the work will be rough and lumpy rather than smooth and mirrorlike. It will be difficult or impossible to heat the work up to the melting point of solder (and this is very important), and consequently a poor bond will form between the solder and the metal.

On the other hand, it is a common mistake of beginners to overheat the iron and burn off the tinning. In this condition, the iron

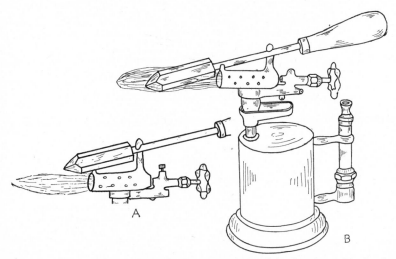

Fig. 357.—The gasoline blowtorch is commonly used for heating soldering irons. Once an iron is heated, it may be kept at working temperature without overheating by pulling it back out of the flame as at A. If the point of the iron changes from a silver to a yellowish color, it is too hot.

is practically worthless for soldering, and it must be retinned. If the tinning begins to turn from a silverish to a yellowish color, it is getting too hot and should be removed from the flame or placed in a cooler part of the fire. The use of an overheated iron may burn some of the solder, or the zinc coating in the case of soldering galvanized iron, forming small particles of cinder or ash. These impurities will then be incorporated in the joint, making it weak and rough and marking it as the work of an inexperienced or careless workman. An iron should therefore be heated until it will readily melt solder, but not until it is so hot that the bright tinning on the point begins to turn yellow.

A good way to judge the temperature of a soldering iron is to note the sound it makes when it is quickly dipped into a cleaning solution. If it is at the right temperature it will make a sharp snap or crack that is easily recognized; if too hot, a gurgling noise; and if too cold, very little or no noise.

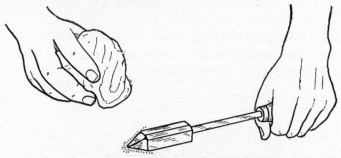

Fig. 358.—Keep the iron clean while soldering by occasionally wiping it quickly with a clean damp cloth.

Keeping Soldering Irons Clean. Soldering irons in use will continually become dirty and will require frequent cleaning, usually immediately after every heating. An iron may be cleaned while hot by (1) wiping it quickly with a damp rag (see Fig. 358), (2) dipping it quickly and only for an instant into a cleaning fluid, or (3) rubbing it on a cake of sal ammoniac and then wiping it with a damp cloth.

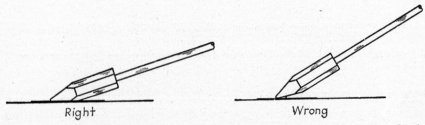

Right   Wrong

Fig. 359.—Press the whole face of the soldering iron—not just the point—firmly against the work.

Using the Soldering Iron. The soldering iron is used primarily for two purposes: (1) to heat the metal being soldered up to soldering temperature and (2) to apply solder to the metal.

To heat the work quickly and easily, press the hot tinned iron *firmly* against it, being sure that both the iron and the work are clean and that a suitable flux has been applied. Hold the flat face of the iron—not just the point—against the metal being soldered (see Fig.

359). Move the iron *slowly* over the work, so as to allow time for the heat to flow from the iron to the metal. Heat may be also applied to the work by the direct flame of a torch (see Fig. 360). This method is especially good when working on large pieces.

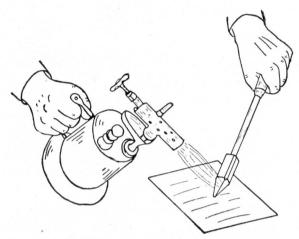

Fig. 360.—Heat may be applied directly to the work while soldering. This method is especially good when soldering heavy or thick pieces.

Fig. 361.—A good way to pick up solder with a hot iron is to bring the iron up under the end of a bar of solder.

Generally the best method of applying solder to small pieces is to pick it up on the iron, a drop or two at a time, and transfer it to the work. To get the solder from a bar to the iron, allow the bar to project over the edge of the bench, or a block or brick on the bench, and bring the hot iron up against the end of the bar *from beneath*, melting one or two drops off onto the iron (see Fig. 361).

When using flux-core wire solder, put the hot iron in place on the metal to be soldered, raise the heel very slightly, and feed the wire under it (see Fig. 362). In this way, the flux does not evaporate or waste away before reaching the work.

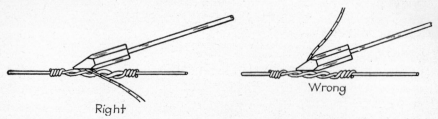

Right   Wrong

Fig. 362.—In using flux-cored solder, apply it to the work and to the face of the iron in contact with the work. In this way the flux does not evaporate or waste away before reaching the work.

Keeping the Work Clean. After every application of the iron, give the solder a moment to cool, and then wipe the work with a damp rag to remove the dirt that always accumulates (see Fig. 363). If the iron is to be applied again, then put another light coat of flux on

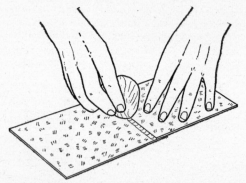

Fig. 363.—Keep the work clean while soldering by wiping it frequently with a damp cloth. When the job is done, wipe off all excess flux to promote cleanliness and prevent corrosion.

the work. When the job is completed, wipe off all excess flux. This is especially important if acid or other corrosive flux has been used.

5. Soldering Different Metals

Various metals respond differently to soldering operations. Some metals oxidize or tarnish more rapidly than others, or oxides may be more difficult to remove, thus requiring special fluxes or special methods for their removal. A flux that works well on one metal may not be suitable for another. Also, metals are affected differently by heat. To do good soldering, as pointed out previously, the work must be heated somewhat above the melting temperature of solder. Some metals, like lead, melt at low temperatures, and special care must be used not to melt them, yet they must be heated above the melting point of solder. Some metals are good conductors of heat and conduct the heat rapidly away from the soldering iron, making it difficult to get the surfaces being soldered above the melting point of solder. In such cases, it is advisable to use a large soldering iron and to heat it as hot as possible without burning the tinning on the point, and possibly also to apply heat direct from a blowtorch or some other means.

Soldering Tin Plate. The metal usually called "tin," such as is used in "tin" cans, is really tin-plated iron or steel sheet. It is the tin coating that must be considered, however, when soldering it. Tin-plated iron or steel is very easy to solder. Several different fluxes work well on it. Commercial soldering pastes are usually quite effective and may be less corrosive than zinc chloride. If it is important that no corrosion take place, use rosin as a flux. It may be sprinkled on the parts to be soldered and then melted with the soldering iron, or a little may be dissolved in gasoline or alcohol and then applied with a small brush or swab.

Use a clean well-tinned soldering iron. Be careful not to heat the metal so hot as to burn off the tin plating. The metal is usually thin, and it heats easily and rapidly. Apply flux sparingly and only to those parts to be soldered. Flow the solder well into the joint or seam, and do not pile it up on the outside. After the work has cooled, wipe it with a damp rag to remove dirt and excess flux.

Soldering Zinc and Galvanized Iron. Galvanized iron is zinc-coated iron. The best flux for soldering zinc and zinc-coated metals is muriatic acid, or raw acid, diluted with about equal parts of water. The muriatic acid acts with the zinc and forms zinc chloride. The melting temperature of zinc is rather low, and it is therefore important not to use a soldering iron so hot as to melt or burn the zinc. If too hot an iron is used, the zinc will burn and leave the work rough and grainy. In case the zinc coating becomes burned and the work is rough and grainy, melt the solder with a well-heated soldering iron and then quickly wipe the joint with a damp rag. Then reflux and proceed with soldering. Since muriatic acid and zinc chloride are corrosive, be sure to wipe off any excess flux when the job is done. It may be advisable to wash the parts with laundry soap, and then rinse and dry.

Soldering Copper and Copper Alloys. Copper and brass and most other copper alloys are easy to solder. Zinc chloride is probably the best flux, although most commercial soldering pastes work well and may be less corrosive and messy to use. Copper and copper alloys conduct heat very readily, and therefore may be a little difficult to heat to proper soldering temperatures. This being the case, it is important to use a hot, well-tinned soldering iron, to press it firmly against the work, and to move it along slowly. A large iron is usually better than a small one. In some cases, it works well to apply heat direct to the work with a blowtorch. As in all soldering, keep the work clean, apply flux only to those parts to be soldered, flow the solder well into the joints, and wash or wipe off excess flux when the job is done.

Soldering Lead. To solder lead, first remove the oxide or tarnish by scraping with a knife or scraper, leaving the surfaces to be soldered perfectly bright. Tallow, rosin, zinc chloride, or soldering paste may be used for flux. Apply the flux immediately after scraping. Since lead melts at a very low temperature, be careful not to melt it with the hot soldering iron. Heat the iron just hot enough to melt solder readily, and then do not hold it in contact with the lead too long at a time. The secret of success in soldering lead is to control the heating, raising the temperature of the lead parts above the melting point of solder, but not to the melting point of lead.

Soldering Iron or Steel. The first step in soldering iron or steel is to remove thoroughly the oxide or scale from the surfaces to be joined. This may be done by grinding, filing, or scraping. Sometimes it helps to apply muriatic acid after filing, grinding, or scraping. After the surfaces are thoroughly cleaned, then apply zinc chloride and tin them. Use a soldering iron that is well heated. On large pieces, a blowtorch may be used also. After the surfaces are tinned, then solder the parts much the same as with other metals.

Soldering Stainless Steel. Since stainless steel resists corrosion, it also resists the action of fluxes. For this reason, it is recommended that the surfaces of stainless steel parts to be soldered, first be etched with muriatic acid. Apply the acid with a small brush or swab, being careful to get it on only those parts to be soldered.   If by accident acid is applied to other parts, wash it off with laundry soap and rinse.   After etching, wipe off the acid with a damp rag and apply zinc chloride for flux.   Soldering is then completed much the same as for other metals.   Wash off excess flux when the work is finished.

Soldering Aluminum and Aluminum Alloys.   Aluminum and aluminum alloys conduct heat away rapidly and also oxidize rapidly. They are therefore difficult to solder.   Once such a metal is tinned, however, it can then be soldered much the same as other metals except that no flux is used.   To tin aluminum or aluminum alloy, heat it with a blowtorch or other means until it will melt solder.   Then brush it vigorously with a wire brush in the presence of molten solder. This may have to be repeated several times, being careful not to overheat and burn parts already tinned.   Another way recommended for tinning aluminum is to heat it, then melt some solder onto it, and then rub back and forth with a clean hot soldering iron.   Before the solder solidifies, wipe it off quickly with a small piece of steel wool. Apply more solder, and continue to rub with the iron, being sure the iron stays clean and hot.

Special aluminum solders and fluxes are on the market and are usually available in small packages with directions for use.   For the occasional repair job, it is probably best to buy some such solder from a reputable company and then carefully follow directions for its use.

6. Repairing Small Holes

To repair a small hole in sheet metal, first thoroughly clean the metal around the hole, on both the top and bottom sides where practical.   Then apply a suitable flux, usually a paste flux, except on galvanized surfaces which are best treated with muriatic acid.

Next apply a drop of solder to the hole, using a clean well-tinned hot soldering iron.   Use firm pressure on the iron, and move it about slowly so as thoroughly to heat the metal around the hole.   Sometimes it works well to put the point of the iron straight down into the hole and rotate it back and forth slowly (see Fig. 364).

When the solder is thoroughly melted and spread out evenly around the hole, remove the iron and allow the solder to cool.

Turn the work over, if possible, and inspect the other side.   If the solder has not spread out evenly and smoothly around the hole, wipe with a damp cloth, place a little flux around the hole, and then apply a clean well-tinned hot iron. After the job is done, wipe it with a damp cloth to clean it and remove any excess flux.

Repairing a Hole with a Rivet and Solder. A hole that is slightly too large to be stopped with a drop of solder may be plugged with a rivet, and then solder can be applied over the rivet (see Fig.

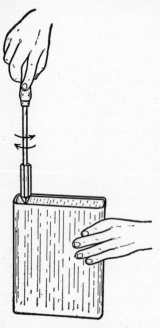

365). To make such a repair, first clean thoroughly around the hole, and then insert a short copper or galvanized rivet, and hammer it down. Do not use a long rivet. If necessary, cut it off. The rivet, of course, should be clean.

Next apply a suitable flux, and flow solder over the rivet and around the hole, using a clean well-tinned hot iron. Solder both the top and bottom sides of the hole. By careful work, a smooth, neat job results.

Fig. 364.—A small hole may be soldered by cleaning and fluxing and then rotating a well-tinned hot iron back and forth slowly with the point in the hole.

7. Patching Large Holes

A hole too large to be stopped with a rivet may be repaired by sweating a patch of metal over it. To make such a repair, first clean the metal around the hole, flux it, and coat it with solder. Likewise, clean the patch itself, and then flux it, and coat it with solder. Be sure to keep the work clean, wiping with a damp cloth between heats as may be required.

Place the patch over the hole, apply flux, and then heat with a clean well-tinned hot iron. When the solder is well melted, hold the patch in place with the tang of an old file, or a piece of scrap iron or wood, until the solder cools. *Never use a tempered tool, like a screw driver or an awl, for this purpose,* as the heat will draw the temper. Also, the flux will likely cause the tool to rust.

It may be necessary to add a little solder around the edges of the patch to make a smooth job. In this case, be careful not to melt the whole patch loose. Use a well-tinned hot iron, and do not hold it in one place too long. Work fast and first on one edge of the patch and then on the other. In this way, there is less danger of melting the patch loose.

It is always a good plan to solder both the inside and the outside of the patch whenever possible, so as to make a smooth, neat job. This requires careful work to avoid melting the patch loose. The main precautions are to use a clean well-tinned hot iron—possibly a little hotter than for average work—and to work rapidly. If the iron is left in contact with the work too long, there is a tendency for heat to spread too far, possibly melting the patch loose.

8. Soldering a Seam or Joint

To solder two pieces together with a plain lap joint, first clean the parts to be joined, and apply a suitable flux. Next, put the pieces

Fig. 365.—An effective way of repairing a medium-sized hole is to clean the metal around the hole, insert a rivet, and then apply solder.

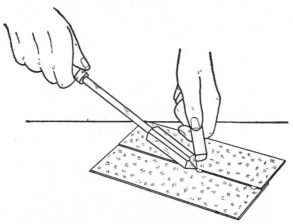

Fig. 366.—In soldering a lap joint, it is best first to "tack" the pieces at intervals and then solder the remainder of the seam.

together and "tack" them in place with a drop of solder at intervals along the seam (see Fig. 366). Move a clean hot iron along, keeping it pressed firmly against the work, and feed solder into the seam by touching it to the iron (see Fig. 367). If there is a tendency for the pieces to melt apart, hold them together with the end of an old file or a piece of scrap iron while the solder cools.

9. Repairing Tubing

A leaky metal tube, such as a fuel or oil line on an engine, can usually be repaired by soldering. If the hole is small, simply file, scrape, or otherwise clean the metal around the hole, apply flux and then solder it. Copper and brass tubes are very easily soldered.

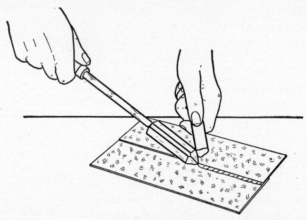

Fig. 367.—Soldering a lap joint after "tacking." When considerable solder is required, hold the end of the bar against the point of the iron as it is drawn along slowly over the work.

If there is a large crack in the tube, it may be advisable to reinforce the joint by wrapping a piece of sheet metal snugly around the break and then soldering it in place. Another method that is often effective is to wrap copper or brass wire tightly and closely around the tube a short distance each way from the hole, and then solder it in place (see Fig. 368). It is necessary, of course, first to clean the tube thoroughly and tin it before wrapping either with sheet metal or wire.

Fig. 368.—A leak in a tube maybe repaired easily by (1) cleaning, fluxing, and tinning; (2) wrapping tightly with clean wire; and (3) applying solder over the wire.

The sheet metal or wire must also be clean. Be sure to flow the solder evenly and smoothly over the whole wrapping.

Points on Soldering

1. Thoroughly clean all parts to be soldered.
2. Use a suitable flux.
3. Heat parts being soldered somewhat above the melting point of solder.
4. Be sure the soldering iron is clean, well-tinned, and properly heated.
5. A general-purpose soldering paste is an excellent flux for most common metals.
6. Zinc chloride, or cut acid, is a good flux for most metals.
7. Muriatic acid is a good flux for soldering zinc and galvanized iron.

8. Wipe off excess flux when the soldering is done. Many fluxes are corrosive.

9. A good way to tin an iron is to heat it, and then rub it in 2 or 3 drops of molten solder on a cake of sal ammoniac.

10. A soldering iron may be cleaned by wiping it while hot with a damp rag, or by dipping it quickly and for just an instant into a cleaning solution or dip.

11. A teaspoon of powdered sal ammoniac dissolved in a pint of water makes a good cleaning solution or dip for soldering irons.

12. Keep the iron at a good working temperature. If it is too cold, it will not readily melt solder. If it becomes overheated, the tinning will be burned from the point.

13. If the bright tinning on the point of an iron begins to turn yellow, it is getting too hot.

14. An iron at a good working temperature makes a sharp snap or crack when it is dipped into a cleaning solution. An iron that is too cold makes very little or no noise; an iron that is too hot makes a gurgling noise.

15. Hold the iron firmly in contact with the parts to be soldered to make it easier for the heat to flow from the iron to the work.

16. Move the iron slowly over the work, advancing it along the seam or joint as the solder melts.

17. To melt solder from a bar onto the iron, allow the end of the bar to project over the edge of a brick on the bench, and bring the iron up against the end of the bar from beneath.

18. As soon as the solder cools, wipe the work with a damp rag to remove dirt which accumulates.

19. Use precaution against accidental fires when filling and operating a blowtorch.

20. In turning out a blowtorch, turn the control valve just tight enough to stop the flow of gas. Screwing it too tight may damage the valve seat or needle.

21. Keep fluxes away from tools.

10. Laying Out Sheet-metal Work

In making appliances of sheet metal, it is very important first to mark out the pattern accurately on the metal, so that it may be cut and bent properly. Marking is best done with a sharp-pointed instrument such as an old saw file that has been ground to a needle point on the end, or an awl. A pencil makes a line that is too wide and indistinct. Also, it is easily erased or smeared by handling. Mark circles and arcs with dividers.

Always use a square to ensure accurate marking of lines at right angles. In marking out patterns for many appliances, it is best first to lay out two base lines at right angles to each other and to do all measuring and squaring from these; or to straighten one edge of the stock and square one end with it, and then use this edge and this end as base lines.

11. Cutting Sheet Metal

The best tool for the average job of sheet-metal cutting is a pair of tinner's snips (see Fig. 369). These tools are made in various styles, as straight snips, curved snips, or hawk-bill snips. The straight snips are used for straight cutting and for cutting on outside curves. They are quite adequate for most farm shopwork. Curved snips and hawk-bill snips are used for cutting inside curves and in close quarters where straight snips would be awkward or difficult to use. For an

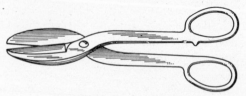

Fig. 369.—Tinner's snips.

occasional job of cutting light sheet metal, even a pair of old scissors may be used.

Using the Snips. To use a pair of snips for straight cutting, open the blades wide and insert the sheet metal all the way back in the throat. Be sure the cutting edge of the upper blade is exactly over the line of cutting, and then squeeze the handles together. It is best not to cut all the way out to the tips of the blades, but to stop and take a new cut. This avoids small nicks and burrs made by the ends of the blades when full-length cuts are made.

When cutting a large sheet, allow the right part to bend down somewhat and pull the left part up a little to allow room for the hand to operate the snips (see Fig. 370). When trimming the edge of a sheet, allow the part trimmed off to curl or roll up (see Fig. 371). Always be careful to keep the cutting edge of the top blade directly over the line to be cut.

When cutting a short distance into the edge of a piece of metal, as in cutting notches, it is usually best to open the blades only a little,

Fig. 370.—Straight cutting with tinner's snips. Keep the cutting edge of the upper blade exactly over the line to be cut.

Fig. 371.—Trimming the edge of a piece of sheet metal.

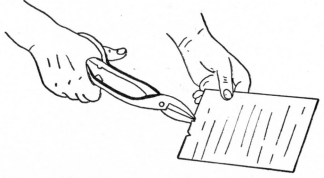

Fig. 372.—When cutting small notches with tinner's snips, use only the tip ends of the blades.

allowing the metal to go back between the blades just the distance to be cut, and then to do the cutting with the tips of the blades (see Fig. 372). When cutting is done well back on the blades, it may not be possible to stop the cut at exactly the point desired.

Cutting Heavy Sheet Metal. Heavy sheet metal may be cut by clamping it in a vise, or between bars or angle irons held in a vise, and then shearing it with a cold chisel (see Fig. 373). In cutting sheet metal in this manner, the following points are important:

1. Clamp the metal with the line of cutting just even with the top of the vise jaws (or the bars or angle irons held in the vise).
2. Keep the bevel of the cold chisel flat against the vise jaw or the bar.
3. Hold the chisel to one side to give an angling shear cut.
4. Use a sharp chisel. It may be advisable to grind the cutting edge keener than for average cutting with a cold chisel.

This same method may be used for cutting thin sheet metal also, but it is difficult to make a smooth, even cut with a cold chisel if the metal is too thin.

12. Folding and Forming Joints

Frequently a hook or lock joint, as illustrated in Fig. 374, can be used to advantage in making appliances of sheet metal. The joint

Fig. 373.—A good way to cut heavy sheet metal with a cold chisel. Use a sharp chisel and keep the bevel of the chisel flat against the angle iron (see also Fig. 383).

is strong, easily made, and gives a neat, finished appearance to a piece of work.

To start such a joint, extend the edge of the metal over the edge of the bench and bend it down with a hammer or a mallet (see Fig. 375*A*); or clamp the metal in a vise or between bars in a vise and

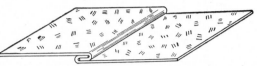

Fig. 374.—The hook joint is a good method of joining pieces of sheet metal. It is strong and it is easily soldered.

bend it (see Fig. 375*B*). The metal may also be bent over the edge of an anvil or over a bar of iron or a piece of hard wood clamped in a vise. After the edges of the two pieces to be joined are bent down about square, then bend them on farther back, forming hooked edges.

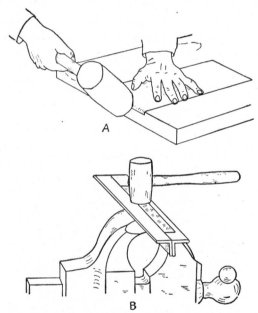

A

B

Fig. 375.—Two good ways of starting the bends for the hook joint. Extend the metal over the edge of the bench and bend with a hammer or mallet as at A; or clamp the metal between irons in a vise and bend as at B.

Then hook the two pieces together, and hammer the joint tight with a mallet or a hammer. The joint is then ready to be cleaned and fluxed if it is to be soldered.

In hammering sheet metal with a hammer, be careful not to stretch or deform the metal with heavy blows. A wooden mallet is usually better than a hammer, because there is less danger of beating the metal out of shape with the mallet.

13. Riveting Sheet Metal

Riveting can often be done to advantage in repairing and making sheet-metal appliances. A good way to rivet two pieces of sheet metal

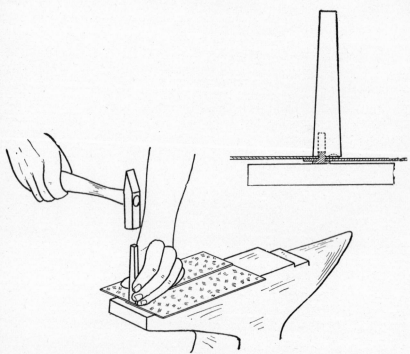

Fig. 376.—A good way to put a rivet through sheet metal is to place the rivet on an anvil or other solid support and then drive the sheet metal down over the rivet with a rivet set thus making the rivet cut its own hole.

together is as follows: Place a rivet on an anvil or other solid support such as a bar held in a heavy vise, and then drive the pieces of sheet metal down over the rivet with a rivet set. The rivet thus cuts its own hole (see Fig. 376). A rivet set is essentially a small bar of steel with a hole drilled up into it to receive the end of the rivets and with a cup-shaped depression for use in forming rounded heads on rivets.

After a rivet is in place in the hole, then hammer down the end and form a head on it. Use a few straight medium-weight hammer blows (see Fig. 377A). Heavy blows or too many blows will cause the metal to stretch and buckle around the rivet.   If a rivet starts to bend, cut it off, remove it, and insert a new one.   Finish the job of heading the rivet with the cuplike hollow in the rivet set if a neat appearance is important (see Fig. 377*B*).

Holes for rivets in sheet metal may be punched or drilled.   Where extreme accuracy is necessary, marking with a center punch and then drilling or punching is better than making rivets cut their own holes.

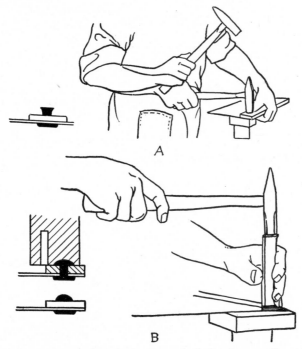

Fig. 377.—A, In hammering down the end of the rivet, strike a few straight, medium-weight blows. Heavy blows or too many blows will cause the metal to stretch and buckle around the rivet. B. Finish the job with the cuplike hollow in the rivet set. (Courtesy of Stanley Tools.)

A good way to punch holes in sheet metal is to use a solid punch over end-grain wood or over a block of lead.

14. Fastening Sheet Metal with Self-tapping Screws

A good way to fasten pieces of sheet metal together is to place the pieces together, make a small hole with an awl or sharp-pointed tool, and then insert a sharp-pointed self-tapping sheet-metal screw.   Such a screw (see Fig. 378) will enlarge the hole slightly and cut threads in it as it is driven up with a screw driver. Screws of this type are available at tinners' shops and hardware stores.

Fig. 378.—Sheet-metal screws are often used for fastening sheet-metal parts together.

This method of fastening is not so tight and secure as riveting, but is much faster and easier, and for many jobs it is entirely satisfactory. It is particularly good for fastening joints of sheet-metal pipe, like stovepipe, together. Pieces fastened with these screws are easily taken apart.

Jobs and Projects

1. Fill, generate, and light a blowtorch; and then heat and tin a soldering iron.

2. Cut small pieces of metal, and practice soldering on them. Stop a few nail holes with solder, leaving a smooth surface on both sides of the metal. Also stop a hole with a rivet soldered in place, and sweat a patch over a larger hole. Practice making both plain lap joints and hook joints.

    After you have practiced repairing different sizes of holes, making joints, etc., cut some other pieces of metal neatly to size, and make samples of work. The work outlined in the following drawing is suggestive.

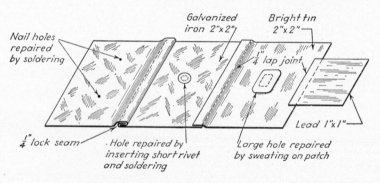

3. Clean and tin a small piece of copper tubing, wrap it tightly with *clean* wire for a distance of about a half inch, and then solder the wrapping in place.

4. Look through shop manuals, books, bulletins, etc., and make a list of small appliances made of sheet metal that you could make in the shop. Select one or two of these that you could use at home and that would give you valuable experience in working and soldering sheet metal, and make them. If the available designs or plans could be modified slightly to suit your needs better, make such changes.

5. A check list of small appliances made of sheet metal includes

Small pan	Lamp shade
Funnel	Book ends
Feed scoop	Dustpan

6. Gather up from around your home, articles made of sheet metal, such as pails, chick water fountains, funnels, and oil cans, that need repairing. Take them to the shop, clean them thoroughly, solder all holes or leaks, and do such other repairing as may be needed to put them in good serviceable condition.

10. Cold-metal Work

COLD-metal work constitutes one of the most important phases of farm shopwork. Most farm machines and many small appliances used on the farm are made of metal. Many valuable repair jobs can be done with the use of only a few simple hand tools, such as a vise, a hack saw, a hammer, cold chisels and punches, and a few files, drills, and threading tools. Every farm shop should have a fair assortment of such tools.

MAJOR ACTIVITIES

1. Distinguishing between Different Kinds of Iron and Steel

2. Laying Out and Marking Metal

3. Cutting with a Cold Chisel

4. Filing

5. Hack Sawing

6. Drilling Holes in Metal

7. Bending Cold Metal

8. Riveting

9. Threading

1. Distinguishing between Different Kinds of Iron and Steel

There are many different kinds of iron and steel which have different uses and properties. The mechanic who works with iron and steel should, therefore, be able to distinguish between these different kinds and know something of their different properties so he can use the best methods of cutting, filing, shaping, etc. A knowledge of the different kinds of steel and iron is also essential for effective work in forging, tempering, and welding.

Pig Iron. The first step in the manufacture of iron and steel is to extract the iron from the iron ore, which is mined in various parts of the world. This is done by means of the modern blast furnace.

The molten iron accumulates at the bottom of the furnace and is drawn off and taken to other furnaces for further refining, or it is cast into short thick bars known as *pig iron*. Pig iron is then used as the source from which other kinds of iron and steel are made.

Cast Iron. To make castings, the pig iron is remelted, together with small amounts of scrap steel and iron, and poured into molds of the desired shape and then allowed to cool and solidify. Cast iron is used extensively because it is cheap and can be readily molded into complicated shapes. It is hard and brittle and cannot be bent. It cannot be forged or welded in the forge, but it can be welded with the electric arc welder or with the oxyacetylene torch. It crumbles when it is heated to a white heat in the forge. It can be drilled and sawed easily and also filed easily after the hard outer shell is removed. Cast iron has a rather coarse-grained texture.

Chilled Iron. Chilled iron is cast iron that has been made in special molds, sometimes water-cooled molds, that cool the outer portions of the casting rapidly, thus making the surface of the casting very hard and wear resistant. Chilled iron is used for bearings on many farm machines and for shares and moldboards of plows designed for use in sandy, gravelly, or rocky soils.

Malleable Iron. Malleable iron is cast iron that has been treated after casting by heating for a long period. This prolonged heating changes or transforms some of the carbon in the outer portions of the casting and reduces its brittleness. Malleable castings are softer and tougher than plain castings and can be bent a certain amount without breaking. They are also more shock resistant. One of the most practical ways of repairing malleable castings is by bronze welding (see page 386).

Wrought Iron. Wrought iron is practically pure iron with a small amount of carbon and slag mixed with it. It is made from pig iron. It is fibrous or stringy rather than grainy. It is easily bent, cut, shaped, and welded. It cannot be hardened or tempered.

Mild Steel. Mild steel, also known variously as *machine steel, low-carbon steel, soft steel*, and *blacksmith iron*, contains about 0.1 to 0.3 per cent carbon. It cannot be hardened appreciably. It can be bent and hammered cold to some extent, and it can be easily sawed, filed, or ground. It can be forged and welded in the forge, but it is a little more difficult to weld than wrought iron. It is easily welded with the oxyacetylene torch or the electric arc welder. It is the kind of steel most commonly used in the farm shop for repair of farm machinery and equipment and the construction of appliances made of steel. It is available in bars and rods and also as angle irons of various sizes.

Tool Steel. Tool steel is made from pig iron, usually by refining in the electric furnace. It is practically free from impurities and contains about 0.5 to 1.5 per cent carbon. The higher the percentage of carbon, the harder the steel may be hardened, and the more difficult it is to weld. Tools like hammers and cold chisels are commonly made of steel having 0.5 to 0.9 per cent carbon. Tools like taps and dies are made from steel that contains 1 to 1.25 per cent carbon. The carbon content of steel is designated by points, one point being one-hundredth of 1 per cent of carbon. Thus a 50-point carbon steel contains $\frac{50}{100}$, or one-half, of 1 per cent of carbon. Tool steel is fine-grained in texture rather than fibrous or stringy. It must be in the annealed or softened state to be readily cut with files, saws, or drills.

Soft-center Steel. Soft-center steel is used in moldboards of plows and in cultivator shovels where it is desired to have a very hard wearing surface combined with high strength and toughness. It consists of a layer of mild steel welded between two layers of high-carbon steel. The outside layers can therefore be hardened while the center remains comparatively soft and tough.

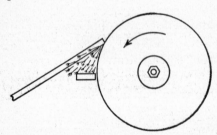

Fig. 379.—Different grades of iron and steel may be distinguished by the sparks produced when ground on a grinding wheel. The higher the carbon content of the steel, the brighter and more explosive are the sparks.

Alloy Steels. Small amounts of one or more other metals, such as tungsten, nickel, chromium, silicon, and vanadium, are commonly mixed with steel to form alloy steels. These metals are used in steel to give it certain desirable properties, such as great strength, resistance to corrosion, toughness, and resistance to shock.

Making Grinding-wheel Tests. A good way to distinguish between many kinds of steel and iron is to grind them on a grinding wheel and note the kind of sparks given off (see Fig. 379). Cast iron gives off a small volume of numerous small sparks. They are dull red as they leave the wheel but then change to a yellow. Sparks from wrought iron are larger in total volume, are light yellow or red, and follow straight lines. Sparks from mild steel are similar but are more explosive or forked. Tool steel gives off a moderately large stream of sparks that are lighter in color and still more explosive. The higher the percentage of carbon in steel, the brighter and the more explosive are the sparks.

2. Laying Out and Marking Metal

Careful measuring and marking of the work before cutting and shaping usually save time and ensure a better job. Measuring and marking on metal are done in much the same manner as on wood, except that a marking awl or scriber is recommended for marking metal (see Fig. 380). An old saw file ground to a needle point makes a very good scriber. A center punch, or a prick punch which is

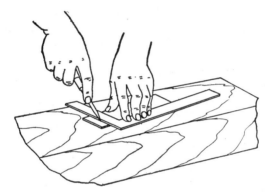

Fig. 380.—Marking metal with a scriber or marking awl. Time spent in careful measuring and marking is time well spent.

ground to a sharper point than a center punch, is also valuable for marking locations for bends, drill holes, saw cuts, etc.

3. Cutting with a Cold Chisel

The cold chisel is an inexpensive tool that has a wide variety of uses in cutting cold metal. And like most other tools, its usefulness is greatly increased when it is kept well sharpened and used properly. (See page 171 for methods of grinding the cold chisel.) Good cold chisels may be bought, or, if a blacksmith's forge is included in the shop equipment, they may be made and tempered in the shop.

Choose a chisel of a size suitable for the work being done. Use heavy chisels for heavy cutting and smaller chisels for light cutting. If the chisel is too small for the work, there is not only danger of breaking it, but it may vibrate and sting the hands when struck, and of course it will not cut so fast as a larger one.

Holding the Chisel; Striking. Hold the chisel firmly enough to guide it, yet loosely enough to ease the shock of hammer blows and keep the hands from becoming tired. Always hold the hammer handle near the end and strike blows in accordance with the kind of cutting being done—heavy blows for heavy cutting and light ones for light work. Strike light blows mostly with motion from the wrist; medium blows with motion from both wrist and elbow; and heavy

Fig. 381.—In cutting with the cold chisel at the anvil, always work over the chipping block—not over the face of the anvil. Hold the chisel firmly, yet loosely enough to ease the shock of hammer blows.

blows with motion not only from the wrist and elbow, but also from the shoulder.

Cutting on the Anvil. When cutting with the chisel at the anvil, be sure to cut on the chipping block, the small depressed surface at the base of the horn (see Fig. 381). The chipping block is soft, while the face of the anvil is hardened. Cutting through a piece on the face of the anvil would not only dull the chisel, but it would damage the face, which should be kept smooth for good blacksmithing.

Nick Deeply, Then Break. Bars and rods can generally be cut most easily by nicking them deeply on two or more sides and then breaking them by bending back and forth. Rods may be held in a vise for bending, or if small enough they may be inserted in one of the punch holes in the anvil.

Cutting in the Vise. If a heavy vise is available, small and medium-size rods or bars may be clamped in the vise and nicked deeply by chiseling close to the jaws. The rods are then easily broken by bending. Always hammer so that the force of the blows will come against the stationary jaw of the vise and not against the movable one (see Fig. 382).

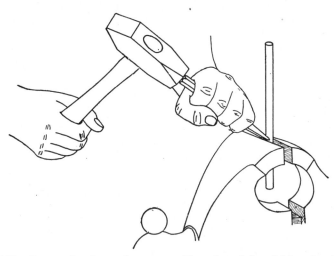

Fig. 382.—Bars and rods may be cut roughly to length by nicking deeply on two or more sides with the cold chisel and then bending back and forth. When cutting in the vise, always cut as close to the vise jaws as possible and always strike so that the force of the blow is against the stationary jaw—not the movable one.

Shearing Thin Bars. Thin bars and band iron up to about $\frac{3}{16}$ in. thick can usually be more easily cut by clamping in a vise and shearing with a cold chisel than by sawing. Clamp the bar or band iron securely in the vise with the cutting line just even with the top of the jaws. Place the chisel at one edge of the piece, with one bevel of the cutting end flat against the top of the vise jaws, and with the handle at an angle of about 60 deg. to the line of cutting (see Fig. 383). The chisel then acts as one blade of a pair of shears, and the stationary vise jaw as the other blade.

When the chisel is properly placed, tap it lightly once or twice to get the proper direction for striking, and then strike firm, well-directed blows. It is important to keep the chisel placed so that it cuts close to the vise jaws, and yet does not cut into them and thus damage the vise and dull the chisel. Driving too straight against either the flat surface or against the edge will cause the work to slip in the vise.

When properly done, the metal cuts fast and easily, leaving a surface that is smooth enough for most work. Where smoother work is required, the surface is readily dressed with a file.

Sheet metal that is too thick for easy cutting with snips can be easily cut with a cold chisel and vise in the same manner as thin bars.

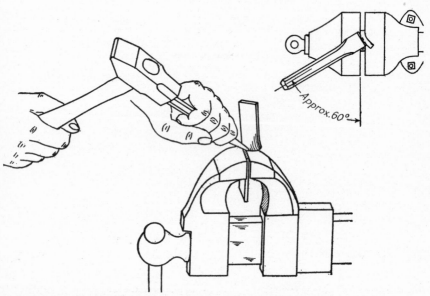

Fig. 383.—Thin bars are easily sheared with a cold chisel and vise. Keep the lower bevel of the chisel flat against the vise and the chisel handle at an angle of about 60 deg. with the line of cutting.

Cutting Soft Metals or Thin Sheet Metal. Soft metals like brass, lead, babbitt, or thin sheet metal are easily cut with the cold chisel. If much cutting of this type is to be done, it is best to use a chisel that has been ground to a very keen cutting edge of possibly 30 to 45 deg., instead of the usual angle of about 70 deg. The chisel will then cut both faster and smoother.

Using a Slitting Chisel. A special kind of cold chisel, known as a slitting chisel, is very useful for fast cutting of thick sheet iron, as cutting out the head of an old oil drum or cutting it in two in the middle (see Fig. 384). The slitting chisel is ground to a blunt square end, instead of being ground with a beveled cutting edge like a regular cold chisel. The slitting chisel is slightly thicker at the cutting end than a little further back. It shears out a ribbon of steel just as wide as the cutting end.

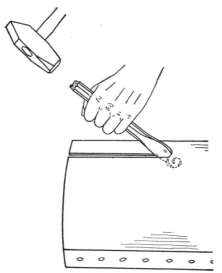

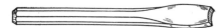

Fig. 384.—Heavy sheet metal, oil drums, water tanks, etc., are easily cut with a slitting chisel.

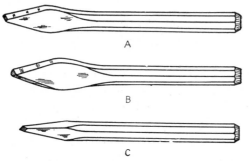

Fig. 385.—Types of special grooving chisels that may be made in the farm shop: A, cape chisel for cutting keyways; B, round-nosed chisel for cutting round grooves; C, diamond-point chisel for cutting grooves.

To start a slitting chisel in the middle of a piece of sheet iron, first drill a hole. A little experimenting will quickly indicate the best angle of cutting. An angle of about 45 deg. to the surface of the metal being cut is usually about right.

Cutting Slots and Grooves. Other special chisels, such as those illustrated in Fig. 385, will occasionally be found useful in cutting grooves and slots, such as oil grooves in a bearing or a keyway in a shaft. If a forge is a part of the shop equipment, such chisels are easily made in the shop.

Points on Cutting with the Cold Chisel

1. Always use a sharp chisel and one of a size suited to the cutting to be done.
2. Hold the chisel firmly enough to guide it, yet loosely enough to ease the shock of the hammer blows.
3. Grasp the hammer handle near the end.
4. Tap the chisel once or twice to get the direction of striking, and then use firm, well-directed blows.
5. For light chiseling, strike blows with wrist motion only; for heavier work. use both wrist and elbow action; and for very heavy work, use motion from the shoulder as well as wrist and elbow.
6. In cutting at the anvil, always work over the chipping block and not the face of the anvil.
7. Nicking a bar deeply and then breaking it by bending is usually easier than cutting it all the way through with a chisel.
8. In hammering in a vise, always strike so the force of the blow comes against the stationary jaw and not against the movable jaw.
9. For shearing in a vise, clamp the work tight and place the chisel so as to get a good shearing cut. Keep the bevel on the end of the chisel flat against the top of the vise jaws. Hold the chisel at an angle of about 60 to 70 deg. to the line of cutting.
10. For cutting brass and similar soft materials or thin sheet metals, the chisel cuts smoother and faster if ground to an angle of 30 to 45 deg. instead of the usual angle of about 70 deg.
11. A slitting chisel is very useful for fast cutting of thick sheet iron, such as cutting an old oil drum in two.

4. Filing

A file is a most valuable cutting tool. Its real value is rarely fully appreciated by the beginner. Unless a file is properly cared for and used, it will do only moderately satisfactory work, however, and filing with it will prove tedious and laborious. On the other hand, when a file is properly cared for and used, a good workman can often do faster and better cutting with it than he can with a grinding wheel.

Taking Care of Files. A file is a hardened-steel tool that has a series of small sharp cutting edges, or points, on its surface. Files should therefore not be thrown around with wrenches and other tools, nor should they be kept on shelves or in drawers where they will scrape against each other or against other tools. A good method of keeping them when not in use is to hang them on hooks or on a rack by the handles, as shown in Fig. 386.

Keep file handles tight on the tangs. If a handle becomes loose, it usually can be tightened by ramming the file lightly, handle end down, against the bench top.

It is also important that files be kept clean, that is, free from filings or chips, rust, grease, and grime.

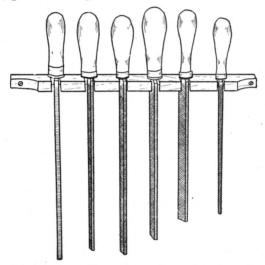

Fig. 386.—A good way to keep files. Do not keep them in a drawer or on a shelf where the sharp cutting edges might be dulled by contact with other files or tools.

Selecting the Right File for the Job. There are hundreds of styles, kinds, and sizes of files, and for best results, the file should be selected to fit the job at hand. Fortunately, a rather small assortment of files will meet ordinary requirements in farm shopwork.

Files may be classified according to (1) size; (2) kind of teeth; (3) shape, style, or use; and (4) degree of coarseness or fineness.

Size of Files. The size of a file is designated by its length, measured exclusive of the tang.

Kinds of Teeth. A file with one series of chisellike teeth running at an angle across the face is known as a *single-cut file.* A *double-cut file* has a second series crossing the first at an angle. A third kind, used on rasps, consists of raised points on the surface, rather than chisellike teeth.

Shape, Style, or Use. Files are commonly named to indicate (1) their general style or shape, as *flat, square, round, half-round, three-cornered (triangular);* or (2) their particular use, as *mill* and *auger bit.*

A particular kind of file is always made in just one kind of teeth. A *flat* file, for example, cannot be obtained with either single-cut or with double-cut teeth. It is made in double-cut only. Likewise, a *mill* file is made in single-cut only. A mill file, so called because of its use in woodworking mills for sharpening saws and planer knives, is very much like a flat file except that it is somewhat thinner, is tapered less in width near the point, and is made with single-cut teeth.

Fineness or Coarseness of Cut. The fineness or coarseness of files is commonly designated by the following series of terms, which, arranged in the order of coarsest first, are: *rough, coarse, bastard, second cut, smooth,* and *dead smooth.* These terms are relative, however, and vary with the kind or style of file and with the length or size of file. For example, a mill bastard file is finer than a flat bastard file of the same size; and an 8-in. bastard file is finer than a 10-in. bastard file.

For rough, fast cutting, use a flat bastard (double-cut) file, 10, 12, or 14 in. long. For finer work, a mill bastard (single-cut) file 8 or 10 in. long is usually satisfactory. For still finer work, a mill smooth file would be better.

For miscellaneous filing jobs and for filing small pieces or working in close quarters, a small assortment of half-round, round (rattail), triangular, and square files is an asset to any shop.

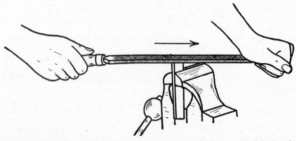

Fig. 387.—A good way to hold a file for heavy filing. Hold the handle firmly, yet not too tightly, with the right hand. Keep the thumb on top. Apply moderate to heavy pressure with the left hand.

Holding the File. For heavy filing, grasp the file handle firmly in the right hand, *thumb on top.* Do not squeeze too tightly. Hold the end of the file between the ends of the first two fingers and the base of the thumb of the left hand (see Fig. 387). Apply moderate to heavy pressure.

For light filing, hold the end of the file between the *end of the thumb* and the fingers of the left hand. Use light to moderate pressure (see Fig. 388).

Wherever possible, clamp the work to be filed securely in a vise. Elbow height, or possibly a little lower, is best for rough heavy filing. A little higher is better for light filing.

Using the File. Probably the most important points to observe in filing are (1) *to use rather slow full-length cutting strokes, and* (2) *to lift the file on the backstrokes.*

In heavy filing, push the file with a combination slow rhythmic swing from both the body and the arms. Stand in front of the vise,

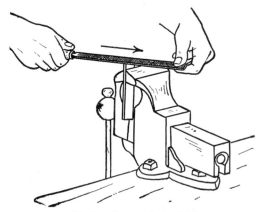

Fig. 388.—A good way to hold the file for light filing. Hold the end of the file between the thumb and finger of the left hand.

with the right foot about 10 to 12 in. back of the left and with the body bending forward slightly at the hips (see Fig. 389).

Start the forward stroke by gradually leaning the body forward and at the same time pushing with the arms a little faster than the body moves. Finish the forward stroke with motion from the arms only, while the body swings back into position for the next cutting stroke. Then lift the file slightly or release the pressure, and quickly draw it back into position for the next cutting stroke.

Use the Right Amount of Pressure. One of the quickest ways to ruin a file is to use too much or too little pressure on the forward or cutting stroke. Different materials require different pressures. In general, use just enough pressure to keep the file cutting. If too little pressure is used and the file is allowed to slide over the work without cutting, the teeth will rapidly become dull, particularly in filing hard metals.

If too much pressure is used, the file will "overload" with cuttings. This is likely to chip or clog the teeth and also scratch and score the work.

On the backstroke, there should be very little or no pressure. It is usually best to lift the file clear off the work.

Never Use Short, Jerky, or Seesaw Strokes. Only a poor workman or an amateur would use short, jerky, or seesaw strokes. When

Fig. 389.—In filing, stand with the left foot 10 or 12 in. ahead of the right, and with the body leaning forward slightly at the hips. Use rather slow, full-length strokes, and lift the file slightly or release the pressure on the backstrokes.

used in such a manner, the file cuts slowly, does poor work, and soon becomes dull.

When a file is pushed too fast, it slides over the metal without properly engaging it, which causes slow cutting and quick dulling of the teeth. Also, the work is likely to vibrate, causing screetching. *Always push a file slowly enough for the teeth to "take hold" and cut.*

Filing a Surface Flat and Straight. When filing a flat surface, use more pressure on the front end of the file than on the handle during the first part of the stroke. Then as the file is pushed forward, gradually ease the pressure on the front end and place more on the handle. In this way, the file may be kept cutting straight and flat all the way across. Unless the pressure on the file is shifted in this manner, the edges of the work will be filed more than the middle portion, and the surface will be rounded instead of flat and straight.

Drawfiling. Drawfiling is a quick, easy method of filing long, narrow surfaces or round rods. To drawfile, grasp the file on each end as shown in Fig. 390 and push it sidewise. Use pressure on the forward stroke, and lift the file slightly or release the pressure on the backstroke.

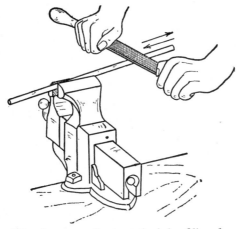

Fig. 390.—Drawfiling is an excellent method for filing long, narrow surfaces. Push the file back and forth sideways, applying pressure on the forward strokes and releasing it on the backstrokes. Use moderately slow strokes.

In blacksmithing, drawfiling can often be used to smooth up a round rod or other piece of work better than hammering. If the work is to be filed while hot, use an old file that can be kept for this purpose, as the heat would soon damage a good one.

Cleaning the File. Small particles of metal will often tear out and lodge between the file teeth and scratch the work. This is more likely to occur with a new file than with an old one, or when filing on narrow work, especially if too much pressure is used. Rubbing the file with a piece of chalk will help prevent clogging or scratching. A file is also likely to become clogged and slick when filing soft metals or dirty, greasy pieces.

After every few strokes, tap the end of the file on the bench to shake the filings from the teeth. Also, clean the teeth frequently by using a file brush or a file card (a small fine wire brush) as shown in Fig. 391. A piece of soft iron wire sharpened to a point and known as a *scorer* may also be used to remove cuttings that become lodged

between file teeth. Most file cards and brushes have scorers attached. Another way to clean lodged cuttings from file teeth is to drag the edge of a thin piece of soft brass or copper along the teeth.

Filing Soft Metal. It is difficult to file soft metal with an ordinary file, because of the tendency of the teeth to clog. Drawfiling, however, usually works reasonably well and better than straight filing. Special files for brass, aluminum, and lead work much better than ordinary files.

Fig. 391.—When a file becomes clogged with dirt, grease, or metal cuttings, it may be cleaned with a small wire brush known as a file card.

Filing Cast Iron. Cast iron has a hard outer surface that would quickly damage the teeth of a good file. In filing cast iron, it is therefore good practice to use an old file for cutting through this outer surface before using a good file.

Points on Filing

1. Clamp the work to be filed firmly in the vise to prevent chattering.
2. The work should be at about elbow height for average filing, possibly a little lower for heavy filing, and a little higher for light filing.
3. Exert just enough pressure on the file to keep the teeth cutting.
4. Use moderately slow, long, full-length strokes.
5. Always lift the file slightly or release the pressure on the backstroke.
6. Never use short, jerky strokes.
7. Do not allow the file to slip over the work, as this dulls the teeth.
8. Do not allow files to be thrown around against tools or against each other, as this will damage the teeth.
9. A good way to keep files is to hang them up by their handles.
10. If a file tears and scratches the surface of the work, rub the teeth with chalk.
11. Keep the teeth of the file clean by means of a file card or brush.
12. Drawfiling works well on long narrow surfaces.
13. Drawfiling on soft metals gives rapid cutting with a minimum of clogging of the teeth.

14. In filing cast iron, first remove the hard outer surface with an old file before using a good one.

15. An 8- or 10-in. mill bastard file is good for fine filing, and a 12-in. flat bastard file is good for fast, rough filing.

5. Hack Sawing

The hand hack saw is one of the most useful tools for cutting metal. Like the file, however, it is often not used properly. Although the hack saw can be used with fair satisfaction by an inexperienced workman, a little thought and study given to its proper use will result in faster and better work and less dulling and breaking of blades.

Selecting the Right Saw Blade for the Job. Good work with a hack saw depends not only upon the proper use of the saw, but also upon proper selection of the blades for the work to be done. There are three general kinds of blades available: (1) all hard, (2) flexible, and (3) high-speed steel. All-hard blades have the whole blade tempered. They are suitable for general use where the work to be sawed can be held securely. The flexible blades have only the teeth hardened without hardening the back. They are used where there is danger of cramping and breaking the blade, as in sawing in awkward positions, where the work cannot be held securely, or for sawing flexible material like armored electric cable. High-speed steel blades are made of a special steel and will cut faster and last many times longer than regular blades, provided they are carefully used and not broken.

Use of saw blades with the wrong size of teeth is a common cause of breaking blades or of stripping teeth from the blade. *In general, use larger teeth for sawing thick pieces and soft materials and smaller teeth for thin pieces and hard materials.* Fine teeth lessen the danger of stripping teeth from the blade or of breaking the blade.

Hand hack-saw blades are commonly available with 14, 18, 24, and 32 teeth per inch. A blade with 14 teeth per inch is best for sawing thick bars and soft materials, although a blade with 18 teeth per inch is usually satisfactory for the heavier sawing. For general use, a blade with 18 teeth per inch is usually best for the experienced workman, while a blade having 24 teeth per inch is best for beginners. A blade with 32 teeth per inch is best for sawing thin metal and tubes with walls thinner than about $\frac{1}{16}$ in. A general rule is to use a blade that will always have at least two teeth cutting at the same time.

This lessens the tendency of teeth to catch on the work and break out of the blade.

Inserting and Tightening the Blade in the Frame.   Insert the blade in the frame with the teeth pointing away from the handle,

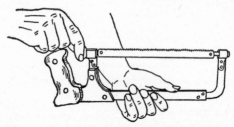

Fig. 392.—Tighten a hack-saw blade until it gives a humming note when picked with the thumb.   After a few strokes, a new blade will stretch and need to be retightened a little.

and tighten it until it gives a humming note when picked with the thumb (see Fig. 392).   After a little sawing, it may be necessary to retighten the blade.   It is important to keep the blade tight but not overstrained.   A slack blade is likely to drift or cut at an angle instead of straight, and it is likely to buckle and break.

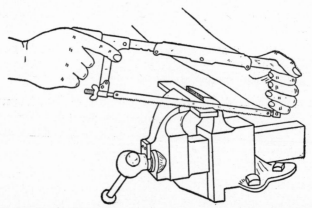

Fig. 393.—Grip the saw with the first finger of the right hand alongside the handle.   Hold the front end of the frame with the left hand to help guide the saw and apply pressure.

Holding and Using the Saw.   Hold the handle of a hack saw in much the same manner as holding a hand wood saw.   Hold the front of the frame with the left hand to help guide it and apply pressure (see Fig. 393).   Stand at ease with the right foot 10 or 12 in. back of the left and with the right forearm, elbow, and shoulder in line with the saw (see Fig. 394). Push the saw with a slight sway of the body as well as with motion from the arms, much as a file is pushed (see page 313).

Use Long, Slow Even Strokes. Do not work too fast. Sixty cutting strokes per minute should be the maximum—40 to 50 are usually better.

Exert pressure on the forward strokes, and release the pressure, or lift the blade slightly, on the backstrokes. Use just enough pressure to make the teeth cut and keep them from slipping over the work. If they slip, little or no cutting is done, and the blade is quickly dulled. Too much pressure increases the danger of breaking the blade.

Sawing too fast is usually accompanied by short strokes and heavy dragging on the backstrokes, with consequent quick dulling and excessive wear in the middle portion of the blade. Also, with fast sawing there is more danger of catching and breaking the blade.

If a blade starts to cut to one side, it is best to turn the stock a quarter or half turn and start a new cut.

In case a blade is broken and

Fig. 394.—In sawing with a hack saw, stand at ease with the right foot 10 or 12 in. back of the left, and with the right forearm, elbow, and shoulder in line with the saw. Use long, even strokes, swaying the body back and forth slightly in rhythm with the arm strokes.

a new one must be used, always start in a new place, possibly on the opposite side of the stock. A new blade would bind, and there would be danger of breaking it if it were used in a cut started by an old blade.

Holding the Work. A piece to be sawed should be held securely. If possible, clamp it in a vise with *the sawing line close to the vise jaws.* Beginners often make the mistake of sawing too far from the vise, thus allowing the work to spring and vibrate, causing poor work and increasing the danger of catching and breaking the saw blade. Where a vise is not available, sometimes two nails can be driven in the bench top and the work held by cramping it between the nails. Sawing can best be done with the work at about elbow height.

In fastening irregular-shaped pieces, as angle irons or channel irons, in a vise, clamp them so the saw will make an angling cut and enough teeth will engage at a time to prevent catching and breaking

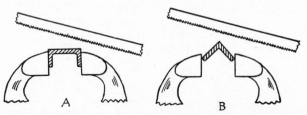

Fig. 395.—Suggested methods of holding irregular work for hack sawing. In cutting channels and angle irons, keep several teeth in contact with the work.

(see Fig. 395). At least two teeth, preferably more, should be in contact with the work. In sawing thin bars, ⅛ to ¼ in. thick, clamp them flatwise in the vise, rather than on edge, and saw with the front end of the saw slightly lowered (see Fig. 396). Sawing across the wide surface instead of the edge not only lessens the danger of stripping teeth from the blade, but also prevents vibration.

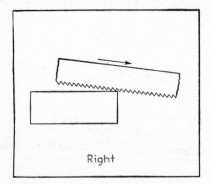
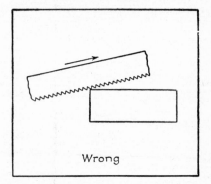

Right   Wrong

Iig. 396.—In sawing thin bars, clamp them flatwise in the vise (not on edge), and saw with the front end of the saw slightly lowered. (See also Fig. 393.)

To hold a piece of metal in a vise without marring it, it may be clamped between two pieces of wood.

Starting the Saw in a File Notch. It is a good plan to file a notch for starting the saw, as shown in Fig. 397, particularly in the case of beginners or when starting on the corner of a bar. The notch helps get the saw started in the right place and decreases the danger of breaking teeth from the blade.

Sawing Tool Steel. Be sure tool steel is in the softened or annealed state before sawing. Otherwise, the saw blade will be quickly dulled. Although tool steel may be clamped in a vise and sawed as ordinary steel, it is generally best to saw only a deep nick

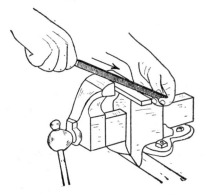

Fig. 397.—A file notch makes it easy to start the hack saw exactly on a line and decreases the danger of breaking teeth from the blade.

(about one-fourth or one-fifth the way through the bar) and then to break the bar by clamping in the vise and striking a sharp blow with the hammer (see Fig. 398). Clamp the bar with the saw nick on the side to be struck with the hammer, and just even with the mov-

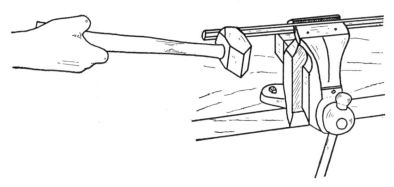

Fig. 398.—A practical method of cutting tool steel is to saw a deep nick with the hack saw, clamp it securely in a vise with the nick even with the movable vise jaw, and then break it with a sharp hammer blow.

able vise jaw. In heavy hammering in a vise, always strike so the force of the hammer blows will come against the stationary jaw.

Sawing Thin Metal. A small piece of thin sheet metal may be easily sawed by clamping it between two pieces of wood in a vise and sawing through both wood and metal (see Fig. 399).   Larger pieces of sheet metal that cannot be held in a vise may be clamped to the bench top with the part to be sawed projecting over the edge.

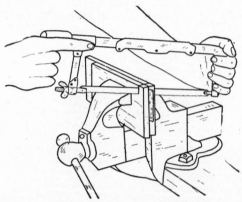

Fig. 399.—A small piece of thin metal may be sawed easily by clamping it between pieces of wood in a vise and sawing through both wood and metal.

Sawing Wide Slots.   It is sometimes desirable to make a wide slot in a bolthead so that it may be held or turned with a screw driver.

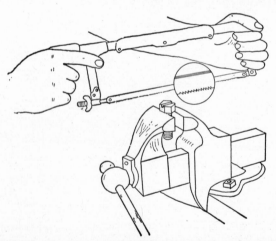

Fig. 400.—Wide slots are easily made with the hack saw by using two or more blades instead of one.

Such a slot is easily made by using two saw blades in the hack saw instead of one (see Fig. 400).   If for some reason a still wider slot is needed, three blades might be used.

Points on Hack Sawing

1. Clamp the work in a vise if possible.
2. Saw as close to the vise jaws as possible in order to keep the work from vibrating.
3. Use a flexible blade for sawing flexible material, or where the work cannot be securely held.
4. It is a good plan to start the saw in a notch made with a file.
5. Use a blade with a suitable number of teeth per inch. Twenty-four is about right for general use, especially for beginners. Eighteen teeth per inch might be better for an experienced workman.
6. Use a blade with 32 teeth per inch for sawing thin metal.
7. At least two teeth should be kept in contact with the work when sawing.
8. Coarser teeth saw faster; finer teeth lessen the danger of breakage.
9. Keep the blade tight yet not overstrained. If it is properly stretched, it will give a humming note when picked with the thumb.
10. Use just enough pressure to make the teeth cut.
11. Release the pressure on the backstroke.
12. Use long, even, slow strokes—not over 60 cutting strokes per minute; 40 or 50 are usually better.
13. Never allow the saw to rub or slip instead of cutting. Rubbing or slipping dulls the teeth.
14. To saw irregular-shaped pieces, clamp them cornerwise in the vise, so as to allow plenty of teeth in contact with the work.
15. Clamp thin bars flatwise rather than on edge. This prevents vibration and lessens danger of stripping teeth.
16. Tool steel is readily cut by sawing a deep nick and then clamping in the vise and breaking with a hammer.
17. Thin metal may be sawed by clamping it between pieces of wood and then sawing through both wood and metal.
18. Two or more blades may be used in a hack saw for sawing wide slots.

6. Drilling Holes in Metal

A few tools for drilling holes in metal will enable a farmer to make many handy appliances and to make many repairs on his machinery and equipment that would otherwise be impossible.

A properly sharpened drill is the first requirement for satisfactory drilling. A drill that is not properly ground will do poor work; it will cut slowly; it will require excessive effort to turn it; and there will be danger of breaking it. Drills require frequent grinding if they are used much. Anyone who expects to use drills with satisfaction or profit to himself should therefore become proficient in grind-

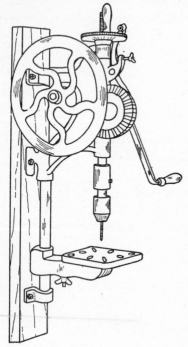

ing them. See pages 165 to 171 for information on drill sharpening.

Selecting Drilling Equipment for the Shop. *The post drill* (see Fig. 401) is probably the best drilling machine for general work inside the shop. It is especially good for drilling holes of the larger sizes—above ¼ in. in diameter. A power-driven post or bench drill press may be desirable, although it is not hard work to turn a drill by hand, especially if the drill bits are kept sharp.

The carpenter's brace can be used for drilling holes in metal. For holes about ¼ in. in diameter, it works very well (with sharp drill bits), but it is somewhat slow for holes of larger size. Small drills are easily broken if the workman is not careful.

Fig. 401.—The post drill is an excellent piece of equipment for the farm shop.

The chain drill (Fig. 402), used with a carpenter's brace, makes a very effective drilling combination. Holes can be drilled much faster and easier than with the brace alone.

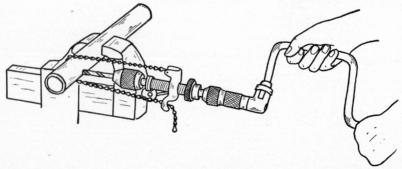

Fig. 402.—The chain drill is good for drilling holes in parts of machines and implements when it is not convenient to take these parts to the shop, as well as for drilling in the shop.

With a chain drill, the workman does not have to push hard against the brace, as the pressure for drilling is supplied by simply turning a

knob.   The chain drill can be used in the shop, or it can be taken to
the machine shed or the field.   Thus a hole may be drilled in a
machine or implement without having to take it, or a part of it, to the
shop.   Chain drills are available with either automatic or plain hand

Fig. 403.—The hand drill is one of the most useful tools for drilling small holes, either
in metal or in wood.

feeds.   The plain-feed drills are cheaper and for the farm shop are
practically, if not altogether, as good.

The hand drill (Fig. 403) is one of the most useful tools for drilling
small holes in lightweight metal and in wood.   For the farm shop,
one that takes drills up to $\frac{1}{4}$ in. in diameter is large enough (see also
page 90).

Fig. 404.—
The breast drill
is similar to the
hand drill but
is larger.

Fig. 405.—The electric drill
is an excellent tool where electricity is available.

The breast drill (Fig. 404) is very much like the hand drill, except
that it is larger and pressure is applied by leaning against it.   Although
a useful tool, it is not greatly needed in the farm shop.   A hand drill
works just as well or even better for drilling small holes, and a chain
drill or a post drill works better on drills larger than $\frac{5}{16}$ in. in size.

The electric drill (Fig. 405) is an excellent tool where electricity is available, although in the larger sizes it is somewhat expensive. An attachment or stand for converting an electric drill into a bench drill greatly increases its usefulness.

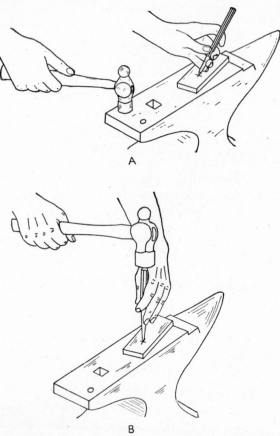

Fig. 406.—Making a center-punch mark in which to start a drill. **A.** Steadying the hand against the work makes it easy to place the point of the punch exactly. **B.** First strike one or two light taps with the hammer, and then stop to make sure the punch is correctly placed. When it is correctly placed, follow with one or two heavier blows.

Center Punching. Always start a twist drill in a *deep* center-punch mark—one that is big enough to take the point of the drill. Otherwise, the drill will likely "wander" and drill the hole off center or not exactly in the right place. In doing accurate work, first mark the location for the hole by the intersection of two scratch lines. Make the lines with a scriber or scratch awl and a square (see Fig.

380).   To locate the center of a small rectangular piece, simply draw diagonals from the opposite corners.   The intersection of the diagonals will be the center.   A very satisfactory scriber can be made by grinding the end of an old saw file to a needle point.

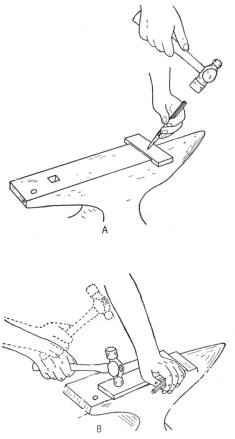

Fig. 407.—In the event that a punch mark is not properly located, it may be changed by driving the punch at an angle, as at A, or it may be hammered out with ball-peen hammer, as at B, and a new one made.   Still another method is to turn the bar over and make a punch mark and drill from the other side.

After making the cross lines, place a large deep center-punch mark at their intersection.   Carefully place the point of the center punch exactly at the intersection.   This may be done easily by steadying the hand against the work, while the point is lowered and moved about into position (see Fig. 406).   First strike a light tap with the hammer, and then inspect to see if the mark is started in exactly the right place. If so, replace the punch and strike one or two heavy blows to make a large, deep mark; and then proceed with the drilling. In case the punch mark is not started in the right place, however, shift it by driving the punch at an angle; or hammer it out with a ball-peen hammer, and make a new mark (see Fig. 407).

Selecting the Proper Size Drill Bit. Be sure to use a drill of the right size. The size of a new drill may be determined from the numbers stamped on the shank. After long use, these numbers may not be legible, particularly if the drill has been allowed to slip in the drill

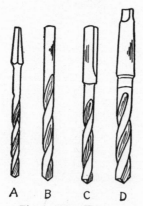

Fig. 408.—Types of drill bits classified according to kind of shanks: A, bit-stock drill; B, straight round shank drill; C, blacksmith's drill (straight round shank with flat side); D, morse taper drill.

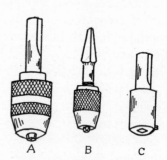

Fig. 409.—Types of adapting chucks: A, three-jaw chuck to fit into a post drill and hold straight round-shank drills; B, three-jaw chuck to fit into a carpenter's brace and hold straight round-shank drills; C, chuck to fit into a post drill and hold bit-stock drills.

chuck. In such a case, the size of the drill may be quickly determined by measuring with a caliper rule (Fig. 431B) or by trying it in a drill gage. Such a gage is simply a piece of steel with various sizes of holes drilled in it and marked. A gage of this type that will serve for ordinary rough work can be made in the shop. Simply select a small piece of steel about ¼ in. thick, and drill holes in it with various sizes of drills and mark them. Drills most commonly used in farm shopwork range from ⅛ in. up to ½ or ⅝ in.

Fitting the Drill into the Drill Chuck. Twist drills are made in different styles of shanks (see Fig. 408), and drilling tools or machines are equipped with different kinds of chucks to hold drills. It is important that the drill and the chuck match, or fit each other. Do not try to put a bit-stock drill (with a square tapered shank) into a chuck made to hold blacksmith's drills, for example, or into a three-jaw chuck made for straight round-shank drills. Attempting to use a drill in the wrong kind of chuck will most likely result in damage to the drill or the chuck or both.

Always fasten the drill in the chuck securely. The drill should not be allowed to slip, for slipping will mar the drill shank and may damage the chuck.

Adapting chucks of various kinds (see Fig. 409) may be used to hold drill bits in drilling machines when the bits do not have the right kind of shanks for the chucks on the machines.

Holding Work on the Drill Table; Safety in Drilling. When using a power drill, always clamp the piece or in some other manner hold it securely on the drill table. Otherwise, the drill may catch and throw the work off the table, possibly breaking the drill or injuring the operator or both. A method sometimes used to hold small pieces is to bolt a board to the drill table and then drive nails in the board to keep the piece from turning.

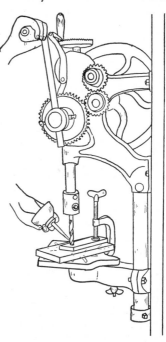

Fig. 410.—It is a good plan to clamp the work to the drill table. Lubricate the drill with lard oil or threading oil when drilling in steel.

When drilling with a hand-driven post drill, the work may be clamped, as shown in Fig. 410, or held by hand if it is long and a firm hold can be secured. Holding small pieces by hand, even on a hand-driven drill, is not safe.

When a hole is to be drilled completely through a piece, always put a block of scrap wood on the drill table to prevent the drill from drilling into the table.

Drilling Pressure and Speeds. Use enough pressure in forcing the drill into the work to give moderate to fast cutting, yet not so much as to cause the drill to gouge or catch in the work

When using a power drill with a variety of speeds, select one fast enough to give rapid cutting, but not so fast that the drill squeaks and fails to bite into the metal, nor so fast that the drill overheats. In general, turn small drills fast and large ones slowly. Woodworking drill presses are usually geared too high for satisfactory drilling in metal.

Lubricating the Drill. When drilling mild steel, apply lard oil or a threading oil as soon as the drill is centered in the center-punch mark and starts to cut. Keep oil on the cutting end of the drill, applying more as required as the drilling proceeds. Turpentine or kerosene is recommended for drilling hard steel or other very hard materials. No lubricant is required for drilling in cast iron or brass, although some mechanics prefer to use lard oil. Ordinary lard, when melted, makes a good lubricant for drills and other cutting tools.

A lubricant helps cool the drill, and it makes it cut easier and smoother. Also, if a good lubricant is used, a drill will stay sharp much longer. Always keep a squirt can of lard oil or threading oil handy when drilling.

Preventing Drill Breakage. Most drill breakage occurs just as the drill goes through the piece being drilled. To prevent such breakage, be sure the work is held securely, and release the pressure slightly on the drill just as it goes through. Turning the drill at a higher speed will also help.

In case a drill gouges and catches just as it starts through and there is difficulty getting the hole finished, use a center punch to smooth the gouges in the bottom of the hole and proceed with the drilling. Another method that may be used is to turn the work over and drill back through from the other side. Whenever there is danger that a drill may gouge or catch, use as high a drilling speed as practical, and use a very light pressure or feed.

Drilling Holes through Round Rods or Pipes. In drilling a hole in a round rod or a pipe, there is a tendency for it to roll on the drill table, increasing the danger of drill breakage, and making it difficult to get the hole drilled straight through on a diameter. To keep the rod or pipe from rolling, it may be placed in a V notch sawed in the surface of a 2-by-4 or a 2-by-6 block (see Fig. 411). Always start the drill in a deep center-punch mark.

Drilling Large Holes; Using Reamers. Where a large hole is to be drilled, it is generally easier to drill through first with a smaller

drill and then to follow it with a drill of the desired size.   Where
extreme accuracy is required in locating a large hole, a small hole is
sometimes drilled through first to serve as a pilot hole for keeping the
large drill centered.

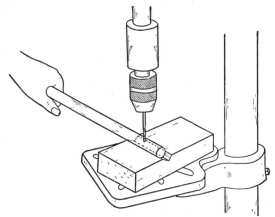

Fig. 411.—A V notch sawed in a block of wood is useful to keep round rods or pipes
from rolling on the drill table while holes are being drilled in them.

If a large drill of the desired size is not available and if a slightly
tapered hole will do, a hole may be drilled first somewhat smaller
than the desired size and then enlarged with a repairman's taper
reamer (see Fig. 412).   Such a reamer is also valuable in assembling

Fig. 412.—Repairman's taper reamers are valuable for enlarging holes in metals.

machinery when the bolt holes in iron or steel parts do not quite line
up.   A few turns with the reamer will usually allow the bolt to go in.
This type of reamer is commonly made with a square taper shank and
is used in a carpenter's brace.

Drilling Holes in Thin Metal. There is a tendency for a drill to catch and gouge when drilling in thin metal. Turning the drill a little faster than normally and using a light feed pressure will help prevent gouging and catching. It is sometimes possible to avoid this difficulty by clamping the metal between two pieces of hardwood and drilling through wood, metal, and all.

Points on Drilling

1. A properly sharpened drill is the first requirement for satisfactory work.
2. Always insert the drill carefully in the chuck. Be sure the drill and the chuck match.
3. Do not allow the drill to slip in the chuck, as it will damage both the drill and the chuck.
4. Start the drill in a large, deep, center-punch mark accurately located.
5. Use lard oil or a threading oil when drilling in steel.
6. Securely fasten the work to the drill table when using a power drill.
7. A block of wood on the drill table prevents drilling holes into the table.
8. Use lighter pressure just as the drill goes through and run the drill faster if possible. This lessens the danger of drill breakage.
9. Large drills can be turned too fast for good work.
10. Small drills work best when turned fast.
11. To keep round rods or pipes from rolling on the drill table, lay them in a V notch cut in a wooden block.
12. A taper reamer is valuable in enlarging holes and in reaming out holes in machine parts that do not line up.
13. To prevent catching and gouging when drilling holes in thin metal, clamp the metal between pieces of hardwood and drill through both wood and metal.

7. Bending Cold Metal

Although it may be necessary to heat large pieces of iron or steel to bend them satisfactorily, many small pieces can be bent better and more easily cold.

Bending in the Vise. With a good machinist's or blacksmith's vise, rods and bars up to $\frac{5}{16}$ or $\frac{3}{8}$ in. thick can be readily bent. Many appliances and repairs can be made with stock of this size and smaller. Care should be exercised, of course, not to do too heavy hammering or bending in a small vise, lest the vise be damaged or broken.

An easy way of making a small bend in strap iron or a thin bar is to bend the stock around a pipe or rod of suitable size. Simply clamp

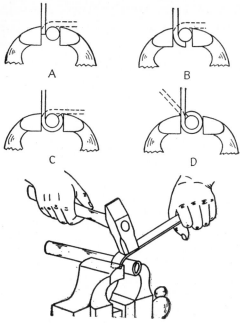

Fig. 413.—To make a short bend or a small eye in a thin bar, clamp it against a pipe or rod held in a vise, and then bend it around the pipe or rod.

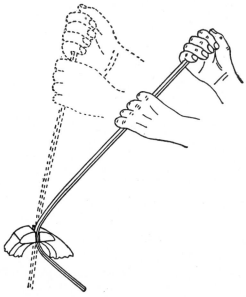

Fig. 414.—To make a smooth, gradual bend of large size, clamp the bar loosely in a vise and bend it a little; then slip the bar through a little farther and bend more, continuing until the desired curve is made.

the stock beside a pipe or rod in a vise and bend by pulling and hammering (see Fig. 413).

To make a uniform bend of large size, slip the end of the stock between the jaws of a vise that are loosely adjusted, and then pull. After the bar is bent slightly, release the pull and slip the bar through the jaws about ½ in. further, and then pull again (see Fig. 414). In this manner make a series of short bends, thus giving a reasonably smooth bend of rather large curvature.

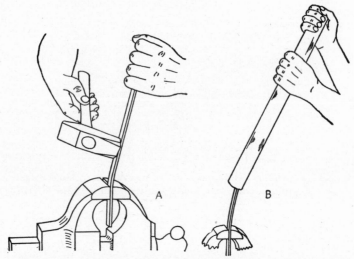

Fig. 415.—Short bars or bars too heavy to be bent by hand may be bent by hammering, as at A, or by pulling on a pipe slipped over the end, as at B.

A bar that is too heavy or too short to be bent by hand may be clamped in a vise and bent by hammering and pulling or by pulling on a pipe slipped over the end (see Fig. 415).

Twisting Cold Metal. Small bars of iron or other metals are easily twisted cold, using the vise and a wrench, or a pair of wrenches, or a pair of tongs. Methods for twisting cold metal are much the same as for twisting hot metal, except for the heating. See pages 359 to 362.

8. Riveting

Riveting offers a convenient and easy method of securely holding metal parts together. Many repairs can be made on machinery and equipment by riveting.

The procedure in riveting is simply to drill holes through the parts to be held together, insert a soft steel rivet, and then hammer it tightly in place. For some work, it is better to heat the rivets and then insert them and hammer them down while hot; but for general farm shopwork, cold riveting is quite satisfactory

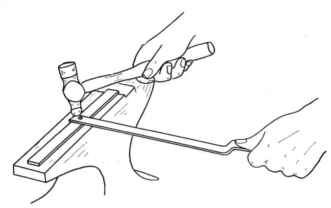

Fig. 416.—A neatly rounded head may be formed on a rivet by first striking a few medium-heavy blows and then finishing by light peening with a small ball-peen hammer

To hammer down a rivet and form a head on it, first strike one or two heavy blows with the flat face of the hammer, and then finish with lighter blows, preferably with the ball peen of the hammer (see Fig. 416). Where available, a tool called a *rivet set* may be used to form a neat round head on a rivet.

A	B	C	D

Fig. 417.—Methods of heading rivets: A, flat head made by straight hammering; B, head rounded with rivet set; C, head rounded with small ball-peen hammer; D, ends of rivet hammered into countersunk holes with ball-peen hammer.

If it is desired to fasten pieces together and not have rivet heads protruding from the surface, countersink the holes as indicated in Fig. 417D. Then use rivets without heads, and peen the ends into the countersunk holes.

Cutting Rivets to Length. A good way to cut a rivet to the desired length is to clamp it in a vise with the head up and then shear it off with a cold chisel. To keep the rivet from flying off and becoming lost as it is cut, hold the chisel near the cutting end with the thumb and first fingers, allowing the palm of the hand to cover the

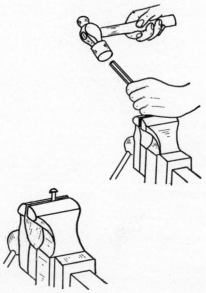

Fig. 418.—Rivets may be cut to length easily by clamping in a vise and shearing with a sharp cold chisel. Placing the hand over the top of the rivet keeps it from flying away and being lost as it is cut off.

rivet (see Fig. 418). Rivets may also be cut with heavy pincers, with small bolt cutters, or with a hack saw.

9. Threading

Bolts and threaded fittings are widely used on all kind of machinery. Equipment for threading bolts, nuts, rods, and pipes, and ability to use it, will greatly increase the scope of repair work that can be done in a shop. Such equipment will also make possible the construction of many handy and useful appliances.

Threads are cut on a rod or bolt by a small hardened-steel tool called a *die*. The die proper, which is usually adjustable, is held in a handle called a *stock*. The tool used for cutting threads inside a hole, as in a nut that screws on a bolt, is called a *tap* (see Fig. 419). A set of taps and dies is called a *screw plate* (see Fig. 420).

Kinds of Threads. *National Coarse* (N.C.) threads are most commonly used on large machinery. *National Fine* (N.F.) threads are used on automobiles, trucks, tractors, and engines (although not exclusively) and on other machines where extra strong steel bolts

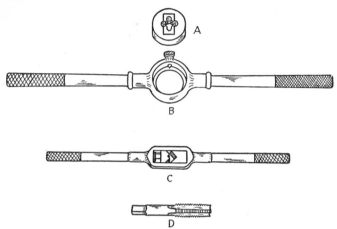

Fig. 419.—Threading tools: A, die; B, stock for holding die; C, tap wrench; D, tap.

are used. The fine threads are of the same general shape as the coarse ones but are smaller and there are more of them to the inch. Nuts with fine threads may be drawn up tighter than those with coarse threads, and they will not shake loose so easily. The coarse

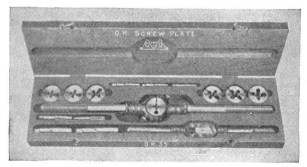

Fig. 420.—A set of taps and dies, known as a screw plate, is a valuable asset in the farm shop.

threads were formerly called U.S.S. (United States Standard) and the fine ones S.A.E. (Society of Automotive Engineers).

Pipe threads, known as Briggs standard threads, are used on pipes and on pipe and tube connections on engines, motors, and machines.

These threads are tapered $\frac{1}{16}$ in. per inch of length.   Since the threaded end of a pipe is tapered, the farther it is screwed into a fitting, the tighter the joint becomes.   This makes it possible to make a tight joint without having any tension or end pull on the pipe.   Bolt threads are straight, that is, they are cut on a cylinder, and a nut is made tight by drawing it up against the piece being held in place by the bolt.

Pipe threads are much larger than the corresponding size of bolt threads, owing to the system of indicating pipe sizes.   The size of a pipe is designated by its inside diameter—not its outside diameter. A $\frac{1}{2}$-in. bolt die cuts threads on the outside of a rod or bolt whose *outside* diameter is $\frac{1}{2}$ in.; and a $\frac{1}{2}$-in. pipe die cuts threads on the outside of a pipe whose *inside* diameter is $\frac{1}{2}$ in.   Therefore, the pipe die is much larger.

Selecting the Proper Size and Kind of Die.   In threading a rod to make a bolt, or in rethreading a bolt, be sure to select the proper

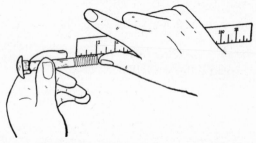

Fig. 421.—The number of threads per inch on a bolt may be determined by measuring with a rule and counting, or by reading the markings on a tap or die that cuts the same thread.

size and kind of die.   The size of rod or bolt may be easily and accurately measured with a caliper rule, or by careful work, with an ordinary rule.   With a little practice, the kind of threads on a bolt (whether fine or coarse) can be determined at a glance.

To determine the number of threads per inch on a bolt, it may be held against a tap from a set of taps and dies, to see if the threads on the bolt and tap correspond.   If they do, then the number of threads per inch may be read from the markings on the tap.   Likewise, dies from the set may be tried on the bolt until one is found that fits perfectly and the threads per inch and the size of bolt determined from the markings on the die.   The number of threads per inch can also be determined by placing a rule against the threaded end and counting the threads for 1 in. (see Fig. 421). In the better equipped shops, thread-pitch gages are used to determine the number of threads per inch on bolts.

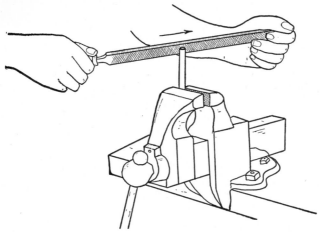

Fig. 422.—To enable the die to start easily, taper the end of the rod slightly by filing, grinding, or hammering. Do not make the taper too blunt.

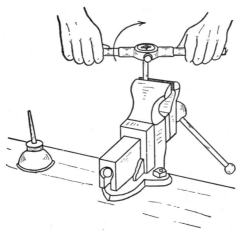

Fig. 423.—Be sure to start the die straight, exerting equal pressure on the two handles.

Threading a Rod or a Bolt. To thread the end of a rod, first taper it slightly by filing, hammering, or grinding so that the die will start on easily (see Fig. 422). Do not make the taper too short and blunt. Be sure the proper size and kind of die has been selected, and then start it on the rod by exerting equal pressure on the two handles of the stock and turning.   Be careful to start the die on straight (see Fig. 423).

As soon as the die starts to cut and feed itself onto the rod, lubricate the die by applying lard oil or threading oil (see Fig. 424). This will make the die cut easier, cut a smoother thread, and the die will stay sharp and last much longer.

Turn the die round and round in the forward direction without backing up until the thread is finished.   Apply more oil as may be required.   If chips and cuttings tend to collect in the die and clog it, stop and punch them out with a small nail, a wire, or a small piece of wood.   If allowed to accumulate, the chips may cause rough or

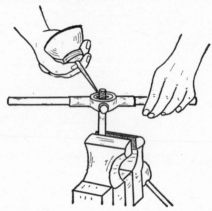

Fig. 424.—When threading steel rods or bolts, always lubricate the die with lard oil or a threading oil.   Oil makes the die turn easier, cut smoother threads, and last longer.

torn threads.   When the thread is cut as far as desired, simply screw the die back off.   Shake the cuttings from the die by tapping the end of the stock against the bench top.   Before putting the die away, be sure it is well cleaned and that any excess oil is wiped off.

Tapping Threads in a Hole or Nut.   If a threaded hole is to be made in a piece of metal, a hole of suitable size must first be drilled. The hole must be somewhat smaller than the size of the bolt to be screwed into it, usually $\frac{1}{16}$ in. smaller for bolts $\frac{1}{4}$ to $\frac{1}{2}$ in. in size.

Probably the best method to determine the exact size of drill to use is to refer to a table, such as Table 9, 10, or 11, page 342.   In the absence of such a table, a hole somewhat smaller than the bolt may be drilled in a piece of scrap material and the tap tried in it.   If the hole does not prove to be the proper size, try a size larger or smaller as may be required.

After a hole of the proper size has been drilled, start the tap in the hole by applying pressure and turning it slowly. A good way is to use only one hand, pushing firmly with the palm against the center of the tap wrench, and giving it a slow twist with the wrist (see Fig. 425*A*). Be particularly careful to start the tap straight.

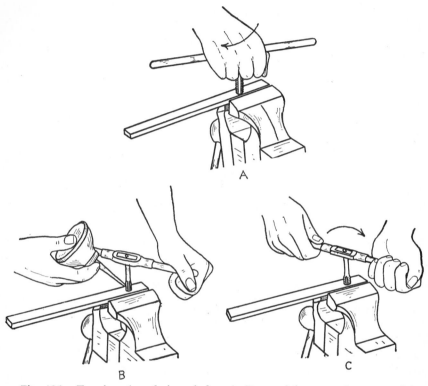

Fig. 425.—Tapping threads in a hole. A. Be careful to start the tap straight. Use one hand, pressing against the top of the tap, and turn slowly with a wrist motion. B. Once the tap is well started, stop and oil it with lard oil or a threading oil. C. Then turn steadily, applying equal pressure with the two hands.

After the tap starts, turn it by pulling with equal force on the two ends of the tap wrench (see Fig. 425*C*). Always use lard oil or threading oil on the tap when cutting threads in steel. Turn the tap until the tapered part is all the way through the hole and the tap turns easily. Then stop and unscrew it from the hole.

Making a Bolt. A very simple way to make a bolt is to cut off a rod to the desired length, thread both ends, and screw on nuts. With a small assortment of nuts, and a few different sizes of rods, one can easily make a bolt for almost any occasion or emergency. Moreover, the bolt is practically as good as a bolt that is bought and usually just as cheap or cheaper. If desired, one of the nuts may be riveted onto the rod to prevent loosening.

TABLE 9. THREADS PER INCH AND TAP DRILL SIZES FOR BOLT THREADS, ¼ TO ¾ IN.

Size of tap	National Coarse		National Fine	
	Threads per inch	Tap drill size	Threads per inch	Tap drill size
¼	20	$\frac{3}{16}$	28	$\frac{7}{32}$
$\frac{5}{16}$	18	¼	24	¼
$\frac{3}{8}$	16	$\frac{5}{16}$	24	$\frac{5}{16}$
$\frac{7}{16}$	14	$1\frac{1}{32}$	20	$2\frac{5}{64}$
½	13	$1\frac{3}{32}$	20	$2\frac{9}{64}$
$\frac{9}{16}$	12	$1\frac{5}{32}$	18	$3\frac{3}{64}$
$\frac{5}{8}$	11	$1\frac{7}{32}$	18	$3\frac{7}{64}$
¾	10	$2\frac{1}{32}$	16	$1\frac{1}{16}$

TABLE 10. THREADS PER INCH AND TAP DRILL SIZES FOR BOLT (OR MACHINE SCREW) THREADS SMALLER THAN ¼ IN.

Size of tap	Outside diameter, in.	National Coarse		National Fine	
		Threads per inch	Tap drill size, No.	Threads per inch	Tap drill size, No.
0	0.060	..	..	80	56
1	0.073	64	53	72	53
2	0.086	56	50	64	50
3	0.099	48	47	56	45
4	0.112	40	43	48	42
5	0.125	40	38	44	37
6	0.138	32	36	40	33
8	0.164	32	29	36	29
10	0.190	24	25	32	21
12	0.216	24	16	28	14

TABLE 11. TAP DRILL SIZES FOR PIPE TAPS
(Briggs Standard)

Size of Tap	Size of Drill
$\frac{1}{8}$	$2\frac{1}{64}$
¼	$\frac{7}{16}$
$\frac{3}{8}$	$\frac{9}{16}$
½	$2\frac{3}{32}$

Rethreading a Bolt.   Threads on bolts often become battered, and it is then difficult or impossible to screw nuts onto them.   Such battered threads can usually be renewed by simply running a regular die down over them in the same manner as in cutting new threads. The use of a rethreading die is usually simpler, however, if such a tool is at hand.   Rethreading dies, which usually come in sets, are six-sided one-piece dies and may be screwed onto bolts with a wrench in the same manner as ordinary nuts.   To use such a die, simply start it onto the threaded end of the bolt with the fingers, apply lard oil or a threading oil, and screw it on with a wrench.   When it has been screwed down to the end of the threads, screw it back off the bolt.

Points on Cutting Threads

1. Taper the end of the rod slightly before starting the die.   Do not make the taper too blunt.   Use a file, a hammer, or a grinder.
2. Start a tap or a die slowly and carefully so as to get it started straight.
3. Use equal pressure on the handles of the stock or tap wrench.
4. Always use lard oil or a threading oil when threading steel.   It will make the tool cut easier and smoother and last longer.
5. Do not back up or reverse the taps and dies until the thread is complete.
6. If chips and cuttings tend to clog the tool, remove them with a small nail or wire or piece of wood.
7. Be sure to drill the proper size of hole before tapping.   A hole that is too small can be easily and quickly enlarged with a taper reamer.
8. An easy way to make a bolt is to thread both ends of a rod and use nuts on both ends.
9. Remove the cuttings and wipe off excess oil before putting taps and dies away.
10. Battered threads on a bolt may be renewed by using either a small rethreading die or a regular die.

Jobs and Projects

1. Make a scriber or metal-marking tool by grinding the end of an old saw file to needle point.
2. If not already done, assemble the cold-metal working tools about your shop or home, and arrange a small cabinet or tool rack for keeping them. Sharpen the cold chisels and punches and other cutting tools if they need sharpening.
3. Look through shop manuals, books, and bulletins, and make a list of small appliances and jobs, involving cold-metal work, that would be useful around your home or farm.

Select a few of these and make them in the shop. Plan your work carefully before starting a job.

4. A check list of small articles and appliances involving cold-metal work includes:

Hasp

Chain repair link

Tool-grinding gage

Endgate rod washer

Shoe scraper

Wire splicer

Shelf brackets

Endgate rod nut

Small clevis

Corner irons for window screens

Small gate hook

Bolt and nut

Flower-pot holder or stand

Lamp bracket

Door antisag rod

5. Look about your home or farm, and make a list of repair jobs to machines or equipment that could be done with only such work as drilling, riveting, hack sawing, filing, punching, and cutting with a cold chisel. Bring two or three such jobs to the shop and do them.

6. Make a trash burner out of an old oil drum as follows: Cut out one end. Cut a draft opening about 4 by 6 in. in the side and near the other end. Use a slitting chisel or a cold chisel. Set the drum with the open end up, and it is ready for use.

11. Forging, Tempering, and Welding

WHETHER a farmer can afford a forge and anvil will depend upon the distance to a blacksmith shop, the amount of forging and other smithing work he needs to have done, and his ability as a mechanic. Although not every farmer can profitably own black-smithing equipment, many farmers can. If a farmer cannot, he should remember that a great variety of repairs can be made with the use of only a few simple cold-metal working tools.

Although blacksmithing is generally more difficult than woodwork, most farm boys with average mechanical ability can soon learn to do simple blacksmithing and will feel well repaid for their efforts. In all mechanical work, much more rapid and satisfactory progress can be made by a study of theory and principles along with practice. This is particularly true of blacksmithing.

MAJOR ACTIVITIES

1. Selecting Forging Equipment for the Shop

2. Building and Maintaining a Forge Fire

3. Heating Irons in a Forge

4. Cutting with the Hardy

5. Bending and Straightening Iron

6. Drawing and Upsetting Iron

7. Punching Holes in Hot Iron

8. Working Tool Steel

9. Welding in the Forge

10. Welding with the Oxyacetylene Torch

11. Welding with the Electric-arc Welder

1. Selecting Forging Equipment for the Shop

The Forge.   The forge for the farm shop should have a gear-driven blower operated by a crank, and it should have a hearth at least 18 in. wide, preferably somewhat larger.   A good way of providing a good forge is to get a good blower and make a hearth or stand of concrete or brick, or of pipe or angle iron.   A plate with holes drilled in it can be mounted in the bottom of the hearth for admitting the air blast.   The forge should have a hood and smoke pipe for taking away the smoke.

The Anvil.   Anvils are of two general kinds: cast iron and steel.   Steel anvils are much better and should be used if they can be

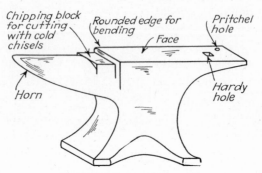

Fig. 426.—Parts of the anvil.

afforded.   The two kinds can be readily distinguished by striking with a hammer.   A cast anvil has a dead sound, while a steel one has a clear ring.

Anvils are commonly available in sizes ranging from 50 to 200 lb. An anvil weighing 100 or 125 lb. is quite satisfactory for the farm shop.

A piece of railroad iron or rail 24 to 30 in. long and mounted on a block or stand (see Fig. 8) makes a good anvil for light hammering and riveting, although a much greater variety of work can be done on a regular anvil.

Place the anvil on a substantial mounting, preferably of wood. Locate it in front of the forge, so that the workman can take irons from the fire and place them on the anvil by simply making a short turn and without the necessity of taking even a full step.   Place the horn of the anvil to the left for a right-handed workman and to the right for a left-handed man.   Mount the anvil at such a height that the top can be just touched with the knuckles of the clenched fist when standing erect and swinging the arm straight down (see Fig. 427).

Tongs. A few pairs of tongs of different sizes and kinds should be provided for the farm shop. Various types are available, the most common ones being flat-jawed tongs and hollow-bit curved-lip tongs (see Fig. 428). The hollow-bit curved-lip bolt tongs are probably best for general work. They can hold flat bars as well as round rods and bolts, and the curved part back of the tip makes it possible to reshape them easily to fit various sizes of stock. By filing or grinding a groove crosswise in the end of each lip, they can be made to hold links and rings practically as well as special link tongs. Tongs 18 to 20 in. long are a good size for average work.

Hammers. A blacksmith's hand hammer weighing 1½ or 2 lb. and another weighing 3 or 3½ lb. will handle all ordinary work very satisfactorily.

Fig. 427.—Mount the anvil on a substantial block at such a height that it can just be reached with the knuckles of the clenched fist, when standing erect.

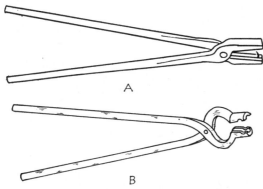

Fig. 428.—Types of tongs: A, flat-jawed hollow-bit tongs; B, hollow-bit curved-lip tongs. This style is very good for the farm shop. Flat bars as well as round rods and bolts can be held in them.

Hardy, Chisels, Punches. There should be at least one hardy of a size to fit the hole in the anvil, and it may be advisable to have two, one that is thin and keen for cutting hot metal, and one that is thick, heavy, and tempered for cutting cold metal. It is a good plan to have a fair assortment of various sizes of chisels and punches. These

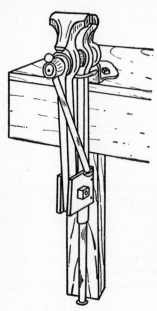

can be made easily in the shop. If considerable blacksmithing is to be done, it would be well to have a hot cutter and a cold cutter (simply large chisels with handles on them) for heavy cutting with a sledge hammer. It would be well also to have one or two large punches with handles on them for punching holes in hot metal. Punches for making holes $\frac{3}{8}$ in. and $\frac{1}{2}$ in. in diameter are probably most useful.

Vise. One vise can well serve for all metalwork in the farm shop, including blacksmithing if it is heavy and strong. A blacksmith's steel leg vise with jaws 4 to 5 in. wide is generally preferred as an all-purpose metalworking vise. A leg vise is one that has one leg extending down to be anchored or fastened into the floor (see Fig.

Fig. 429.—A heavy black-smith's steel-leg vise is a good type of vise for the farm shop.

429). Such a vise can be used for heavy hammering and bending better than other types. If there is a strong steel machinist's vise in the shop, it can be used for blacksmithing work if care is used not to do too heavy hammering or bending with it.

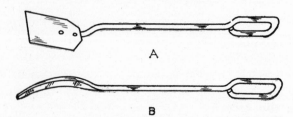

Fig. 430.—Homemade forge fire tools: A, shovel; B, poker.

Fire Tools. A small shovel and poker or rake will be needed for use on the forge fire. These can easily be made in the shop. A flat piece of heavy sheet iron about 3 or 4 in. wide by 4 or 5 in. long, riveted to a bar or rod for a handle, makes a good shovel (see Fig. 430A). A $\frac{1}{2}$-in. round rod, with an oblong eye in one end to serve as a handle and the other end flattened and curved, makes a good combination poker and rake (see Fig. 430*B*).

Measuring Tools.   Some kind of metal rule will be needed for measuring and checking pieces being forged.   A small steel square is very good for both measuring lengths and checking angles and bends. A caliper rule (see Fig. 431*B*) for measuring the diameter of rods and thickness of parts, although not a necessity, will be found convenient.

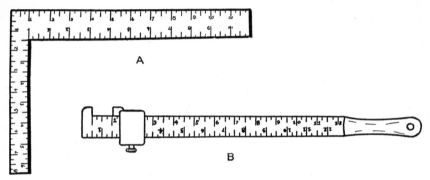

Fig. 431.—Use metal measuring tools in blacksmithing.   A. The small steel square is useful for checking bends and angles as well as for measuring.   B. The caliper rule is especially good for measuring the diameter or thickness of bolts, rods, and bars; as well as for general measuring.

2. Building and Maintaining a Forge Fire

A good fire is the first requirement for good blacksmithing.   Many beginners do poor work simply because they do not know the importance of a good fire.   Two students sometimes attempt to use the same fire.   This is usually a waste of time as it is difficult or impossible to heat irons properly when someone else is poking in the fire and trying to heat other pieces at the same time.

A good fire has three characteristics.   *It is clean*, that is, free from clinkers, cinders, etc.   *It is deep*, with a big, deep center of live burning coke.   And *it is compact*, being well-banked with dampened coal.

Starting the Fire.   To start a fire, first clean the fire bowl with the hands, pushing all coal and coke back on the hearth and *throwing out all clinkers*.   Clinkers are heavy and metallic and have sharp hard corners or projections and are easily distinguished from coke.   Coke is light in weight and is easily crumbled in the hands.   After removing the clinkers, shake the fine cinders and ashes through the grate into the ashpit below the tuyère (that part in the bottom of the hearth through which the blast comes).   Then dump the ashpit and try the blower, making sure that a good strong blast comes through. Sometimes ashes work back into the blower pipe and obstruct the blast.

After the fire bowl is cleaned, light a handful of shavings or kindling *from the bottom* and drop them onto the tuyère. Turn the blower gently and rake fuel, preferably coke left from the previous fire, onto the burning kindling. Once the fire is burning well, rake on more coke, and pack dampened coal *on both sides and on the back* of the fire. This forms a mound of burning coke at the center. The dampened coal on the outside concentrates the heat in the center. In a little while gases will be driven off this dampened coal, sometimes called *green coal*, and it will change to coke.

If the fire tends to smoke excessively, stick the poker down through the center, opening up a hole and allowing the air to come through more freely. Most of the smoke will then usually catch fire and burn.

Always use *blacksmithing coal* in the forge. It is a high-quality soft coal that is practically free from sulphur, phosphorus, and other objectionable impurities. When dampened and packed around the fire, it readily cakes and changes to coke, which burns with a clean, intense flame. Ordinary stove or furnace coal will not work satisfactorily in a forge.

Keeping the Fire in Good Condition. When the coke at the center of the fire burns up, replace it with more coke. This may be done by mashing the fire down in the center with the shovel or rake, or by raking coke in from around the edges of the hearth, or both. Always keep the fire deep. Do not let it become hollowed out and shallow.

Add green coal to the outer parts of the center mound from time to time, as may be required to keep the fire well banked and confined to the center.

Do not continually poke at the fire; simply keep the center well supplied with coke and the outside packed down with dampened coal.

If the fire tends to spread too much, or if it becomes open and loose, throw or sprinkle water on the outer edges and pack it down with the shovel. Use only a moderate blast of air. Excessive air makes for slow heating and scaling of the irons. If a moderate speed of the blower fails to produce enough air, check for leaks in the pipe between the blower and the tuyère, or for a partial stoppage with ashes.

Cleaning the Fire. From time to time—usually every half hour when welding—remove the clinkers and cinders that accumulate over the tuyère. To do this, simply pass the shovel along on the

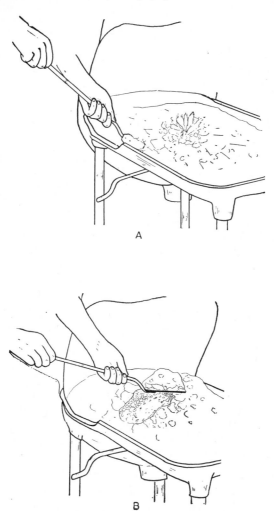

Fig. 432.—Cleaning the forge fire. First, push the shovel along the bottom of the hearth to the center of the fire, as at A; and then lift straight up, as at B. Clinkers and ashes, if any, are then easily seen and removed.

bottom of the hearth to the center of the fire, and then raise it straight up through the fire (see Fig. 432). The clinkers can then be easily seen and removed. Most of them will stay on the shovel.

After the cinders and clinkers are removed, rake coke back into the center and pack it down. Add fresh coal to the back and sides of the fire if needed.

Lining a Forge. Lining a forge hearth with clay, although generally not practiced, will protect it against overheating and rusting and will increase its useful life. A forge hearth may be lined as follows:

1. Make a thin wash of clay, preferably fire clay, and spread it over the inside of the hearth and allow it to dry.
2. Then mix fine sieved coal ash and clay together, using 1 part of coal ash to 3 parts of clay. Make the mixture about as stiff as putty, and apply it about one inch thick around the tuyère opening and at least 8-in. on the sides.
3. Allow it to dry over night.
4. Then build a wood fire and bake slowly for at least an hour, turning the blower a little every 10 or 15 min. to keep the fire going.

In using the forge, be careful not to break up the clay lining with the poker or shovel.

3. Heating Irons in a Forge

To heat irons in a forge fire, place them in the fire in a *horizontal position*, not pointing down (see Fig. 433). Be sure there is burning

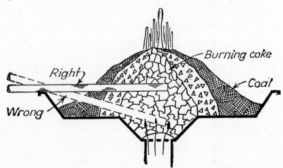

Fig. 433.—In heating irons in the forge, keep them pointed straight in—not down. Keep burning coke below them, on all sides of them, and on top of them.

coke *below the irons, on both sides of them, and on top of them.* Keep the fire deep and compact. Irons heat much more rapidly and oxidize or scale off much less when heated in a deep compact fire than in a shallow, burned-out fire. Some scale will form in spite of a good fire, but the scale should be kept to a minimum.

Small thin parts heat much more rapidly than heavier and thicker parts. To prevent burning the thinner parts, push them on through the fire to a cooler place, or otherwise change the position of the irons to make all parts heat uniformly.

For forging mild steel or blacksmith iron, heat it to a good bright-red heat. Do not allow it to get white hot and sparkle. Sparkling indicates burning. Tool steel must not be heated as hot as mild steel. A bright-red or low-orange heat is hot enough for tool steel.

Fitting Tongs; Holding the Work. If tongs cannot be found to fit the work, reshape a pair by heating and hammering the jaws over the piece to be held. Poorly fitting tongs are a source of continual trouble and should not be used.

A considerable amount of work can be done without the use of tongs. An eyebolt, for example, can be made on the end of a rod 20 to 30 in. long and then can be cut off when finished.

4. Cutting with the Hardy

A blacksmith does much of his cutting of iron and steel on the hardy. Although the hardy does not leave quite as smooth a cut

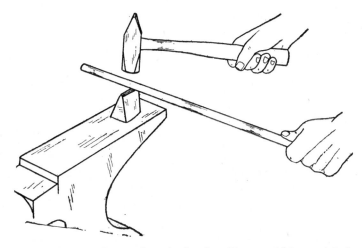

Fig. 434.—Cutting an iron rod on the hardy. To cut cold iron, nick deeply on two or more sides and then break by bending. In cutting hot iron, cut all the way through from one side, being careful to strike overhanging blows at the last to prevent striking the cutting edge with the hammer.

as a hack saw, it is quite satisfactory for most blacksmith work. It cuts faster and easier than a saw and is less expensive, as there are no blades to wear out or break.

To use a hardy, place it in the hardy hole (the square hole) in the anvil, lay the bar or rod on it at the point to be cut, and hammer it down against the sharp edge (see Fig. 434). Hardies may be used for either hot or cold cutting. Some smiths prefer to keep two hardies, one that is thick and stocky and tempered for cutting cold iron, and one that is thin for cutting hot iron. The hardy, like other cutting tools, works better if kept sharp. It may be ground like a cold chisel.

In cutting cold iron, the bar may be nicked deeply on two or more sides and then broken off by bending. In cutting hot iron, it is common practice to cut clear through from one side. Care must be taken, of course, not to let the hammer strike the cutting edge of the hardy, or else both the hammer and the hardy may be damaged. In finishing a cut on a hardy, strike the last two or three blows just beyond the cutting edge and not directly over it.

Cutting Tool Steel. Never attempt to cut tool steel in the hardened state. To cut it on the hardy, cut it hot—not cold—and handle it just like other iron or steel (except, of course, do not heat it above a cherry-red or low-orange heat).

When it is important to have a smooth cut, it is better to cut tool steel by sawing part way through with a hack saw, clamping in a heavy vise at the sawing line, and then breaking by hammering (see page 321).

Estimating Amount of Stock Required. Often before cutting off a piece of stock, it is necessary to estimate the amount required for bends and curves. To determine the amount required for a piece of irregular shape, a small wire may be bent into the desired shape and then straightened out and measured. To estimate the length of stock required for pieces of regular shape, like circles and parts of circles, estimate the length of the center line. For example, suppose it is desired to estimate the length of stock required for a ring $3\frac{1}{2}$ in. inside diameter, the stock itself to be $\frac{1}{2}$ in. thick. The length needed will be the length of the mid-line, halfway between the inside and the outside edges. This length will be equal to the mid-diameter, 4 in., times 3.1416, or approximately $12\frac{1}{2}$ in.

5. Bending and Straightening

In bending iron at the anvil, two points are most important:

1. Heat the iron to a good bright-red heat, almost but not quite white hot, throughout the section to be bent.
2. Use bending or leverage blows—not mashing blows.

Place the iron on the anvil so that it can bend under the hammer blows without being forced down against the anvil and mashed (see

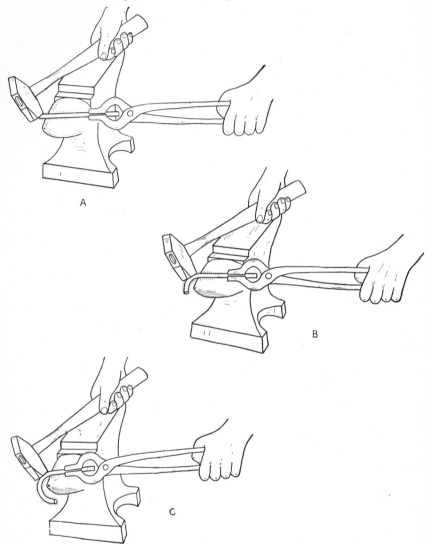

Fig. 435.—To make a uniform bend in the end of a rod, strike the part that projects beyond the horn and keep feeding the rod forward with the tongs as the bending progresses. Keep the iron at a good working heat and do not strike the rod where it rests on the horn.

Fig. 435). If an iron is struck at a point where it is resting firmly against the anvil, it will be mashed instead of bent. A few moderately sharp blows are better than several lighter blows.

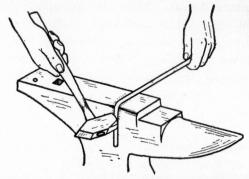

Fig. 436.—Make abrupt square bends over the face of the anvil near the chipping block. Here, the corner of the anvil is rounded to prevent galling the iron.

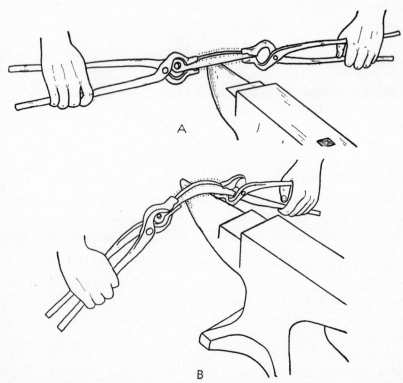

A

B

Fig. 437.—Flat iron may be bent edgeways by heating to nearly a white heat and bending slowly with tongs. This method is good in making flat chain hooks.

Making Square Bends. Make abrupt square bends over the face of the anvil near the chipping block where the corner of the anvil is rounded to prevent marring or galling the iron (see Fig. 436).

As soon as the iron falls below a good bending heat, put it back in the fire and reheat it. To bend an iron at a certain point without bending the adjacent section, heat it to a high red heat, and quickly cool it up to the place of bending by dipping into water. Then bend quickly by hammering or other suitable methods.

Bending Flat Bars Edgeways. A flat bar can usually be bent edgeways by heating and placing over the horn and bending the two ends down slowly, using the hands if the piece is long enough, or two pairs of tongs in case of short pieces (see Fig. 437). Sometimes the bending can be done easily by simply putting one end of the piece in the hardy hole and pulling on the other end (see Fig. 438). If the iron starts to buckle, stop bending at once and lay it on the anvil and straighten it. Hammering the outside edge of the iron when laid flat will tend to stretch it and therefore help with the bending. Once the bend is well started, hammering on edge around the horn is not so difficult. Be sure to always hold the stock firmly, either with the hands or with tongs, and to keep the part to be bent at a high red heat.

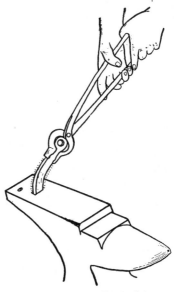

Fig. 438.—Bending of heavy pieces can sometimes be best accomplished in the hardy hole.

Straightening. Straightening can usually best be done on the face of the anvil. Be sure to hold the stock firmly, and to strike at points where it does not touch the anvil face. Sighting along the stock is the best way to test for straightness and to locate high points that need striking.

Striking with the Hammer. Success in blacksmithing depends largely upon ability to strike effectively with the hammer. Most blacksmithing requires heavy, well-directed blows. Where light blows are better, however, they should be used.

Strike light blows mostly with motion from the wrist (see Fig. 439A). Use both wrist and elbow action for medium blows (see

Fig. 439B). For heavy blows, use shoulder action as well as elbow and wrist motion (see Fig. 439C).

To direct hammer blows accurately, strike one or two light taps first, to get the proper direction and feel of the hammer, and then follow with quick sharp blows of appropriate force or strength. It

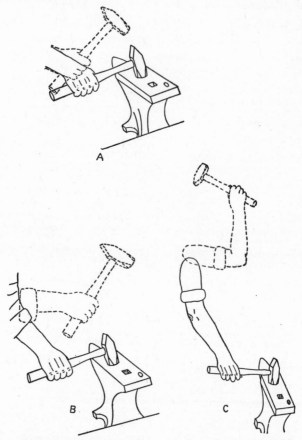

Fig. 439.—Striking with the hammer. Strike light blows largely with wrist motion, as at A; moderate blows with elbow and wrist motion, as at B; and heavy blows with shoulder action, as well as with wrist and elbow motion, as at C.

is also important to use a hammer of appropriate size. A heavy hammer on light work is awkward, and blows cannot be accurately placed. And using a light hammer on heavy work is very slow and tedious.

Bending and Forming an Eye. One of the most common bending jobs in the blacksmith shop is that of forming an eye on the end of a rod. The following is a good method of making such an eye:

1. Heat the rod to a good red heat back for a distance of about 5 to 8 in., depending on the size of the eye.
2. Quickly place the rod across the face of the anvil with just enough of the heated end projecting beyond the edge of the anvil to form the eye. For exact work, the length of hot iron that is to project over may be quickly measured with a metal rule (see Fig. 440A and B). Place the iron across the anvil well up near the horn where the edge is rounded.
3. Bend the end down, forming a square bend, with a few well-directed blows (see Fig. 440C). Work rapidly before the iron cools. In finishing the bend, strike alternately on the top of the anvil and then on the front vertical side.
4. Heat the end of the stock, and start bending the tip end around the horn. Work from the tip back toward the stem (see Fig. 440D, E, and F). Keep the iron hot throughout the part being bent; otherwise the bending will be slow and difficult, and the iron will not bend at just the places desired. If the square bend at the juncture of the stem and eye tends to straighten out, it is an indication that the end of the stock is not being kept hot enough while being bent.
5. Round the eye by driving it back over the point of the horn, noting carefully where it does not rest against the horn and striking down lightly in these places (see Fig. 440G). Keep the iron well heated.
6. Center the eye on the stem, if necessary, by placing the stem flat on the anvil face with the eye projecting over the edge, and strike the eye. Have the stock well heated at the juncture of the stem and eye, but have the eye itself practically cold. Such a condition can be produced by heating the whole eye and then quickly cooling most of the rounded part by dipping in water.

Blacking. After forging a piece of iron, it is a good plan to black it by heating it slightly and rubbing it with an oily rag (see Fig. 441). Blacking gives the piece a better appearance and provides some protection against rusting. In blacking an iron, heat it just hot enough to burn off the oil that is rubbed on. It should not be red hot. If the iron is not hot enough, it will have a greasy appearance after rubbing with an oily rag.

Do not black tempered tools in this manner, as heating would draw the temper.

Twisting. Twisting is really a form of bending. Small pieces may be twisted by heating, clamping a pair of tongs on each end of the section to be twisted, and applying a turning or twisting force. To twist larger pieces (say more than about $\frac{1}{4}$ in. thick by 1 in. wide), clamp them in a vise and twist with a pair of tongs or a monkey wrench

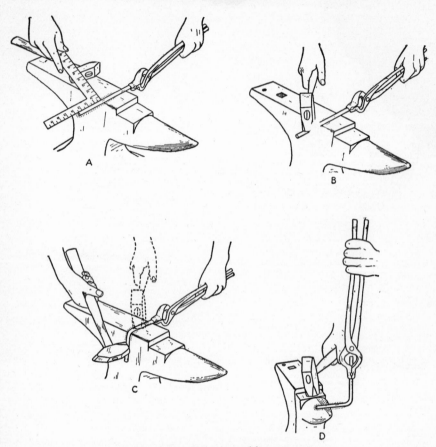

Fig. 440.—Steps in making an eye.

A. Place a well-heated iron across the anvil with enough stock projecting over to form the eye. Where the eye must be made accurately to size, use a metal rule or square for measuring. Work rapidly.

B. Bending the projecting portion down, forming a right angle.

C. Finish the right angle bend by striking alternately on top and on the side, keeping the iron at a good working heat all the while.

D. Start bending the tip end around the horn, being careful to strike "overhanging" or bending blows.

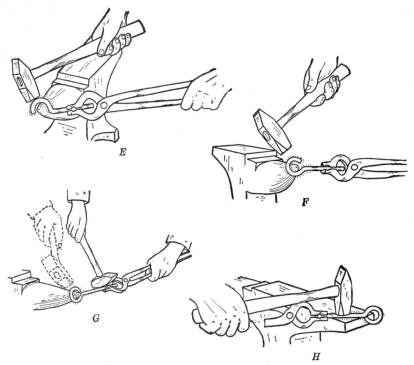

Fig. 440.—Steps in making an eye (continued).

E. Gradually work back from the end to the square bend.

F. Turn the eye over and close it up. Exert considerable back pull on the tongs to keep the upper part of the eye up off the horn. In this position the hammer can strike bending blows instead of flattening or mashing blows.

G. Round the eye by driving it back over the point of the horn. Carefully note where the eye does not touch the horn and strike down lightly in these places.

H. To straighten the stem of an eyebolt, place it across the corner of the anvil face and strike the high points while the iron is at a good working heat.

(see Fig. 442). Be careful to clamp the vise and the tongs or wrench at exactly the ends of the section to be twisted.

In twisting, work rapidly before the iron cools. For a uniform twist, the iron must be at a uniform temperature. If the twist is

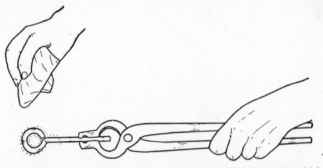

Fig. 441.—An iron may be blacked by heating it slightly and rubbing it with an oily rag. Have the iron just hot enough to make the rag smoke. Blacking improves the appearance and affords some protection against rust.

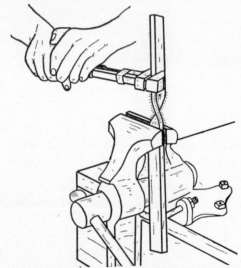

Fig. 442.—Heavy bars may be twisted by heating to a good working heat, clamping in a vise, and twisting with a wrench or pair of tongs.

to be confined to a definite section of the stock, it is a good plan to place center-punch marks at the ends of the section before it is heated.

6. Drawing and Upsetting Iron

Drawing is the process of making a piece longer and thinner Two points are particularly important in drawing:

1. Keep the iron at a good forging heat, a high red or nearly white heat.
2. Use *heavy*, *straight-down*, *square* blows.

Many beginners make the mistake of striking a combination down-and-forward pushing blow, thinking that the pushing helps to stretch the metal.

Drawing may be done more rapidly over the horn than on the face of the anvil, as the round horn wedges up into the metal and lengthens it and there is less tendency for it to stretch in all directions. If a piece tends to get too wide, place it on edge and hammer it.

Keep the iron hot. Hammering after the red heat leaves is hard work and accomplishes little. Also, the iron is apt to split or crack if hammered too cold.

Drawing Round Rods. To make a round rod smaller, the following steps should be carefully followed:

1. Make it four-sided, or square in cross section.
2. Draw it to approximately the desired size *while it is square*.
3. Make it *distinctly eight-sided* by hammering on the corners *after it is drawn* sufficiently.
4. Make it round again by rolling it *slowly* on the anvil and hammering *rapidly* with *light* blows or taps.

An attempt to draw round rods without first going to the square section not only requires a lot of extra work, but usually results in a badly distorted and misshaped piece.

Pointing a Rod. To make a round point on a rod, first make a *square* tapered point. Then make it eight-sided, and finally round. In pointing a rod, the following suggestions are important:

1. Work on the *far edge* of the anvil. The toe of the hammer is then not likely to strike the face of the anvil.
2. Raise the back end of the rod.
3. Strike with the toe of the hammer lower than the heel (see Fig. 443).

After the point is drawn, sight to see if it is centered, or roll the rod on a flat surface and see if the point wobbles up and down (see Fig. 444). A few well-placed blows of the hammer are usually all that is required to center the point.

Upsetting. Upsetting is simply the reverse of drawing, or the process of making a piece shorter and thicker. It is done when more metal is needed to give extra strength, as when a hole is to be punched for an eye. There are two main points to be observed in upsetting:

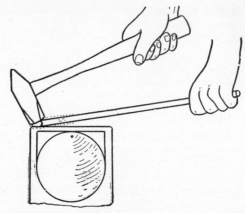

Fig. 443.—In pointing the end of a rod, three things are important: 1. Raise the back end.   2. Tilt the toe of the hammer down.   3. Work on the far edge of the anvil.   To make a round point, make it square first, then eight-sided, and finally round.

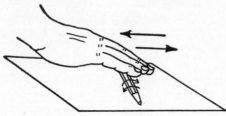

Fig. 444.—To tell if a punch or rod is straight and the point centered, roll it on a flat surface.   If the point wobbles, it is off center.

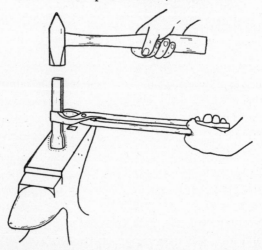

Fig. 445.—To ensure success in upsetting, work the iron just under a white heat and strike tremendously heavy blows.   Light blows simply flare the end without upsetting very far back from the end.

1. Heat the bar or rod to a high red or nearly white heat throughout the section to be upset.

2. Strike *extremely heavy* well-directed blows.

Light blows simply flatten and burr the end instead of upsetting the piece throughout the heated section. The extra-heavy blows needed for upsetting can best be struck by first striking a light blow or two to get the direction of striking and then following with an extra-heavy blow. If the bar starts to bend, stop and straighten it at once. Further hammering will simply bend it more instead of upsetting it.

Probably the best way to upset a short piece is to place the hot end down on the anvil and strike the cold end (see Fig. 445). The

Fig. 446.—When it is desired to have only a portion of an iron hot, and the adjoining parts cool, as in upsetting, it may be necessary to heat a larger portion and then cool back to the desired point by dipping in water.

hot end, of course, may be up, but it is usually easier to upset without bending if the hot end is down.

Usually three heats are enough for upsetting a piece. When more than three heats are required, it is generally an indication that the iron has not been heated enough or that the hammer blows have been too light.

In order to thoroughly heat the part to be upset, and yet confine the heat to this part, heat the work somewhat further than the upsetting is to go and then cool it quickly back to the line of upsetting by dipping in water (see Fig. 446).

The end of a long bar may be upset by laying it on the anvil face, with the hot end projecting beyond the edge, and striking heavy blows endways with the hammer. If the bar is long and heavy enough, it may be upset easily by ramming the hot end against the face or the side of the anvil.

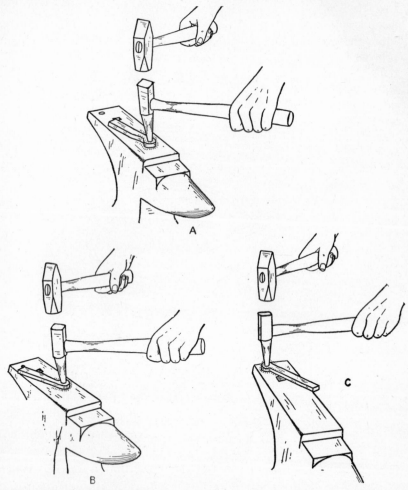

Fig. 447.—In punching holes in hot iron, work it just under a white heat. **A.** Carefully locate the punch and drive it about two-thirds of the way through. **B.** Then turn the iron over and drive it back through from the other side. **C.** Finally move the piece over the pritchel hole or hardy hole and drive the slug or pellet through.

7. Punching Holes in Hot Iron

It is sometimes easier to punch a hole in a piece of iron than to drill it; and for some purposes a punched hole is better. For example, in forming an eye on the end of a bar in making a hook or clevis, punching makes a stronger eye.   Very little metal is wasted when a hole is punched.   To punch a hole, proceed as follows:

1. Heat the iron to a good working temperature, a high red or nearly white heat.
2. Place the hot iron quickly on the flat face of the anvil—not over the pritchel hole or hardy hole.   Punching over a hole would stretch and bulge the iron.
3. Carefully place the punch where the hole is to be, and drive it straight down into the metal with heavy blows until it is about two-thirds of the way through (see Fig. 447A).
4. Turn the iron over, and drive the punch back through from the other side (see Fig. 447B).   Be sure to locate the punch so that it lines up with the hole punched on the first side.   Reheat the iron and cool the punch as may be needed.
5. Just as the punch is about to go through, move the piece over the pritchel hole or hardy hole to allow the small pellet or slug to be punched out (see Fig. 447C).
6. Enlarge the hole to the desired size by driving the punch through the hole first from one side and then the other.   Always keep the metal at a good working temperature, reheating as may be necessary.

Fig. 448.—When punching hot iron, cool the punch frequently by dipping it in water.

When punching holes, cool the punch frequently by dipping it in water (see Fig. 448).   A little powdered coal dropped into the hole will help keep the punch from sticking.

An easy way to place the punch so as to get the hole centered in a bar is to lean it and twist it slightly until it is in the correct position.

Although a hand punch can be used in punching hot iron, it is much better to use a regular blacksmith's punch with a handle in it.

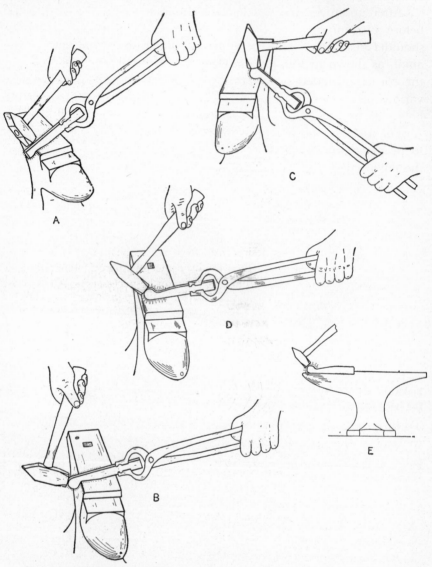

Fig. 449.—Forming a shoulder or neck, preparatory to punching an eye. First drive the iron down against the corner of the anvil, as at A. Then finish shaping as suggested in the various other views.

Forming Punched Eyes. Usually, although not always, when a hole is to be punched for an eye, as in a chain hook or a clevis, it is best to upset the stock first so as to give more metal and make a stronger eye.

After upsetting, the end is shaped and the corners are rounded before punching. This can best be done by forming a neck or shoulder just back of the eye by hammering over the far edge of the anvil, as shown in Fig. 449A. The end is then further shaped and the corners rounded by working over the anvil as suggested in the various other views of Fig. 449. Having the end thus shaped, the hole may be punched in the usual fashion.

For a clevis, the punched holes should be left with straight sides to fit the clevis pin. Holes in chain hooks, however, should have the ends flared and rounded somewhat to fit better in chain links.

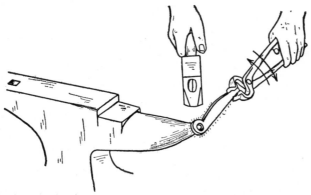

Fig. 450.—Smoothing the inside and the outside edges of a punched eye. First, make the stock approximately eight-sided and then round, by rolling it slowly on the horn and striking fast, light blows.

To thus shape an eye for a chain hook, place the eye at an angle on the end of the anvil horn and make the stock approximately eight-sided and finally round by rolling it slowly while striking light rapid blows (see Fig. 450).

Points on Forging

1. A clean, deep, compact fire is the first requirement for good black-smithing.
2. Put the irons in the fire in a horizontal position—never point them down into the fire.
3. Always work the irons at a good forging heat—a bright red or nearly white heat for mild steel.
4. In bending, use bending or leverage blows—not mashing blows.
5. In drawing, strike square direct blows straight down—not forward-pushing or glancing blows.

6. In drawing round rods, always make them square first and do the drawing while square. When drawn sufficiently, make them eight-sided and finally round.

7. To smooth up a round rod, roll it slowly on the anvil while striking a series of light quick blows.

8. In pointing rods, work on the far edge of the anvil. Raise the back end of the rod, and strike with the toe of the hammer tilted down.

9. In upsetting use a high heat, and strike extra-heavy blows.

10. To make a good twist, have the section to be twisted at a uniform temperature.

11. To punch a hole in a hot iron, start in on the flat face of the anvil. Then turn it over, and drive the punch back from the other side. Move the iron over a hole in the anvil face for finally driving out the pellet.

12. In cutting on the hardy, be careful not to let the hammer strike the cutting edge.

13. Use the chipping block for cutting with the cold chisel—not the flat face of the anvil.

14. Strike light hammer blows with wrist motion only; medium blows with motion from both the wrist and the elbow; and heavy blows with motion from the shoulder, wrist, and elbow.

15. Blacking a forging gives it a better appearance and provides some protection against rust. To black, simply rub the piece with an oily rag when it is just hot enough to make the rag smoke.

8. Working Tool Steel

One of the main advantages of having a forge in the farm shop is to be able to redress and make and temper tools like cold chisels, punches, screw drivers, picks, and wrecking bars. Tool steel for making cold chisels and punches and similar tools may be bought from a blacksmith or ordered through a hardware store; or it may be secured from parts of old machines, such as hay-rake teeth, pitchfork tines, and axles and drive shafts from old automobiles.

Heating Tool Steel. Always heat tool steel slowly in a clean, deep coke fire. Uneven heating, which is usually caused by heating in a poor shallow fire or by too rapid heating, results in unequal expansion, which, in turn, may cause internal flaws or weaknesses in the steel.

Never heat tool steel above a bright-red or low-orange heat, and heat to this temperature only for heavy hammering. Heating higher is likely to cause the steel to become coarse-grained and weak, rather than fine-grained and strong. In case a piece of tool steel is heated a little too hot, the grain size may be restored by (1) allowing it to cool slowly and then reheating, being careful not to overheat it again, or (2) by heavy hammering at a bright-red or low-orange heat.

Forging Tool Steel. It is important to observe the following points when forging tool steel:

1. Do not hammer below a red heat, as this may cause cracking and splitting.
2. Be sure tool steel is uniformly heated before it is hammered. Otherwise, the outside parts, which are hotter, may stretch away from the inside parts, which are colder, and thus cause internal flaws.
3. Avoid very light hammering, because this may draw the outer surface without affecting the inner portions.
4. Do as much forging as possible by heavy hammering at a bright-red or low-orange heat, as this will make the grain size smaller and thus refine and improve the steel.
5. When finishing a piece by moderate blows, do not have the steel above a dark-red heat.

Annealing Tool Steel. It is best to anneal a tool, or soften it, after it has been forged and before it is hardened and tempered. This is to relieve strains that may have been set up by alternate heating and cooling and by hammering. To anneal a tool, heat it to a uniform dark-red heat and place it somewhere out of drafts, as in dry ashes or lime, and allow it to cool very slowly.

Hardening and Tempering Tool Steel. If tool steel is heated to a dark red and then quenched (cooled quickly by dipping it into water or other solution), it will be made very hard. The degree of hardness will depend upon the carbon content of the steel and the rapidity of cooling. The higher the carbon content, the harder it will be; and the more rapid the cooling, the harder it will be.

A tool that is hardened in this manner will be too hard and brittle and must be tempered or softened somewhat. This may be done by reheating the tool to a certain temperature (always lower than the hardening temperature) and quickly cooling it again. The amount of softening (or tempering) accomplished will depend upon the temperature to which the tool is reheated. For practical purposes in the farm shop, these temperatures are judged by the color of the oxide or scale on the steel as it is being reheated. A straw color, for example, indicates a comparatively low temperature, and if the tool is quenched on this color, it will be softened only a little. A blue color, on the other hand, indicates a higher temperature, and if the tool is quenched on this color, it will be made considerably softer.

Different grades of tool steel will have different degrees of hardness when quenched at the same color. Therefore, it may be necessary to experiment a little with the first piece of a new lot of steel in order to secure the desired degree of hardness.

Hardening and Tempering A Cold Chisel. After a cold chisel is forged and annealed, it may be hardened and tempered as follows:

1. Heat the end to a dark red, back 2 or 3 in. from the cutting edge.
2. Cool about half of this heated part by dipping in clean water and moving it about quickly up and down and sideways until the end is cold enough to hold in the hands (see Fig. 451A).

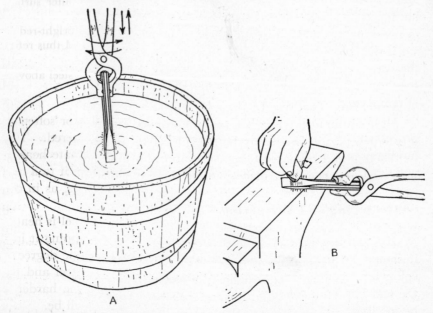

Fig. 451.—Tempering a cold chisel. **A.** First heat about 3 in. on the end to a cherry red. Then cool about half the heated portion, moving the tool about rapidly up and down and sideways, to avoid too sharp demarcation between hot and cold parts. **B.** Then quickly polish the cooled end with emery cloth or other abrasive to enable the colors to be seen. When the dark blue appears at the cutting edge, dip the end again, working it about, keeping the end cold while any heat left up in the shank is allowed to dissipate slowly.

3. Quickly polish one side of the cutting end by rubbing with emery cloth, a piece of an old grinding wheel, a piece of brick, or an old file (see Fig. 451B).
4. Carefully watch the colors pass toward the cutting end. The first color to pass down will be yellow, followed in turn by straw, brown, purple, dark blue, and light blue.

5. When the dark blue reaches the cutting edge, dip the end quickly into water and move it about rapidly.   If much heat is left in the shank above the cutting edge, cool this part slowly so as not to harden the shank and make it brittle.   This is done by simply dipping only the cutting end and keeping it cool while the heat in the shank above slowly dissipates into the air.

6. When all redness has left the shank, drop the tool into the bucket or tub until it is entirely cool.

When the tool is first dipped, it is important that it be moved up and down to prevent the formation of a sharp line between the hardened and unhardened parts, as such a line might cause the tool to break at this point sometime later when in use.

If the colors come down too rapidly, the tool may be dipped into the water and out again quickly to retard their movement.   When they move down slowly, it is easier to watch them and do a good job of tempering.

Dipping the end at the beginning of the hardening and tempering process makes it very hard.   The heat left up in the shank of the tool, however, gradually moves down to the cutting end and softens it; and when it is softened to the desired degree of hardness, as indicated by the color, the tool is then quickly quenched to prevent any further softening.   The various colors are simply indications of different temperatures.

If a tool is tried and found to be too soft, as indicated by denting, it should be retempered and the final quenching made before the colors have gone out quite as far as they did originally, that is, before the end has been softened quite as much.   In case a tool proves to be too hard and the edge chips or crumbles, it should be retempered and the colors allowed to go out a little further.

Tempering Punches, Screw Drivers, and Similar Tools. Tools like punches, screw drivers, and scratch awls may be tempered in the same manner as a cold chisel but may be made harder or softer according to the requirements of the tool.   A scratch awl should be made somewhat harder than a cold chisel, a rock drill somewhat harder, a center punch just a little harder, a punch for lining up holes somewhat softer, a screw driver somewhat softer, etc.

Tempering Knives. Knives and tools with delicate parts are usually hardened and tempered in a manner slightly different from that used for cold chisels, in order to avoid the danger of overheating and warping and to ensure uniform hardening and tempering of the cutting edges.

After a knife blade is forged, anneal it. Then heat it slowly and uniformly to a dark red. Then cool it quickly by dipping the blade edgewise, *thick edge first*, in clean tepid water or oil. This method of cooling helps to ensure uniform cooling and therefore uniform hardening and freedom from warping. Next polish the blade and then reheat it by drawing it back and forth through a flame or by laying it against a large piece of red-hot iron and turning it frequently to ensure uniform heating. When the desired color, usually blue, appears, quickly cool the blade again by dipping edgewise in the water or oil.

Another method of heating knives and similar tools for hardening and tempering is to draw them slowly back and forth inside a pipe in a forge fire. The pipe must first be uniformly heated in a big fire and then turned frequently to keep it uniformly heated on all sides. Do not allow the knife to touch the pipe.

Points on Working Tool Steel

1. Use a clean deep coke fire for heating tool steel, and heat it slowly and evenly.
2. Heating in a poor shallow fire, or heating too rapidly, is likely to cause uneven heating, which results in unequal expansion, which in turn may cause internal flaws or cracks.
3. Proper hammering of tool steel at the proper temperature refines it, making the grain size smaller.
4. Do not hammer tool steel unless it is at least at a dark-red heat and heated uniformly clear through.
5. Hammering below a red heat is likely to cause cracking and splitting.
6. Hammering when not heated clear through may cause the outer parts to stretch away from the inner parts and cause internal flaws or cracks.
7. Avoid light hammering even when the steel is well heated, because of danger of drawing the outer surface without affecting the inner parts.
8. Never heat tool steel above a bright-red or low-orange heat, and then only for heavy hammering.
9. For moderate hammering, as in finishing and smoothing a job, do not heat above a dark red.
10. Tool steel is ruined if it gets white hot.
11. In case tool steel is accidently overheated somewhat, allow it to cool slowly and then reheat, being careful not to overheat it again; or heat it to a bright-red or low-orange heat and forge by heavy hammering to restore the fine grain size.

12. After a tool is forged, anneal it by heating to a uniform low red and placing it in dry ashes or similar material to cool slowly.

13. In quenching a tool like a cold chisel, move it about rapidly up and down and around—to prevent a sharp line of demarcation between the hot and cold parts.

14. Tempering colors should move slowly so they may be easily seen. If they move too fast, dip the tool quickly into water for an instant.

15. In the final quenching of a tool like a cold chisel, cool the end quickly but dissipate any heat left in the shank very slowly. Otherwise, the shank may be hard and brittle.

16. In case a tool is found to be too hard, retemper it and allow the temper colors to go out a little further before final quenching.

17. In case the tool is too soft, quench before the colors go so far.

9. Welding in the Forge

Although welding in the forge is somewhat more difficult than ordinary forge work, welding of links, rings, and bars and rods is

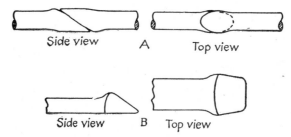

Side view   A   Top view

Side view   B   Top view

Fig. 452.—A, round rods upset, scarfed, and in position for welding. B, flat bar upset and scarfed for welding.

not particularly difficult if careful attention is given to the fire and to a few simple precautions.

A good fire is the first requirement for welding in the forge. It is important for any blacksmithing work, but for welding it is indispensable. Poor fires account for most of the difficulties experienced by beginners. Clean the fire about every half hour when welding.

Scarfing the Irons. Ends to be welded should first be properly shaped or scarfed. Scarfed ends should be short, usually not over $1\frac{1}{2}$ times the thickness of the stock. They should also have rounded or convex surfaces (see Fig. 452), so that when they are welded together any slag or impurities will be squeezed out rather than trapped in the weld. Avoid long, thin, tapering scarfs, because they are easily burnt in the fire, and because they cool and lose their welding heat very rapidly when removed from the fire, thus making welding exceedingly difficult. The ends of bars and rods should usually be upset before they are scarfed.

Heating the Irons. Heat the irons slowly at first so they will heat thoroughly and uniformly. Turn them over in the fire once or twice during heating to ensure equal heating of all sides and parts.

After the irons reach a bright-red heat, remove them and sprinkle welding compound or flux on the scarfed ends (see Fig. 453). Then replace the irons in the fire, and quickly bring them to the welding temperature. At this point, it is especially important that the fire be deep and compact. Otherwise, it will be difficult or impossible to heat the irons to the welding temperature. If one iron heats

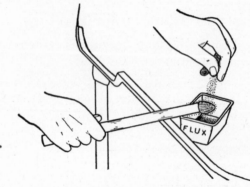

Fig. 453.—Apply welding flux with irons at a red heat, and just before the welding heat is taken.

faster than the other, pull it back into the edge of the fire for a few seconds. They should reach the welding heat at the same time. During the last part of the heating period, have the scarfed sides of both irons *down* so that they will be fully up to the welding temperature when removed from the fire.

When the irons reach the welding temperature, they will be a brilliant, dazzling white; their surfaces will appear molten; and a few explosive sparks will be given off. It is then time to remove them from the fire, get them in place quickly on the anvil, and weld them together.

Getting the Irons Together. When the irons reach the welding heat or temperature, remove them from the fire, quickly rap them over the edge of the anvil to shake off any slag, place them together, and hammer them in place (see Fig. 454). Strike medium blows at first, because the iron is soft and is easily mashed out of shape. Follow with heavier blows. Fast, accurate work is required, or the irons will lose their welding heat before they can be joined.

In placing the irons together on the anvil, put the right-hand one down first, and then put the left-hand one down on top of it. The pieces can thus be held together with only one hand, leaving the right hand free to use the hammer. Steadying the pieces over the edge of the anvil will help get them accurately and quickly placed together

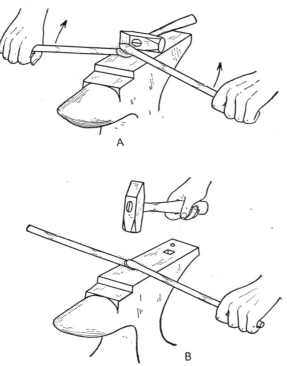

Fig. 454.—Placing irons together for welding. Steady the irons over the edges of the anvil, the one in the left hand being on top, as at A. Gradually raise the hands until the iron in the left hand holds the other one against the anvil, as at B, while the right hand strikes with the hammer.

(see Fig. 454). It is a good plan for the beginner to practice bringing the irons out of the fire and placing them together a few times before taking the welding heat. Pieces that are long enough to be held in the hands without tongs are handled more easily than short pieces.

If the irons do not stick at the first attempt, do not continue hammering, but reshape the scarfs and try again, being sure that the scarfs are properly shaped, that the fire is clean, and that it is deep and compact. Irons will not stick if there is clinker in the fire or if it is burned low and hollow. It is generally not possible to make irons stick after two or three unsuccessful attempts because they will most likely be burned somewhat, and burned irons are difficult or impossible to weld. In such cases, cut off the burned ends before rescarfing.

Fig. 455.—Steps in making a link.

Finishing the Weld. If, as is often the case, the irons stick but the lap is not completely welded down on the first heat, simply reapply flux and take another welding heat. In taking an extra heat, be sure to have the lap down in the fire just before removing for placing on the anvil and hammering.

After a weld is completed in a round rod, the welded section may be easily smoothed and drawn to size by first making it square; draw-

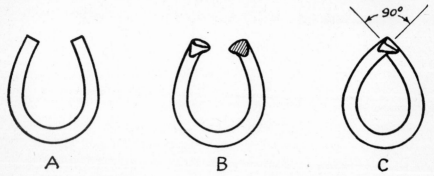

Fig. 456.—Steps in making a ring. A. Bend to horseshoe shape. B. Scarf same as chain link. C. Bend egg-shaped and weld.

ing as may be required while it is square; and then making it eight-sided and finally round, in the same manner as drawing round rods.

Welding a Link or Ring. To make a link or ring, heat the stock and bend it into a horseshoe or U shape (see Figs. 455 and 456). Then heat and scarf the ends in the following manner:

1. Place an end diagonally across the shoulder between the anvil face and the chipping block. Keep the other end against the vertical side of the anvil (see Fig. 457).

2. Strike a medium blow to the end on the shoulder. Swing the tongs a little, and strike another blow. In a similar manner, swing the tongs and strike again. This makes a short, blunt, angling taper with a slightly roughened surface.

3. Turn the piece over, and scarf the other end in the same manner.

4. Finish the scarfs by striking lightly with the cross peen of the hammer.

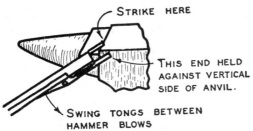

Fig. 457.—A good method of making scarfs for links or rings.

After scarfing the ends, heat and bend them over the horn and lap them together. Be sure the ends meet each other at an angle of about 90 deg. This ensures plenty of material for finishing the link and prevents a thin, weak weld.

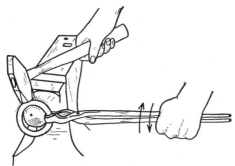

Fig. 458.—Finish the weld on a link or ring by rolling it slowly on the horn while hammering with a series of rapid, light blows. Large rings may be finished by making the stock square, then eight-sided, and finally round.

Next, heat the link, apply flux, and then take a welding heat. When the welding heat is reached, remove the link quickly from the fire, rap it over the edge of the anvil to shake off any slag, place it flat on the anvil (not on the horn), and hammer. Strike one or two medium blows, turn the link over, and strike a few medium blows on the other side. Follow with heavier blows. Take one or two more welding heats if necessary.

After the link or ring is welded, finish it at a good forging heat over the horn. Roll it slowly by twisting the tongs back and forth (see Fig. 458), while hammering rapidly with light blows. In case of a large ring, finish by making the stock square, then eight-sided, and finally round, as in drawing rods.

Points on Welding in the Forge

1. Use a clean, deep, compact coke fire.
2. Clean the fire every half hour.
3. Make the scarfs short and thick, rather than long and thin. Scarfs should not be longer than $1\frac{1}{2}$ times the thickness of the stock.
4. Round the surfaces of scarfs so slag will be squeezed out rather than trapped in the weld.
5. Heat the irons to a good welding heat, yet do not burn them.
6. Bring both irons up to the welding heat at the same time.
7. Before welding the irons together, shake off any slag or impurities by quickly rapping the tongs over the edge of the anvil.
8. Strike light or medium blows when irons are at the welding heat. Simply forcing the parts together is all that is necessary. Heavy blows mash the irons.
9. In case of failure to stick, do not continue hammering. Reshape the scarfs, reflux, and try again, being sure the fire is in good shape and that you heat the irons hot enough.

10. Welding with the Oxyacetylene Torch

Steel, cast iron, malleable iron, and many other metals can be effectively welded with the oxyacetylene torch. Although a little more skill is required for oxyacetylene welding than for most other kinds of farm shopwork, it is not difficult to learn. Those who can weld in a forge can quickly learn to do oxyacetylene welding. Oxyacetylene welding is also usually easier for beginners than electric-arc welding.

Because of the expense of the equipment required, and also because of the expense and inconvenience of returning empty oxygen and acetylene cylinders and getting full ones, oxyacetylene welding has not been widely done in home farm shops. With the increased use of community shops, however, and with the greater need of repair services on farms, the use of oxyacetylene welding in farm shopwork may increase.

Oxygen and acetylene when mixed in the correct proportions and burned produce an extremely hot flame that will readily melt metals.

If two pieces of steel are brought together and heated at the joint with the oxyacetylene flame, the edges will fuse or melt and run together. This is known as *fusion welding*. Another type of welding commonly done with the oxyacetylene flame is *bronze welding* (also called *braze welding*). It is used on cast iron, malleable iron, brass, copper, and also on steel. In bronze welding, the surfaces to be joined are not fused or melted, but are simply brought together and heated to a dull cherry red, and then coated with molten bronze from a bronze welding rod with the aid of a flux. Additional bronze is then added to increase the size and the strength of the weld. When the bronze weld metal is allowed to solidify, a solid joint is formed.

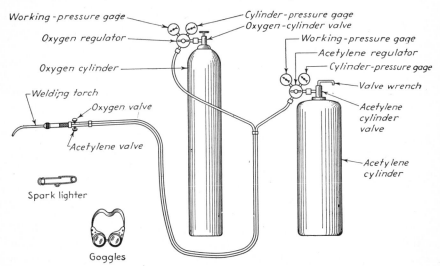

Fig. 459.—An oxyacetylene welding outfit.

Oxyacetylene Welding Equipment. The oxyacetylene welding outfit, as commonly used, consists of a supply of oxygen and acetylene under pressure in cylinders, pressure regulators, a blowpipe or torch, hoses, connections, and accessories like goggles and a lighter (see Fig. 459). The oxygen and acetylene supply cylinders are connected to the torch or blowpipe through pressure regulators and hoses. The torch mixes the two gases in proper proportions and controls and directs the flame. Goggles with colored lenses are used to protect the eyes against glare and against flying bits of hot metal.

Handling and Using Oxyacetylene Equipment Safely. Acetylene is a fuel gas and will burn readily; and oxygen supports or accelerates burning and will cause oil and grease and such materials to burn with great intensity. It is evident, therefore, that great care must be used in handling acetylene and oxygen. Read and study the specific safety instructions furnished with your welding outfit. A few of the more important safety precautions are listed below.

A Few Safety Precautions

1. Keep acetylene and oxygen cylinders away from fire, furnaces, or radiators.
2. Handle cylinders carefully. Rough handling may damage them or cause leaks.
3. Always open cylinder valves *slowly*.
4. Never use oxygen or acetylene direct from cylinders without pressure-reducing regulators.
5. Keep oxygen away from oil or grease. Do not handle oxygen cylinders or apparatus with greasy hands.
6. Do not weld around combustible materials.
7. Do not allow flame or hot metal to come in contact with hoses or other parts of the welding equipment or with clothing.
8. Weld in well-ventilated places, particularly when doing bronze welding or welding of galvanized iron.
9. Never weld without goggles.
10. Never lay a lighted torch down.
11. Do not hang the torch and hoses on the cylinder valves or regulators.

Adjusting and Lighting the Welding Torch. The steps in adjusting and lighting the welding torch are as follows:

1. Attaching the pressure regulators to the cylinders (if not already attached).
2. Selecting the proper size of tip and putting it on the torch.
3. Adjusting the working pressures.
4. Lighting the torch and adjusting the flame.

Attaching Regulators to Cylinders. Before attaching a regulator to a cylinder, open the cylinder valve slightly, and for just an instant, to blow out dust and dirt that may have accumulated. Then wipe off the connections, and attach the regulators to the cylinders.

Selecting Size of Tip. Select a torch tip of suitable size for the thickness of metal to be welded, and put it on the torch. Use small tips for thin metal and larger ones for thick metal. Instructions or tables furnished with the torch give suggestions on the sizes of tips to use.

Adjusting Working Pressures. Be sure the pressure regulators are released (adjusting screws turned to the left until loose), and then open the cylinder valves *slowly*.   Open the oxygen valve fully, but do not open the acetylene valve more than 1 to 1½ turns.   Leave the acetylene cylinder valve wrench in place while working so that the valve may be closed quickly should this be necessary.

Open the torch acetylene needle valve, and then adjust the acetylene pressure regulator to give the desired working pressure.   Then close the torch needle valve.   Adjust the oxygen pressure in the same manner.   The exact working pressures to use are suggested in the instructions furnished with the torch.   Different torches require different working pressures.   In the absence of specific instructions, set the pressures about equal on acetylene and oxygen and at about 1 lb. per sq. in. for welding thin metal, up to ⅛ in. in thickness.   For thicker metals, use higher pressures, up to 5 or 6 lb. per sq. in. for metal ½ in. thick.

Lighting the Torch and Adjusting the Flame.   To light the torch, open the acetylene needle valve at the torch and light it, prefer-

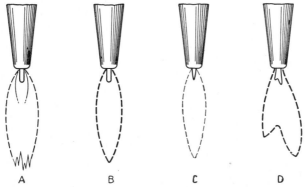

Fig. 460.—Adjusting the oxyacetylene welding flame: **A, an excess acetylene flame; B, a neutral flame; C, an oxidizing flame; D, a flame produced by an obstructed tip.**

ably with a spark lighter.   Using a match may result in burns to the hand.   Then open and adjust the oxygen needle valve to give a neutral flame (see Fig. 460).

When there is not enough oxygen, the flame will have two white inner cones at the tip of the torch.   The innermost one will be a brilliant white, and the outer one not so brilliant (see Fig. 460*A*). This is called a *carburizing* or *reducing* flame, or simply an *excess acetylene* flame.   As more oxygen is slowly turned on, the outer cone becomes smaller and smaller and disappears or merges into the inner cone.   Just as it merges into the inner one, the flame is neutral (see Fig. 460*B*). If more oxygen is turned on, the inner cone becomes smaller and the flame takes on a purplish tinge (see Fig. 460*C*). This is an *oxidizing* flame.

The neutral flame is the one used most in welding, although for some work a slightly excess acetylene flame is better, and for other work a slightly oxidizing flame is needed. The welder should inspect his flame frequently and change the adjustment as may be needed. It is easier to get an exactly neutral flame by adjusting from an excess acetylene flame rather than from an excess oxygen flame.

If the flame is uneven as in Fig. 460*D*, the tip is probably dirty or partly obstructed.

If the inner cone of the flame is sharp and pointed, the pressure of the gases is probably too high for the size of tip being used.

Turning Off the Torch. The following method is a good one to use in turning off the torch and will prevent acetylene and oxygen from becoming mixed in the hoses or the torch.

1. Close the acetylene and the oxygen needle valves on the torch.
2. Close the acetylene and the oxygen cylinder valves.
3. Open the torch acetylene valve, keeping the oxygen valve closed, to drain the hose. Release the adjusting screw on the acetylene pressure regulator, turning it to the left. Then immediately close the torch acetylene valve.
4. Open the torch oxygen valve, keeping the acetylene valve closed, to drain the hose. Release the adjusting screw on the oxygen pressure regulator, turning it to the left. Then immediately close the torch oxygen valve.

Controlling Backfires and Flashbacks. Sometimes when welding, the torch will backfire, or go out with a loud snap. Usually it can be relighted from the hot metal being welded. Backfires may be caused by touching the tip against the metal, overheating the tip, not using the correct gas pressures, or by a loose tip on the torch.

A flashback, or a burning back inside the torch, indicates that something is radically wrong with the torch or the manner in which it is operated. Should a flashback occur, shut off the torch at once, and check it thoroughly. Also, make sure it is being operated in an approved manner.

Preparing Parts for Welding. The preparation of the joints and the placing of the pieces to be welded is important in making strong welds. Edges to be joined, except on metal ⅛ in. thick or less, should usually be ground or otherwise cut out to a V to allow fusion of the parts all the way through (see Fig. 461). Also, when the parts are put in position for welding, it is usually best to leave a small space (about $\frac{1}{16}$ in.) between them to allow the weld to penetrate deep or all the way through the joint. Where possible on parts more than about $\frac{3}{8}$ in. thick, make V grooves on both top and bottom sides. Less grinding and cutting is required and less welding rod will be used for two small V grooves (one on each side of the joint) than for one large one on one side.

Remove scale, rust, and other impurities from edges to be welded to prevent inclusion of such material in the welds, thus weakening them.

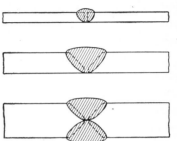

Welding Steel. To weld steel parts, first prepare the edges to be joined and place them in position, laying them on the welding table or otherwise supporting them and holding them in place. Use a suitable size of tip and heat the metal, beginning usually at one end of the weld and working toward the other. Be sure to wear goggles to protect the

Fig. 461.—Time spent in cutting or grinding V grooves and in fitting up work preparatory to welding is usually time well spent. Pieces $\frac{1}{8}$ in. thick and under may be welded without grooving. Wherever possible make a V groove on each side of pieces over $\frac{3}{8}$ in. thick.

eyes from glare and flying bits of metal. Hold the torch so that the inner cone of the flame comes to within about $\frac{1}{8}$ in. from the metal, leaning the torch tip at an angle of about 30 to 45 deg. from vertical (see Fig. 462). Move the torch about a little, using a weaving or zigzag motion or a circular motion. As a puddle of molten metal forms, advance the torch slowly along the joint, adding metal from a welding rod or filler rod as needed to build the weld up to suitable size and strength. To do this, heat the end of a welding rod almost to melting temperature and place the hot end in the puddle of molten metal, allowing the metal from the rod to melt off and accumulate in the joint. Do not hold the rod up too high and allow molten metal to drop onto the weld. *Do not use too much filler rod.* A good welder uses very little filler rod and is careful to fuse thoroughly all parts of the weld together as the torch is moved along. Beginners are inclined to use too much filler rod, piling metal from the rod on top of the work, rather than fusing all parts thoroughly.

Welding Cast Iron. Welding of cast iron is accomplished in much the same way as steel, but with a few important differences. A cast-iron filler rod is used instead of a steel one; and a special cast-iron welding flux is used. The joint is prepared in much the same manner as for steel parts.

In veeing out a joint preparatory to welding, make the V wide and deep, so the weld will be deep and strong. Heat the parts to the melting temperature, and add metal from the end of a cast-iron welding rod that has been fluxed. To flux the rod, simply heat the end to a red heat and dip it in the flux.

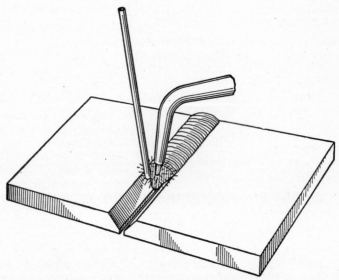

Fig. 462.—In welding with the oxyacetylene torch, hold the tip so that the inner cone of the flame comes to within about ⅛ in. of the metal, and lean the torch tip at an angle of about 30 to 45 deg. from the vertical. Fuse all parts thoroughly.

If welds are to be made in large castings, it is usually necessary to preheat them. This is to avoid strains which might lead to cracks when the hot parts of the casting cools and therefore contracts.

Bronze Welding. Bronze welding can often be done to better advantage than fusion welding. Since bronze welding is done at a red heat, instead of at a melting temperature, the work can be done faster and with less trouble from expansion and contraction of parts. Bronze welding of castings can often be done with little or no preheating. Malleable iron can be effectively welded by bronze welding but not by fusion welding with the oxyacetylene flame. Bronze welding is also often used to build up or *bronze surface* worn parts, like shafts, and missing parts, like broken gear teeth.

For bronze welding of cast iron, it is better to prepare V grooves by chipping than by grinding. It is hard to make bronze stick to a freshly ground or machined cast-iron surface. First searing such surfaces by heating them almost to the melting point with the torch will make it easier to get bronze to stick to them.

For bronze welding, adjust the torch flame so that it is slightly oxidizing. This is important to secure a better bond between the bronze and base metal. Heat the parts to be bronze welded gradually up to a red heat. Then heat the end of a bronze welding rod, and dip it into bronze welding flux. Next melt a few drops from the end of the heated rod onto the work. If the work is at the proper temperature, the molten bronze will flow out and form a thin even layer over the heated portion. If the metal is too hot, the bronze will tend to boil and form into drops which roll around but will not stick. If the metal is too cold, the bronze will not spread but will tend to be sticky and will usually remain in drops. A little experience will indicate the proper temperature. This coating of the parts to be bronze welded with a thin layer of bronze is called *tinning*. After tinning, then add more bronze to the weld to build it up to suitable size and strength.

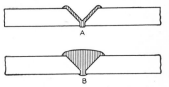

Fig. 463.—In bronze welding a joint, first heat the parts to a red heat and "tin" them with bronze from a bronze welding rod. Then add bronze to build up the weld to suitable size. Use a slightly oxidizing flame.

When a bronze weld is completed, allow it to cool very gradually to room temperature. It may be a good plan to play the torch back and forth over the whole weld for a moment after the weld is completed, particularly for welds in cast iron. This helps to relieve strains set up by uneven expansion and contraction of parts.

Cutting with the Oxyacetylene Cutting Torch. The tip of an oxyacetylene cutting torch has a number of small holes arranged in a circle about a larger central hole. The outer holes supply oxygen and acetylene for preheating flames which heat the metal to a red heat. The central hole supplies oxygen under high pressure which is to be turned on by the operator when the metal is at the desired temperature. When the oxygen strikes the hot metal, the metal is quickly burned or oxidized to a molten slag and blown through, making a cut. As the cutting torch is moved slowly along, a narrow cut or kerf results. Ordinary steels are readily cut by the oxyacetylene cutting process, but cast iron and certain alloy steels are more difficult to cut.

A cutting torch is lighted and adjusted in a manner similar to a welding torch, but with certain differences. For best results, follow any special instructions available for a particular torch. In the absence of special instructions, the following general method may be used. Adjust the oxygen pressure regulator to give the desired pressure with the preheat oxygen valve closed, but with the cutting oxygen valve open. Then close the valve. Next, adjust the acetylene pressure regulator with torch acetylene valve open. Close the valve.

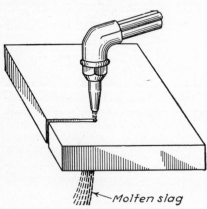

Open the torch acetylene valve and light it. Adjust to a neutral flame with the preheat oxygen valve. Then open the cutting oxygen valve, by depressing the lever or trigger, and see if the flame remains neutral. If not, readjust the preheat oxygen valve until the flame does remain neutral when the cutting oxygen valve is opened.

Cutting Steel. In cutting steel, hold the torch with the tip perpendicular to the metal and with the inner cones of the preheating flames almost but not quite touching the metal. When the metal

Fig. 464.—Cutting steel with the oxyacetylene cutting torch. Hold the cutting tip about perpendicular to the metal with the inner cones of the preheating flames almost touching the metal. When the steel becomes a bright red, open the cutting oxygen valve and move the torch steadily along. (Courtesy of Linde Air Products Company.)

becomes a bright red, open the cutting oxygen valve and move the torch slowly and steadily along the cutting line (see Fig. 464). A straight-edged bar clamped along the cutting line greatly helps to steady the torch and make the cut straight.

Cutting Cast Iron. Cutting of cast iron with the oxyacetylene cutting torch is done in a manner similar to cutting of steel, but with a few important differences. To cut cast iron, adjust the preheating flames to strongly excess acetylene, with the length of the outer streamer or cone approximately equal to the thickness of the metal to be cut, and heat the iron to nearly the melting temperature before turning on the cutting oxygen valve. To start a cut, lean the tip of the torch to point backward somewhat, and gradually straighten it up after the cut gets started until it makes an angle of 75 to 90 deg. with the surface of the work (see Fig. 465). Another difference in cutting of cast iron is that the torch needs to be moved slightly from side to side with a weaving motion as it advances along the line of cutting.

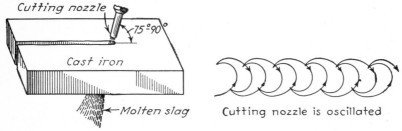

Fig. 465.—In cutting cast iron with the oxyacetylene cutting torch, adjust the preheating flames to strongly excess acetylene, lean the tip slightly, preheat to almost the melting point, and then open the cutting oxygen valve. Move the tip forward with a weaving motion. (Courtesy of Linde Air Products Company.)

Points on Oxyacetylene Welding and Cutting

1. Use care in handling and using oxyacetylene welding equipment, so as to avoid injury to the operator or damage to the equipment.
2. Always wear goggles when welding to protect the eyes from glare and from flying bits of hot metal.
3. Use a size of torch tip suitable to the thickness of metal being welded.
4. Always use an accepted procedure for adjusting the working pressures and lighting and turning out the torch.
5. It is easier to adjust the flame to exactly neutral by adjusting from an excess acetylene flame than from an excess oxygen flame.
6. Inspect the flame frequently, and change the adjustment if needed.
7. Pieces $\frac{1}{8}$ in. thick and under can usually be satisfactorily welded without first veeing out the edges.
8. For welds in parts over $\frac{1}{8}$ in. thick, make the V grooves wide and deep.
9. When placing parts in position for welding (except thin parts), leave a small space between them so that the weld may penetrate deep.
10. Do not use too much welding rod.
11. In adding metal from the welding rod, keep the end of the rod in the puddle of molten metal.
12. Do not allow molten metal to drop from the end of the welding rod and pile up on relatively cold metal. Thoroughly fuse the added metal into the weld.
13. Hold the torch so that the inner cone of the flame comes about $\frac{1}{8}$ in. from the metal.

14. Use a slightly oxidizing flame for bronze welding.

15. It is important to have the temperature of the metal just right for bronze welding. If it is either too cold or too hot, the bronze will not stick properly.

16. In adjusting the cutting torch, be sure that the character of the flame does not change when the cutting oxygen valve is opened.

17. Use a neutral preheating flame for cutting steel and a strongly excess acetylene flame for cutting cast iron.

11. Welding with an Electric-arc Welder

An electric-arc welder is a very valuable piece of equipment in the farm shop. It is true that a welder is somewhat expensive compared with other pieces of shop equipment, and possibly a little more

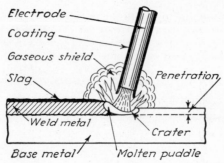

Fig. 466.—Weld metal being deposited by an arc from a shielded or coated type of electrode.

skill is required for its operation. Where electricity is available from a power line, however, and where a large amount of machinery and equipment is to be kept in repair, the purchase of an electric welder should be given careful consideration.

In arc welding, the intense heat of the electric arc is used to fuse the parts being joined. There are three different kinds of arc welding—metallic-arc welding, carbon-arc welding, and arc-torch welding.

Metallic-arc Welding. Metallic-arc welding is the kind of electric welding most commonly done in farm shopwork. A heavy current is made to flow across a small gap between the end of a metal electrode or welding rod, held in a suitable holder, and the metal being welded. The heat of the arc that spans this gap is so intense that it instantly creates a crater, or small puddle, of molten metal on the work and also melts small globules of metal from the end of the electrode. These molten globules pass through the arc and are added to the puddle on the piece being welded (see Fig. 466). As the electrode is slowly moved along the joint, a weld is made.

In the case of heavily coated or shielded electrodes, the molten metal passing from the electrode and the molten base metal are shielded or protected from the atmosphere, and thus oxidation or burning is prevented or kept at a minimum. This protection is provided in three ways: (1) The coating on the electrode burns slower than the electrode melts, leaving a protruding shield on the end (see Fig. 466). (2) The coating when it burns or becomes a gas shields the arc with a protecting shield or envelope of inert gas,

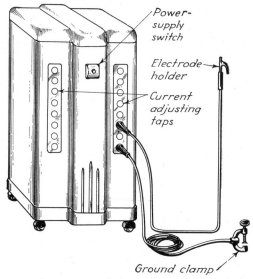

Fig. 467.—The a.c. transformer type of electric-arc welder is the type most commonly used in farm shopwork.

keeping the oxygen and the nitrogen of the air away. (3) A film of slag is deposited on the top of the molten metal, protecting it from the air while it cools.

Carbon-arc and Arc-torch Welding. In carbon-arc welding, a carbon electrode is used instead of a metal one, and a metal welding rod or filler rod is used in much the same manner as in oxyacetylene welding. In welding with an arc torch, current passes between the ends of two carbon electrodes held in the torch. The flame of the arc is played upon the parts being welded and the metal filler rod used, also much as in oxyacetylene welding.

Arc-welding Machines. Special generators or transformers are used to supply the current for electric-arc welding. When generators are used, they are driven by gas engines or electric motors. Both direct-current (d.c.) and alternating-current (a.c.) welders are in common use, each having their special advantages. Alternating-current welders are most commonly used in farm shops, mainly because of their simplicity and their lower cost in the smaller sizes.

Using Electric-arc Welders with Safety. Welding with the electric arc is a comparatively safe process, although accidents may be caused by careless or improper use of the apparatus. Probably the most important safety precaution is to protect the eyes, and other parts of the body also, from spatter of molten bits of metal and from

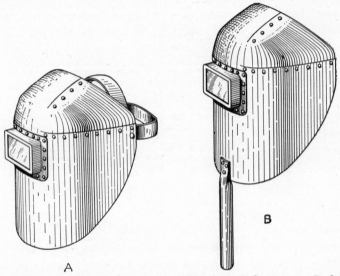

Fig. 468.—Head shields for electric-arc welding: A, helmet type, B, hand type.

the rays of the arc. Always use a helmet or face shield (see Fig. 468) that is in good condition, and *never, under any circumstances, look at the electric arc with the naked eye.* Do not strike an arc or weld without first being certain that those in the vicinity have the necessary protective equipment or will look in the opposite direction. For welding in open spaces, it is a good plan to use portable screens for shielding others from the rays of the arc.

Prevent Eye Burn. Looking at the arc with the naked eye, even for a very brief time, may cause eye burns that are exceedingly painful, although they are not permanent injuries. In the case of eye burn, which feels like "hot sand in the eyes," apply cool boric acid, or a few drops of 5 per cent argyrol every 4 or 5 hr. Aspirin in ordinary doses may be taken to relieve pain and headache.   In severe cases, see a doctor.

Wear Protective Clothing. Always wear suitable clothing when welding, not only to protect the body from flying bits of hot metal, but also from the rays of the arc.   Exposure to the arc usually causes a condition similar to sunburn, but often more severe.   Wear gloves to protect the hands and wrists.   Dark-colored shirts are better than light ones.

Avoid Shock Hazards.   There is very little danger of electric shock from the output of an electric-arc welder.   Electricity to operate welders is commonly supplied at 220 or 440 volts, however, and it is important that the wiring to welders be properly installed and grounded for safety.   When working on any part of the welder, be sure that the main switch is opened and that the circuits are "dead."

To avoid damage to the equipment, and possible burns, do not make adjustments of the welding current when the welder is under load, and do not leave the electrode holder on the welding table or in contact with a grounded metal surface.

Provide Ventilation. Always weld in a well-ventilated place. Exhaust fumes from an engine (in case of an engine-driven welder) are dangerous.   Fumes given off from the arc are unpleasant and may be injurious.   Never breathe them in any appreciable quantity.

A Summary of the More Important Safety Precautions

1. Never look at the welding arc with the naked eye.
2. Always use a helmet or face shield that is in good condition.
3. Replace any cracked or poor-fitting lenses in the helmet or shield.
4. Wear suitable clothing to protect all parts of the body from spatter of hot bits of metal and from arc burn, which may be more painful than sunburn.
5. Do not strike an arc or weld until you are sure those in the vicinity have protective equipment or will look in the other direction.
6. Do not weld around combustible or inflammable materials.
7. Do not pick up hot metal.
8. Do not weld in confined spaces without adequate ventilation.
9. Change the welding current adjustment only when the welding circuit is open.
10. Do not work on live circuits. Always open the main switch when checking over a machine.
11. Do not leave the electrode holder on the welding table or in contact with a grounded metal surface.
12. Do not use worn or frayed cables.

Adjusting the Welding Current. Before starting to weld, adjust the welder to a suitable current value, and select an electrode suitable for the kind of work to be done. Instructions accompanying an arc welder suggest current settings and electrode sizes to use. After some experience, an operator soon learns to set his machine and select electrodes for various kinds of welding. In using a d.c. welder, make the electrode negative and the work positive, or vice versa, depending upon the recommendations for the particular kind of electrode used. (With the electrode negative and the work positive, the machine is said to be set on "straight polarity"; and with the electrode positive and the work negative, on "reversed polarity.")

Fasten the ground clamp securely to the work or the welding table, and get the helmet adjusted or the face shield in readiness.

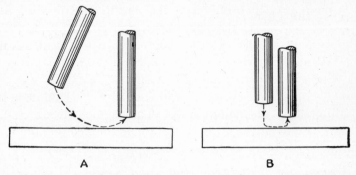

A   B

Fig. 469.—Two methods of striking an arc. A. The scratching method. Move the electrode downward with a sweeping motion. When the end touches the metal, continue a little further, raising the electrode slightly. B. The tapping method. Simply lower the electrode slowly, tapping it or bouncing it lightly on the work.

Striking an Arc and Running a Bead. To strike an arc, be sure the flux on the electrode tip has been knocked off and then touch the end of the electrode to the base metal and then quickly withdraw it a short distance. Two general methods are used to do this. One, called the *scratching method*, is similar to striking a match. Simply move the end of the electrode with a slow, downward-sweeping motion. When the electrode touches the metal, continue a little further, raising the electrode slightly (see Fig. 469A). To use the other method, known as the *tapping method*, keep the electrode perpendicular to the work, lowering it slowly and tapping or bouncing it lightly on the work (see Fig. 469B).

In case the electrode sticks to the work, quickly bend it back and forth, pulling at the same time. If this does not free it, then release the electrode from the holder or stop the welder.

Once the arc is struck, move the electrode along slowly from left to right, assuming the workman is right-handed. (Moving from left to right for a right-handed workman and from right to left for a left-handed workman enables the arc to be seen better.) Lean the top of the electrode about 10 or 15 deg. in the direction of welding. This not only gives a better view of the arc, but also keeps more of the molten metal back out of the crater and gives better penetration.

To break the arc at the end of the bead, hold the electrode still long enough to fill the crater and then gradually lift it straight up from the work.

Do not grip the electrode holder too tightly, as a tight grip not only produces fatigue, but also makes it difficult to control the movement of the electrode.

Holding the Proper Length of Arc. The length of arc greatly affects the strength, smoothness, and uniformity of the weld. The best length of arc varies somewhat with different kinds of electrodes but is usually about equal to the diameter of the electrode. Instructions are often furnished with electrodes or with welding machines regarding the most suitable length of arc to use. Experience will also enable an operator to know quickly when he is using an arc of the correct length. As the end of the electrode melts, it must be continually lowered, of course, to hold the same length of arc.

In general, use a short arc, but not so short that the coating of the electrode touches the molten metal on the work, as this will cause a porous weld. A slightly longer arc is required when welding with alternating current than with direct current.

Whether or not the proper length of arc is being used may be judged by (1) the appearance of the arc, (2) the appearance of the weld produced, and (3) the sound of the arc.

A short arc gives a steady flame that surrounds and protects both the molten metal as it passes from the end of the electrode to the work, and the puddle of molten metal on the work (see Fig. 470). A long arc, on the other hand, gives a flame that whips about and exposes the molten metal passing from the electrode and also the molten metal on the work (see Fig. 471). The hot metal, being unprotected from the atmosphere, burns or oxidizes somewhat and becomes porous, resulting in a rough, weak weld.

A short arc gives good penetration, with little or no overlap of the deposited metal (see Fig. 470). Also, there will be a minimum of spatter (bits of molten metal thrown from the melting electrode to the surrounding work). A long arc gives poor penetration and considerable overlap (see Fig. 471). The deposited bead will not be uniform and smooth, but will be uneven and rough and porous, and there will be excessive spatter.

A short arc, with proper adjustment of the welding current (and polarity in case of a d.c. welder), will give a sharp crackling or "frying" sound. A good length of arc with a.c. electrodes also has a humming sound. A little experimenting will enable the operator to judge the difference in sound of long and short arcs.

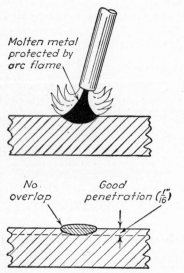

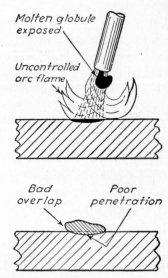

Fig. 470.—Weld with a short arc. It protects the metal being deposited, and gives good penetration with little or no overlap.

Fig. 471.—A long arc gives a flame that whips about and provides insufficient protection for the metal being deposited. The penetration is poor and the overlap is bad.

Weaving the Arc. Although a steady uniform movement of the electrode will produce a satisfactory bead for some purposes, a slight weaving or oscillating motion should be used for most welds. Some gas and slag are usually present in the molten metal of an arc weld. Weaving or oscillating the electrode keeps the metal molten a little longer and allows the gas to escape and the slag to come to the surface, and thus avoids porosity in the weld. Weaving is also often done to produce a wider deposit or bead than would be possible without weaving, and sometimes also to secure better penetration or better building up of the bead along the edges.

Various different patterns of weaving are used by different operators. Some of the more common patterns are illustrated in Fig. 472. The crescent weaves (Fig. 472*A* and *B*) are perhaps the patterns most commonly used. They are also among the easiest to learn. For weaving a narrow bead, a motion that is mostly forward and backward and with very little side motion (Fig. 472*C*) is used. The pattern shown in Fig. 472*D* is used to give slight hesitation at the edges of the bead in order to build up the weld better, or give better penetration, along the edges.

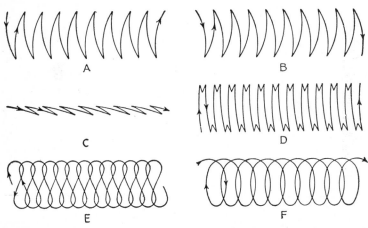

Fig. 472.—Common patterns for weaving or oscillating the electrode: A, a crescent weave; B, another crescent weave; C, a back-and-forth weave with practically no side motion used for making narrow beads; D, a pattern giving slight hesitation at the edges of the bead; E, a figure-8 weave; F, a circular weave.

Current Setting and Speed of Electrode Movement. The current setting and the speed with which the electrode is moved greatly affect the penetration and the strength of the weld. Too high a current penetrates too deeply, makes too wide a crater, and tends to produce excessive spatter; while too low a current gives only shallow penetration with a high bead that is not well fused into the base metal (see Fig. 473).

Moving the electrode too slow builds the bead up too high, while moving it too fast gives a flat, shallow bead of poor penetration (see Fig. 474). In general, move the electrode just fast enough to keep the arc at the forward edge of the crater. In running a straight bead with little side weaving, an electrode should produce a bead about equal in length to the electrode consumed.

Preparing Work for Welding.   Clean the work thoroughly of scale, rust, or other foreign material before welding.   Thin parts (⅛ in. thick and under) may generally be placed together and welded without first beveling the edges, but thicker parts should be beveled or veed out to allow adequate penetration and fusion of all parts of the weld.   On all but thin metal, the parts to be welded should be separated slightly (about ¹⁄₁₆ in.) to allow better penetration of the

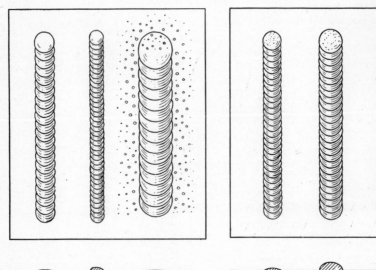

Fig. 473.—Effects of current setting on the weld: A, proper adjustment of current; B, current too low; C, current too high.

Fig. 474.—Effects of speed of electrode movement on the weld: A, proper speed; B, movement too slow; C, movement too fast.

weld.   Where possible on parts more than ⅜ in. thick, make V grooves on both top and bottom sides.

Usually time spent in veeing out joints and fitting them properly together is time well spent.   Careful preparation of the work makes for stronger welds, as well as for faster welding, better appearance, and lower consumption of electrodes.

Making Different Types of Joints and Welds.   The most common kinds of joints are *butt*, *lap*, *tee*, and *corner* joints (see Fig. 475). Welds in general may be classified as *bead*, *groove*, and *fillet* welds.   A bead weld is a deposit of weld metal made on a surface by a single

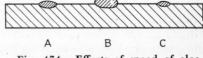

pass of the electrode, as in "building up" or in making a surface weld like a plain closed butt joint.   A groove weld consists of one or more beads deposited in a groove.   When more than one bead is made, the weld is called a *multiple-pass* weld (Fig. 476), a single-bead weld being

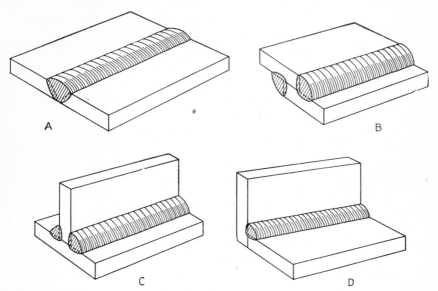

Fig. 475.—Common types of welded joints: A, butt joint; B, lap joint; C, tee joint; D, corner joint.   The welds on joints B, C, and D are known as fillet welds.

called a *single-pass* weld.   A fillet weld consists of one or more beads run in the angle or corner between two surfaces, as in a tee, lap, or corner joint (Fig. 475B, C, and D).

In making a fillet weld, hold the electrode so that it about bisects the angle between the two surfaces.   Also lean the top of the electrode in the direction of welding about 30 deg. (see Fig. 477). Change the angle or position of the electrode slightly if necessary to obtain good fusion on both pieces and good penetration.   Hold a short arc, and use a slight weaving motion. Fillet welds on light material may be made

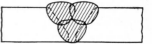

Fig. 476.—A multiple-pass weld made in thick metal.   The first bead is laid in the bottom of the V groove and the other beads are laid on top.

with one pass, while two or more passes may be required on thicker material.

Before laying a bead on top of another in making a multiple-pass weld, scrape or brush the first one vigorously with a wire brush to remove the slag, and then peen it with a hammer.

In making a groove weld, weave the electrode back and forth side-wise as it is advanced in order to fuse all parts thoroughly. In case of a groove weld in thick material, however, make a multiple-pass weld rather than trying to weave the electrode sidewise too far on a single-pass weld.

Place parts to be welded in a flat or horizontal position where pos-sible, as welds are much easier to make in this position than in vertical or overhead positions.

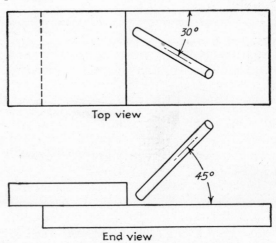

Fig. 477.—In making a fillet weld, hold the electrode so that it about bisects the angle between the two surfaces being welded and lean the top about 30 deg. in the direction of welding.

Points on Electric-arc Welding

1. Never look at the welding arc with the naked eye.
2. Do not strike an arc until you are sure that those in the vicinity have protective equipment or will look in the other direction.
3. Wear suitable clothing to protect all parts of the body from spatter and from arc burn.
4. Clean the work of rust, scale or other foreign material before welding.
5. Bevel the edges to be joined if the parts are over ⅛ in. thick.
6. Before welding parts over ⅜ in. thick, vee out on both top and bottom surfaces where possible.
7. Except with thin pieces, leave a small space (about ¹⁄₁₆ in.) between parts to be welded.
8. Time spent in securing good "fit up" of parts to be welded is usually time well spent.
9. Strive to maintain a suitable length of arc. A short arc, about equal to the diameter of the electrode. is usually best.

10. Be sure to use a suitable current setting to give good penetration.
11. Move the electrode along just fast enough to keep the arc at the forward edge of the crater.
12. Lean the top of the electrode 10 or 15 deg. in the direction of welding; and, for most welds, use a slight weaving motion.
13. Wherever the work will permit, place parts to be welded in a flat or horizontal position.
14. In making a fillet weld, hold the electrode so that it about bisects the angle between the two surfaces being joined, and lean the top about 30 deg. in the direction of welding.
15. In making a groove weld in thick pieces, first lay a bead in the bottom of the groove, and then lay other beads on top.
16. Before laying a bead on top of another, scrape or brush the first one vigorously with a wire brush, and then peen it with a hammer.

Jobs and Projects

1. Make a list of forging equipment you would consider adequate as an initial installment for a shop on your home farm or some typical farm in your community. Consult catalogues and specify makes, model numbers, sizes, prices, etc.
2. Obtain a piece of railroad iron or rail, and make a suitable mounting for it, so that it may serve as a light anvil.
3. Make a list of the principal forging operations or processes, like bending, which are done frequently. Write down beside each operation two or three of the most important points to be observed in performing it.
4. Make a list of several small forged appliances, such as meat hooks, gate hooks, hay hooks, and eye bolts. Consult shop books and manuals for suggestions. Select five or six, or possibly more, of these that involve most of the common forging operations, and that you would like to make and use about your home or farm.
5. Make plans, including lists of material, etc., for the jobs selected in No. 4 and proceed to make them.
6. Make an inspection of a few farm machines, like plows, cultivators, and planters, and make a list of such parts as straps, irons, and clevises that need repairing or replacing. Sketch the parts to be made, or plan the repairs, and make them.
7. Practice welding two or three chain links, and then repair a broken chain, if one is available, by welding in links.
8. Do one or two other jobs involving forge welding, such as making a ring or a singletree clip.
9. Make one or two chain repair links, which may be used for repairing a broken chain without bringing it to the shop.
10. Make a chain hook, and attach it to a chain with a welded link.

11. Reshape a chain hook that has been overstrained and pulled out of shape.

12. If oxyacetylene- or electric-arc-welding equipment is available in the shop, read and study the instructions applying to this particular equipment, as well as the general information in this book.

13. Practice making welds in scrap material until you can do creditable work. Clamp your practice welds in a heavy vise, and bend the pieces, possibly cutting some with a cold chisel. Inspect them for flaws or weaknesses. Then make other practice welds and see if you can improve your work.

14. Locate parts of machines or equipment about your farm or home that need welding. Bring them to the shop, and put them in first-class condition. Do not be satisfied with a weld if you can make it better. Always inspect your own work with a view to making it better next time.

12. Pipework and Simple Plumbing

Nⁿᵒᵗ ᵐᵃⁿʸ tools are required to do a moderate amount of pipework and simple plumbing. Many jobs can be done with only one or two pipe wrenches and the common woodworking and metalworking tools in a shop. If a considerable amount of piping is to be installed or kept in repair, it might be advisable to have some pipe-cutting and threading equipment. For the occasional job, pipe may be bought already cut to the desired length and threaded, so that all that is necessary is to assemble the pipes and fittings.

MAJOR ACTIVITIES

1. Selecting Pipe Tools for the Shop

2. Selecting Pipe and Pipe Fittings for a Job

3. Measuring and Cutting Pipe

4. Reaming Pipe

5. Threading Pipe

6. Assembling Pipe and Pipe Fittings

7. Cutting a Gasket

8. Removing a Section of Defective Pipe

9. Repairing Leaky Valves and Faucets

10. Repairing Pumps

11. Taking Care of an Automatic water System

12. Installing a Kitchen Sink and Drain

13. Installing a Simple Shower Bath

1. Selecting Pipe Tools for the Shop

Wrenches. Every farm shop should have at least one pipe wrench, preferably two, even if very little pipe work is to be done. Pipe wrenches are useful for holding and turning round rods and various machine parts, as well as pipes. Wrenches of 14- and 18-in. size are most useful, and wrenches with all steel handles (see Fig. 478) are usually better for farm work than those having wooden grips.

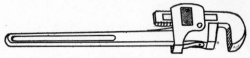

Fig. 478.—Every well-equipped farm shop should have at least one pipe wrench, preferably two.

Vise. Probably the most practical kind of vise for holding pipes in farm shops is a regular blacksmith's or machinist's vise equipped with auxiliary pipe jaws. Where considerable pipe work is to be done, a regular pipe vise may be advisable. For an occasional job, pipe can be held in an ordinary flat-jaw vise with the aid of a pipe

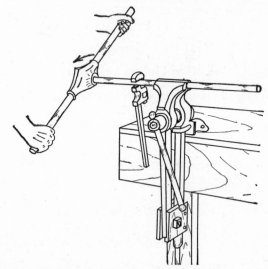

Fig. 479.—For an occasional job of pipework, the pipe can be held in an ordinary vise with the aid of a pipe wrench.

wrench by cramping the handle against the side of the bench (see Fig. 479).

Pipe Cutter. For an occasional job of pipe cutting, a hack saw is quite satisfactory. Where considerable pipe is to be cut, a pipe cutter will be better and more economical.

Dies and Taps. A set of pipe dies will be needed only when a large job of pipe fitting, such as installing a water system, is undertaken, or when considerable pipe work is to be kept in repair.   Pipe dies and taps in the two smallest sizes ($\frac{1}{8}$ and $\frac{1}{4}$ in.) may be justified for such work as repairing grease and oil pipes on machinery and engines.

Sets of pipe-threading tools do not normally include taps, as pipe fittings that screw onto pipes are threaded at the factory, and therefore there is little use for taps.

One-piece dies are generally preferred to two-piece dies, because of the ease with which they may be changed in the stock.   A stock with a ratchet handle is much easier to use than a plain stock for threading any but the smallest sizes of pipe.   The ratchet handle can be worked back and forth through that portion of the turn where force can be best applied.   Ratchet equipment costs more, however,

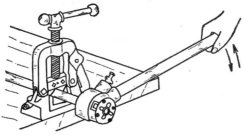

Fig. 480.—The ratchet die, although more expensive than the plain die, is much easier to operate and is preferred by many mechanics.

and for an occasional threading job, the extra expense may not be justified.

2. Selecting Pipe and Pipe Fittings for a Job

Pipe is available in either black or galvanized iron.   Black pipe is used for oil, air, or gas.   Galvanized pipe should be used for a water-supply system.   Galvanized pipe will of course last longer when used under conditions that tend to cause rusting and corrosion.

Pipe fittings are made of cast iron or wrought iron.   Wrought-iron fittings are available in either black or galvanized finish.

In selecting or buying pipe, be sure to get the right size.   The size of a pipe is designated by its *inside* diameter.   The actual inside diameter of a pipe, however, is slightly larger than its nominal or designated size.   For example, a $\frac{1}{2}$-in. pipe measures actually a little more than $\frac{1}{2}$ in. in diameter.   Pipe is made in sizes ranging from $\frac{1}{8}$ to $\frac{1}{2}$ in. by steps of $\frac{1}{8}$ in., and from $\frac{1}{2}$ to $1\frac{1}{2}$ in. by steps of $\frac{1}{4}$ in.

The size of a pipe fitting is designated by the size of the pipe upon which it fits, and not by the diameter of the fitting itself. For example, a ½-in. elbow has an inside diameter of about ¾ in., but it is called a ½-in. elbow because it fits on the outside of a ½-in. pipe.

Selecting Pipe Fittings. There are on the market many different pipe fittings. By careful planning and selection of fittings, a particular job may often be simplified and made easier. The most common

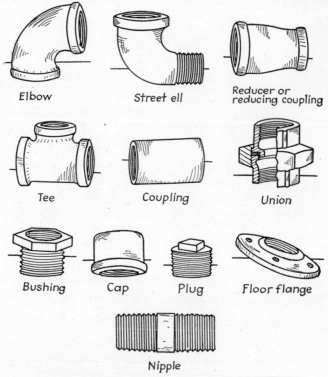

Elbow   Street ell   Reducer or reducing coupling

Tee   Coupling   Union

Bushing   Cap   Plug   Floor flange

Nipple

Fig. 481.—Common pipe fittings.

pipe fittings are illustrated in Fig. 481 and their uses described in the following paragraphs.

The *coupling* is simply a short sleeve threaded on the inside at both ends and is used for joining two pieces of pipe in a straight line, where at least one of the pieces can be turned.

The *union* is used for joining two pipes where neither can be turned. It consists of three pieces, one to screw onto each of the two ends being joined, the third part being a nut for drawing the other two parts tightly together. There are two general kinds of unions, one that

requires a gasket to make a tight joint, and one that does not.   The
parts of a union that requires no gasket fit together much as an
engine valve fits into its seat.

A *nipple* is simply a short piece of pipe threaded on both ends.

The *elbow*, or *ell*, is used for making right-angle turns in a line of
pipe.   A 45-deg. elbow is used for making a turn of 45 deg.

The *street ell* is similar to the ell, except it has one end threaded on
the outside, so that it may be screwed *into* a fitting such as a tee.   It

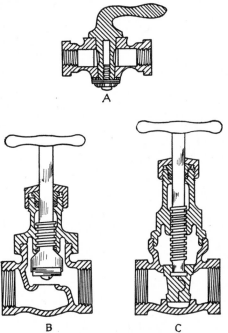

Fig. 482.—Shutoff valves: A, stop-and-waste cock; B, globe valve; C, gate valve.

can be used instead of an ell and a short nipple.   It is also frequently
used in piping to give a certain degree of flexibility to allow a limited
movement of parts without causing undue strain on the joints (see
Fig. 488).

A *tee* is used for joining a side branch to a main line of pipe.

A *reducing coupling* is a coupling with one end made to fit one size
of pipe, and the other end a different size.

A *bushing* is a short sleeve used to reduce the size of a threaded
opening.   It is threaded on the inside, and also on the outside at one
end.   The other end is hexagon shaped to receive a wrench.

A *cap* is used to screw over the threaded end of a pipe, thus stopping it.

A *plug* is used to screw into a threaded opening, such as one outlet of a tee, and thus stop the opening.

A *floor flange* is used for fastening the end of a pipe to a wall or floor, as in stair rails.

Selecting Pipe Valves and Faucets. The most commonly used pipe valves and faucets are as follows:

A *stop-and-waste cock* (Fig. 482*A*) is commonly used in a supply pipe. When it is turned off, it allows the water in the pipes beyond the cock to drain out.

The *globe valve* (Fig. 482*B*) is the most commonly used type of shutoff valve. In passing through it, the water must make two right-

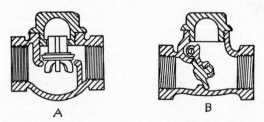

A   B

Fig. 483.—Check valves: A, lift type; B, swing type.

angle turns. It should be installed so that when it is turned off there will be no pressure on the packing around the valve stem. This not only lessens the possibility of leakage around the stem, but it also enables the stem to be repacked without turning off the pressure on the whole line.

The *gate valve* (Fig. 482*C*) offers less resistance to the passage of water through it than does the globe valve; but it is not so easily repaired and is used less. It is used in places where it is important not to impede the flow and where the valve would have to be closed only rarely, such as at a pump or storage tank where the valve would need to be closed only when repairs are made on the system. In a gate valve, the flow is stopped by lowering a wedge-shaped gate into a seat.

A *check valve* (Fig. 483) is used to prevent a backflow in a pipe. Two general styles are in common use, the lift valve and the swing valve.

The *compression bib* (Fig. 484*A*) is the most common type of faucet. In principle it is very similar to the globe valve. When it is closed, a composition disk is held against a seat.   When the disk becomes worn, it is easily replaced or turned over.

The *Fuller's bib* (Fig. 484*B*) is a faucet in which the flow of water is stopped by pulling a rubber ball or valve disk onto a seat.   It is a little more difficult to renew the ball or disk in a Fuller's bib than the disk in a compression bib.   Both the compression bib and the Fuller's bib may be purchased plain or with ends threaded to receive a hose connection.

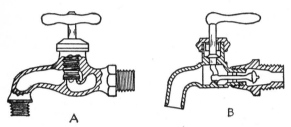

Fig. 484.—Faucets: A, compression bib; B, Fuller's bib.

3. Measuring and Cutting Pipe

To ensure a good job and to prevent waste of materials and time, it is always advisable to take measurements before doing any cutting. If the job is a large one and involves the use of many parts or joints, first make a sketch to show dimensions and kinds of fittings to be used. Careful planning often makes it possible to get along with fewer joints and fittings and consequently with less expense and work. Many small jobs of pipework may be done on the farm even though pipe threading tools are not available in the farm shop.   In such cases, simply take accurate measurement of the lengths of pipe needed and then buy the pipe already cut to length and threaded.   In measuring or cutting pipe to fit into a given place, be sure to take into account the distance which the pipe will screw into the fittings on its ends.

Cutting Pipe with a Hack Saw.   In cutting pipe with a hack saw, careful work is required to avoid catching and breaking out some of the teeth or breaking the blade.   Use a medium- or fine-toothed blade (18 or 24 teeth per inch).   It is a good plan when measuring to indicate the places to be sawed by filing a notch with a file.   Such a notch not only marks the place distinctly, but helps to start the saw exactly at the right place.   Use moderate pressure on the saw blade and make long, moderately slow strokes.   Release the pressure on the backstrokes.   Keep the blade stretched tight in the saw frame.

and be careful to keep the saw cutting square with the pipe. See pages 317 to 323 for further suggestions on the use of hack saws.

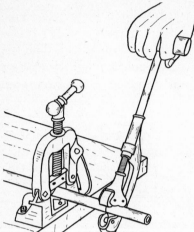

Fig. 485.—Cutting pipe with a pipe cutter. Although a hack saw is satisfactory for an occasional job of pipe cutting, a pipe cutter is recommended where much pipe work is to be done.

Cutting Pipe with a Pipe Cutter. A pipe cutter leaves a smoother end on the pipe than a hack saw, although it forms a burr inside the pipe that should be removed for most pipework.

To use a pipe cutter, place it carefully on the pipe so that the cutting wheel engages the pipe at exactly the place to be cut (see Fig. 485). Tighten the wheel against the pipe by screwing the handle. Apply threading or cutting oil and turn the cutter around the pipe, tightening the handle a little once each turn to force the cutting wheel into the pipe.

4. Reaming Pipe

After cutting pipe with a pipe cutter, it should be reamed to remove the burr left on the inside by the cutter, if it is important that

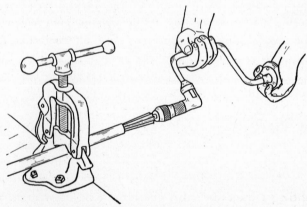

Fig. 486.—Reaming the inside of a pipe to remove the burr left by the pipe cutter. The burr may be removed with a round or half-round file if a reamer is not available.

the carrying capacity of the pipe not be reduced. The best way of removing this burr is to use a reamer in a carpenter's brace (see Fig.

486). For an occasional job when a reamer is not at hand, the burr may be removed with a round or a half-round file.

5. Threading Pipe

To thread a pipe, first clamp it securely in a vise. If a pipe vise is not available, a flat-jaw vise and a pipe wrench may be used (see Fig. 479). Select the proper size of collet (guide) and die, and insert them in the stock. Place the die on the end of the pipe, and exert considerable pressure while the die is slowly turned. As soon as the die starts to cut and thread itself onto the pipe, stop and apply threading or cutting oil. Then continue to turn the die onto the pipe until about one thread projects through the die. If the die is screwed on

Fig. 487.—Threading pipe. Keep the die lubricated with lard oil or threading oil. Stop and back the die off when about one thread projects through.

further, the end of the pipe that projects through will have straight threads instead of tapered threads. (Pipe threads are tapered so they will tighten securely as they are screwed into fittings.)

After the die has been screwed far enough onto the pipe, simply stop and turn it back off. Shake the cuttings from the die, as they may interfere with threading the next pipe. When finished with the die, wipe with a cloth to clean it before putting it away.

6. Assembling Pipe and Pipe Fittings

Assembling Pipes to Avoid Strain. Pipes should be assembled so as to avoid strains and bending at the joints wherever possible. Figure 488 illustrates a method of using a street ell in combination with an ell to allow slight movement of the pipes without placing undue strain on the joints. This method is especially good for aligning pipes and connecting a water tank to a pump or a pipe line, or a gas stove or heater to a pipe line.

Using Thread Compound.   To ensure tight joints, the threads may be coated with some sort of thread compound before they are screwed together.   Thick paint, a mixture of graphite and heavy grease, or a paste of Portland cement and linseed oil may be used for this purpose.   Thick paint is all right for rather permanent work, but not so good if the joints may need to be taken apart frequently.

Pipe compound may be applied to the threads on the outside of the pipe, or to the threads on the inside of the fittings.   On pressure lines, a tighter joint will result if it is applied in the fitting; and on suction lines, if applied to the threads on the pipe.   When applied in the fitting, however, any excess material will be left inside the pipe.

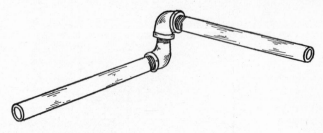

Fig. 488.—A street ell in combination with an ell used to form a sort of "universal joint" to allow slight movement of the pipes without placing undue strain on them.

Where this would be objectionable, as in the case of pipes for drinking water, the material should be applied to the threads on the pipe.

On a large job where many joints are to be assembled, it is best to use a small stiff brush for applying compound to the threads.   On small jobs, where only a few joints are to be made, it may be applied with a small wooden paddle or stick, or simply with the fingers.

Screwing the Joints Tight.   Some judgment is required to know just how tight to screw the joints.   The threads are tapered, and the farther a joint is turned, the tighter it will fit.   Just tight enough to prevent leaking is all that is required.   It is possible, particularly on small pipes, to screw the joints too tight, and possibly deform or unduly mar the fittings.

It is a good plan in assembling pipe work to use two pipe wrenches, keeping one set to fit the pipe and the other to fit the pipe fittings.

Installing Valves.   When screwing a valve onto the end of a pipe, use the wrench on the end of the valve next to the pipe.   This avoids placing strain on the valve which might twist it and thus damage it.   Likewise, when screwing a pipe into one end of a valve, place a wrench on the end of the valve receiving the pipe to avoid damaging it.

In order to avoid marring a valve or faucet with wrench marks, use a monkey wrench or other smooth-jaw wrench, wherever possible, rather than a pipe wrench.

7. Cutting a Gasket

Some unions require gaskets to make tight joints.   Gaskets are also used in various places on water tanks, heaters, and other pieces of equipment.   When fittings are dissembled, gaskets are usually destroyed, and new ones must be used when the parts are reassembled.

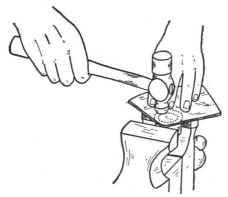

Fig. 489.—**Cutting a gasket for a union.   To cut, place the sheet of gasket material over one end of the union and hammer lightly; or the material may be marked with a few taps and then cut with a knife, scissors, or snips.**

Gaskets are usually available at hardware stores and plumbing shops, but they may be made at home if some suitable material, such as sheet rubber or sheet fiber, is available.

To cut a gasket, place the gasket material over the part it is to fit (see Fig. 489) and cut it to shape with light blows from a hammer; or, if preferred, simply mark the gasket material with a few light hammer taps or with a pencil, and then cut it out with a sharp knife, snips, or scissors.

8. Removing a Section of Defective Pipe

It frequently happens that one length of pipe in a system of pipe work develops a leak, as from freezing, and must be removed.   Probably the simplest way to do this is as follows: Cut the defective section in two, using a hack saw.   Then remove the two parts of the defective pipe by unscrewing them and without disturbing any other parts of the system. Cut and thread two or more pieces of pipe to go in the place of the leaky joint, using a union to make the final connection.

9. Repairing Leaky Valves and Faucets

Leaky faucets and globe valves can usually be repaired by simply replacing the disks or washers that fit down on the seats. To repair such a faucet or valve, turn off the pressure from the pipe line and take the faucet or valve apart. Use smooth-jaw wrenches rather than pipe wrenches or pliers to avoid marring the parts. Once the valve or faucet is apart, the disk or washer can usually be removed by taking out a small screw (see Fig. 490). To complete the job, simply install

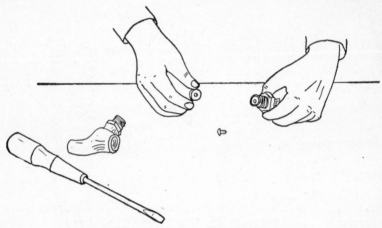

Fig. 490.—Replacing a faucet disk or washer.

a new disk or washer, clean the various parts, and reassemble them. In case no new washer or disk is at hand, it is sometimes possible to turn the old one over, and make it do for a time longer. It is a good plan to keep an assortment of such valve and faucet disks on hand for repairs.

After long use, the seat in a valve or faucet may become pitted or corroded until it will not hold even when a new disk is installed. In such cases, a special valve and faucet seat reamer may be used to smooth the seat. On some types of faucets, it may be possible to remove the old seats and replace them with new ones.

10. Repairing Pumps

There are many different types of pumps used on farms, and the exact method of repairing a particular pump will depend upon the type of pump and the troubles. The first step in repairing a pump is to study it to determine how it should work and to locate the troubles, if possible, before taking it apart.

Most farm pumps are plunger pumps, with cup-shaped leathers on the plungers to make them fit snugly in the cylinders (see Fig. 491). The most common causes of pump trouble are worn leathers and worn valves. The most common pump-repair job, therefore, is the replacement of worn leathers and valves.

Pumps are usually not difficult to repair, once they are taken apart and the valves, leathers, and other working parts are accessible. It usually requires much more work to take a pump apart than it does to install the new leathers and other parts. Therefore, whenever a pump is taken apart, it is a good plan to replace all parts that show any appreciable wear, even though they might last considerably longer.

Before taking a pump apart, be sure to have on hand, if at all possible, all leathers and other parts, including any special gaskets, that may be needed. Repair kits for common makes of small automatic motor-driven pumps are often available from dealers. These kits include all leathers, valve washers, gaskets, etc., that may be needed, and, if they are available, it is best to have one on hand before starting to overhaul such a pump.

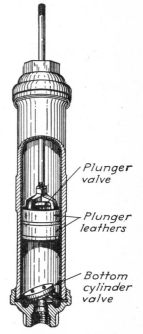

Plunger
valve

Plunger
leathers

Bottom
cylinder
valve

Fig. 491.—A view showing working parts of a pump cylinder. The most common pump repair job is the replacement of worn leathers and parts in the pump cylinder.

Removing a Pump from a Well. Many farm pumps are so made that the whole pump and pipe must be withdrawn from the well in order to remove the working parts from the cylinder. In such cases, be sure to use safe lifting tackles of some sort. A set of blocks and tackle, chains and pry poles, rope hitches, etc., can often be used to advantage. A special safety holder, or a combination lifter and holder, to keep the pipe from slipping back into the well is a valuable piece of equipment when removing a pipe or a pump from a well. Such a holder may be bought or borrowed from a pump dealer or a similar device may be made in the shop.

Some deep-well pumps have cylinders so made that the plunger and valves can all be withdrawn up through the pump pipe without the necessity of removing the whole pump (see Fig. 492). In such cases, disconnect the plunger rod at the top, and allow the plunger to sink gently to the bottom of the cylinder. Then attach the plunger to the bottom cylinder valve, by turning the plunger rod to the right. The bottom cylinder valve can then be withdrawn along with the plunger, by simply pulling up on the plunger rod. The plunger rod is jointed, and sections may be removed as they are lifted from the well.

Fig. 492.—Some deep-well pumps are so made that the plunger and valves may be withdrawn without removing the whole pump from the well. (Courtesy of The Deming Company.)

Taking a Pump Cylinder Apart; Installing New Leathers. Once a pump is removed from the well, proceed to take the cylinder apart. Large strong wrenches may be required, and considerable patience as well, for the joints often become rusted and are difficult to loosen. Striking a few sharp blows with a medium-sized hammer will often help to loosen a tight joint. Once the cylinder is apart, remove the plunger and the valves and renew all leathers or other worn parts. Plungers themselves are usually taken apart by unscrewing the bottom parts.

To complete the pump-repair job, simply clean all parts and reassemble them carefully. It may be advisable to use pipe compound on the pipe joints to avoid the possibility of air leaks. It is particularly important to use good gaskets on cylinder heads, valve chamber covers, etc., of small automatic water pumps. Small leaks in pipes and connections often cause loss of prime and erratic or otherwise unsatisfactory operation.

11. Taking Care of an Automatic Water System

Many pieces of automatic equipment require so little attention that they are often forgotten or neglected. This is sometimes the case with an automatic water system. The main points in the care and maintenance of such a system are (1) lubricating the pump and motor occasionally, (2) tightening the stuffing box, (3) adjusting the air-intake valve (on some systems), and (4) adjusting the belt tension.

Most pumps have oil-level plugs or marks on the side (see Fig. 493). Be sure to keep the oil up to the proper level, adding medium-weight motor oil as may be required. A few drops of oil in each motor bearing once every month or two is usually adequate. For further information on lubricating electric motors, see page 458.

Check the stuffing box (see Fig. 493) occasionally. Keep it tight, but not too tight. A good system is to tighten the stuffing box

Fig. 493.—A small automatic water pump. The main points in taking care of such a system are; lubricating the motor and pump, tightening the stuffing box, adjusting the belt tension, and adjusting the air intake valve as needed. (Courtesy of General Electric Company.)

until it stops leaking, and then unscrew it a half turn or more until it leaks two or three drops a minute. A slight leakage helps to lubricate the piston rod.

For proper operation, a closed water-storage tank must be partly filled with air. The proper water level in the tank is indicated by try cocks or by a water glass on the side. Water absorbs some of the air, and, in time, if the absorbed air is not replaced, the tank will become waterlogged, causing the pump to start often, pump just a little, and then shut off. Some water systems have automatic controls so that the pump pumps a little air along with the water as may be needed to keep the tank properly supplied. On other systems, particularly older models, the operator must open a small air valve on the pump to allow it to replenish the air in the tank and then shut it off after enough air has been pumped. This air valve on the pump is often like an air valve on a tire, and is opened by loosening or removing a cap on the end, and closed by screwing it back on.

Keep the belt tight enough to prevent slippage, but not so tight as to cause undue pressure and wear on the motor and pump bearings. The tension on a V belt is usually about right when the belt can be depressed or pushed in about ¾ in. at a point midway between the two pulleys.

12. Installing a Kitchen Sink and Drain

A kitchen sink solves the problem of disposing of much of the kitchen wastes and eliminates considerable drudgery. If a well or cistern is located near by, and if the low water is not more than 20 ft. below the sink, a pump can also be installed at the sink, and this will save the work of carrying water into the house.

A sink should be at least 18 by 24 in. in size—preferably somewhat larger. A sink with an enameled back attached is much preferred to one without a back. A drainboard likewise is desirable.

Mounting the Sink. The sink should be carefully located so as to be convenient and to save steps in doing the kitchen work. It should be mounted at a height that is best suited to the individual who is to use it. About 36 in. from the floor to the top of the sink rim is an average height. A sink is mounted on hangers which are first attached to the wall. These hangers must be securely screwed to the vertical studs in the wall, or to a board that is nailed or screwed to the studs. In mounting these hangers, be careful to get them level and placed so as to give the desired height to the sink. Use screws of appropriate size and length, and drill suitable size pilot holes for the screws.

Attaching the Trap. Every sink should have a trap or water seal in the outlet to prevent gases from the sewer or drain from coming back into the house. Traps are usually made in an S or U shape.

To install a trap, assemble it loosely and mark the place for the outlet in the floor beneath the sink. Then bore a hole of suitable size through the floor. After the drain is laid beneath the floor, assemble the trap and connect it to the drain.

The exact method of connecting the pipe from the trap to the main drain beneath the floor will depend upon the type of fittings used. Most sink-trap connections are attached directly to the threaded end of a standard-size pipe, the other end of this standard-size pipe being screwed into or otherwise attached to an opening in the larger drain beneath the floor.

Making the Drain. The main drain beneath the floor should be at least 2 in. in inside diameter—preferably larger. Cast-iron soil pipe is most commonly used. Regardless of the type of drain pipe used, a clean-out plug should be provided where the vertical pipe

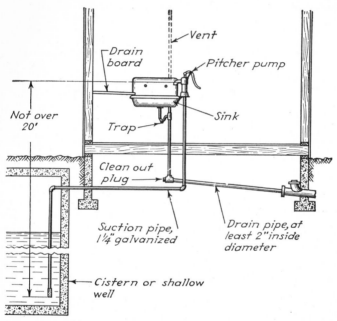

Fig. 494.—A pump, sink, and drain is a practical, simple system. It eliminates much drudgery.

from the sink joins the horizontal drain that goes through the foundation wall. Then, in case the drain becomes clogged, the plug may be removed and the trouble remedied.

Bell-mouthed sewer tile may be used to connect to the cast-iron soil pipe that goes through the foundation wall. The joints in the drain should be made tight to a point 40 or 50 ft. from the house, and at least 100 ft. from any well supplying water for drinking purposes. Give this section of the drain a fall of $\frac{1}{8}$ to $\frac{1}{4}$ in. per ft., and be sure all joints are cemented or otherwise made watertight.

From the end of the tight section of the drain, lay a line of common 4-in. drain tile, giving it a fall of $\frac{1}{4}$ in. per ft. and leaving the joints slightly open or loose.   The required length of the open-joint section will depend upon the porosity of the soil, 50 to 75 ft. usually being sufficient.   If the soil is heavy or tight, it may be advisable to lay the tile in a trench of gravel, the trench being about 2 ft. deep, and the tile being laid not over 18 in. deep.

In case a complete sewage-disposal system is later installed, the sink drain should be connected into it.

13. Installing a Simple Shower Bath

Where a suitable drain can be provided, a simple shower bath can often be installed with very little work or expense.   If hot and cold water under pressure is available, piping connections to a shower-bath sprinkler head are easily made.   Where a pressure water system is not available, it is often possible and practical to mount a small tank or barrel in an elevated place to give pressure; and if a metal tank or barrel is used and placed where the sun can shine on it, water can be pumped up in the morning and allowed to heat during the day.

Where a modern water and plumbing system is already installed in a house, it usually is a simple matter to make a shower bath in the basement.   A floor drain and trap must be installed for the shower and connected to the main sewer drain.   The concrete floor for the shower should have side walls a few inches high to prevent water from running out onto the basement floor.

A sprinkler head is then mounted in the shower booth and connected to the hot- and cold-water lines.   Although it might be desirable to have a special mixing valve for the hot and cold water, a plain globe valve connected in the hot-water line and another in the cold-water line will work satisfactorily.   Simply run the hot water through a valve to one end of a tee, the cold water through another valve to the other end of the tee, and then connect the sprinkler head to the side opening of the tee.

Jobs and Projects

1. Practice cutting and threading a few pieces of pipe until you understand exactly how to do it.   Fit a few pieces of pipe together, and test under water pressure if possible.
2. Make a list of pipe or plumbing repair jobs that need to be done about your home or farm.   Also list any simple extensions to an existing water system that would save time and labor in doing chores or in taking care of livestock.

3. Make plans and actually perform some of these repair jobs, or make an extension if it can be arranged.

4. Renew the disks on some leaky faucets, either about your home or farm, or about the school or shop.

5. Take a pump cylinder apart, and renew the leathers and any valve parts that are worn.

6. Make a plan, including sketches or drawings, for a kitchen pump, sink, and drain for your home, or some typical farm home in your community. Consult catalogues, and make a list of all equipment and materials required, giving sizes, kinds, costs, etc.

7. If possible, install a pump, sink, and drain according to a plan previously made.   (This makes an excellent class project.)

8. Make a plan for a basement shower, listing all materials and supplies needed, together with sizes, kinds, costs, etc.

9. If possible, install a shower according to a plan previously made.

13. Repairing and Reconditioning Machinery

A COMPLETE treatment of the subject of repairing and reconditioning of farm machinery is somewhat beyond the scope of this book. Therefore, only a few of the more important and more common activities in this field will be considered here. It may be observed, however, that repairing machinery involves many of the basic operations treated in the preceding sections of this book, particularly those on metalwork.

MAJOR ACTIVITIES

1. Removing Worn and Broken Machine Parts
2. Repairing and Adjusting the Cutting Parts of a Mower
3. Repairing and Adjusting Sprocket Chains
4. Sharpening Plowshares
5. Sharpening Harrow Teeth
6. Sharpening and Adjusting Ensilage-cutter Knives
7. Sharpening Disks and Coulters
8. Sharpening and Setting Cultivator Shovels
9. Repairing Broken Tongues
10. Protecting Machinery from Rust
11. Lubricating Machinery

1. Removing Worn or Broken Machine Parts

Removing parts of machines generally involves the unscrewing of nuts and bolts, and sometimes also the removal of other fastenings, such as rivets, cotter keys, and setscrews.

Using Wrenches. Nuts and bolts can be removed from many parts of common machines with ordinary adjustable wrenches

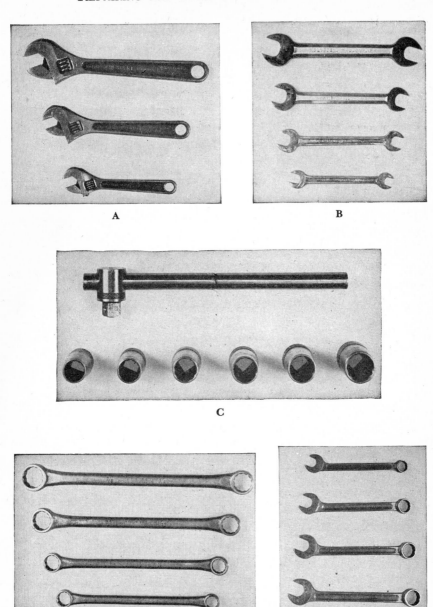

Fig. 495.—Good types of wrenches for the farm shop: A, adjustable wrenches; B, open-end wrenches; C, socket wrenches: D, box-end wrenches; E, combination open-end and box-end wrenches.

(Fig. 495*A*), or with open-end wrenches (Fig. 495*B*). In close quarters, however, it may be necessary to use socket wrenches (Fig. 495*C*) or box-end wrenches (Fig. 495*D*). Where considerable machinery is to be kept in adjustment, a few good open-end, box-end, and socket wrenches will prove to be a profitable investment.

Always use a wrench that exactly fits the nut or bolt. Loose-fitting wrenches are liable to slip and round the corners of the nut or bolt. Also, there is danger of skinning or bruising the hands.

When using an adjustable wrench, always adjust it to fit the nut or bolt tightly, and put it on the nut or bolt so that the pull or push will be toward the open end of the jaws (see Fig. 496). Pulling this way lessens the danger of slippage and also the danger of overstraining or springing the jaws of the wrench.

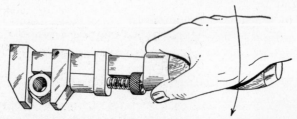

Fig. 496.—Always adjust an adjustable wrench to fit tightly and always pull (or push) toward the open end of the jaws.

Do not use pliers on nuts or bolts, except possibly as a last resort. It is difficult to get a tight grip with pliers, and they will mar the parts and make it difficult or impossible later to use wrenches. Using pliers on brass nuts and fittings is especially bad.

Removing a Rusted Nut. Sometimes, because of rust, it is impossible to turn a nut by ordinary methods. In many such cases, it is best simply to cut the nut off with a sharp cold chisel or a hack saw and to use a new nut and bolt for replacement. Heating with a gasoline blowtorch or applying kerosene or a penetrating oil will sometimes help in loosening a rusted nut.

Extracting Broken Screws and Bolts. The extraction of broken screws or bolts often requires very painstaking work. If the bolt or screw is not screwed too tightly into place and if a portion of it protrudes, it may be removed with pliers or a small pipe wrench. If it is tightly in place with a small portion protruding, say ⅛ to ¼ in., it may be feasible to saw a slot across the top with a hack saw and then remove it with a screw driver. Alternate heating and cooling will sometimes loosen a tight screw. Heating can be done with a gasoline blow torch, and cooling may be done by applying water.

Probably the most satisfactory way of removing large broken screws and bolts is to drill a hole somewhat smaller than the diameter of the bolt, down through the center of the bolt, and then to use a broken-bolt or screw-plug extractor, commonly called an "ezy out" (see Fig. 497). Such an extractor might be described as a tapered tool with coarse left-hand threads on the pointed end. To use it, simply start it into the drilled hole, turning it to the left. It wedges itself tightly in the hole, and continued turning will ordinarily loosen the broken bolt and unscrew it.

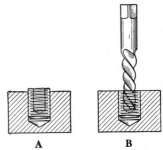

Fig. 497.—Steps in removing a broken screw or stud bolt with a screw extractor. A. Drill a hole in the broken screw. B. Insert a screw extractor, turning it to the left.

2. Repairing and Adjusting the Cutting Parts of a Mower

Heavy draft, ragged cutting, and excessive wear and breakage can often be avoided by a few simple adjustments and replacement of parts of the cutting mechanism of a mower.

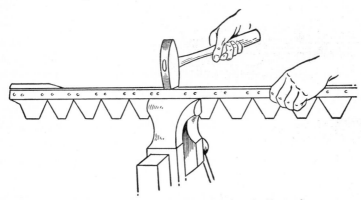

Fig. 498.—Removing a worn or broken knife section.

Replacing Knife Sections. To remove a broken or worn knife or sickle section, support the knife rib or bar firmly, and strike the back of the section one or two sharp blows with a hammer. A good way to support the knife is to clamp it loosely in a vise, pointed end of the sections down (see Fig. 498).

After the broken section is removed, then punch the sheared rivets from the holes in the knife bar, being careful not to enlarge the holes. Then put the new section in place, insert the rivets, and rivet them down. Strike one or two heavy blows first to swell the rivets, and then form the heads by light peening with the ball peen of the hammer, or with a rivet set.

Sharpening a Mower Knife. A dull or improperly ground knife causes ragged cutting, rapid wear, and extremely heavy draft. Three points are very important in grinding a knife:

1. Maintain the original width of bevel (see Fig. 499). A narrow, blunt bevel does not cut easily; and a wide, keen bevel nicks easily.

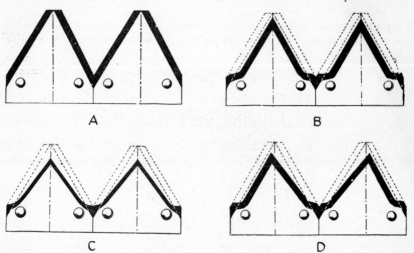

Fig. 499.—Right and wrong ways to grind mower knives. Dotted lines show outlines of new sections. A. New sections with proper width of bevel and angle of shear. B. Sections properly ground. Proper width of bevel and angle of shear are retained. C. Sections improperly ground. The bevels are too narrow, and the angle of shear is wrong. D. Proper width of bevel, but improper angle of shear. (Courtesy of Deere and Company.)

2. Maintain the original angle of shear, or the angle between the cutting edge of the section and the guard plate. Otherwise, the grass will tend to slip away and not be cut.
3. Do not overheat and draw the temper.

Mower knives can best be sharpened on special grinders or grinding wheels made for that purpose (see Figs. 500 and 501). With such a grinder, it is easy to maintain the original bevel and angle of shear. With a little practice and patience, however, mower knives can be ground satisfactorily on regular grinding wheels.

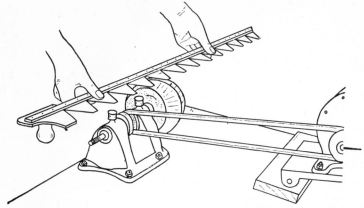

Fig. 500.—Mower sickles are ground easiest on special grinders made for that purpose, although they can be satisfactorily sharpened on an ordinary grinding wheel.

A

B

Fig. 501.—Two types of mower knife grinders.   (Courtesy of Deere and Company.)

Grinding may be done on either the flat vertical side or on the regular curved grinding surface. Motor-driven grinders are much faster and require less work than hand- or foot-operated grinders.

Straightening a Mower Knife. If a knife bar is bent, either edgewise or flatwise, it will bind as it works back and forth in the cutter bar and cause both rapid wear and increased draft. To straighten a knife, sight along the knife bar to locate the bend. Then place it on some straight surface, as a bench top or mower tongue, with the bend or bulge up, and strike with a hammer (see Fig. 502). Sight again and hammer more as may be required. Be sure to check the knife bar for bends both edgewise and flatwise.

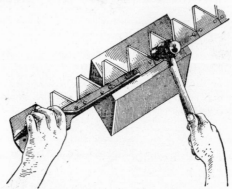

Fig. 502.—A bent knife may be straightened by hammering it over a straight surface. (Courtesy of Deere and Company.)

Replacing Guard Plates. A guard plate, also called *ledger plate*, serves as one blade of the shears and should be kept in good cutting condition just the same as the knife section. When guard plates become worn and dull or nicked or broken, remove them and install new ones.

Guard plates may be removed with the guards either on or off the cutter bar. To remove a guard plate, firmly support the guard from beneath. A special guard-repair anvil (see Fig. 503) is excellent for this purpose. Drive the guard rivet down from the top, using a stout punch to start the rivet and a slim punch to finish removing it. Then insert a new guard plate and rivet, and securely brad the end of the rivet. If there is any part of the rivet projecting above the guard plate, trim it off smooth with a sharp cold chisel. New rivets may be inserted from the bottom or from the top of the guard.

Guards should be aligned frequently, for they often strike stones, sticks, or other obstructions and become bent. Even a nearly new

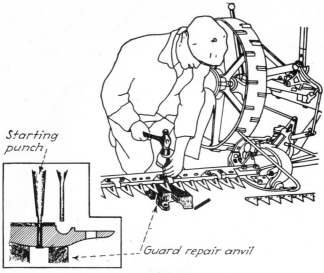

Starting punch

Guard repair anvil

Fig. 503.—Worn or broken guard plates may be removed without taking the guards off the cutter bar. Start to remove the rivets with a stout starting punch and then finish with a slim punch.

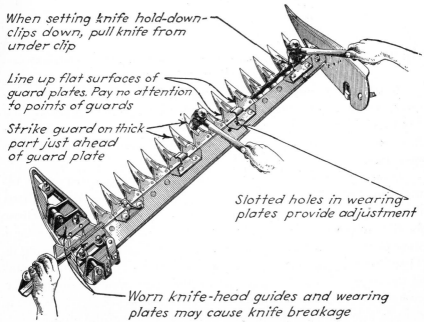

When setting knife hold-down-clips down, pull knife from under clip

Line up flat surfaces of guard plates. Pay no attention to points of guards

Strike guard on thick part just ahead of guard plate

Slotted holes in wearing plates provide adjustment

Worn knife-head guides and wearing plates may cause knife breakage

Fig. 504.—Alignment of guards and adjustment of wearing plates and knife hold down clips are important for clean cutting and light draft.

mower is likely to need some guards bent back in line.   Probably
the best way to align guards is to insert a straight knife and hammer
the guards that are out of line, bending them up or down as may be
required.   Strike on the thick part of the guard just ahead of the
guard plate.   Pound the high ones down first, and then bring the
low ones up (see Fig. 504).   Be sure the guard bolts are tight, and
remember in hammering to make the guard *plates* line up.   It is not
important if the points are somewhat out of line

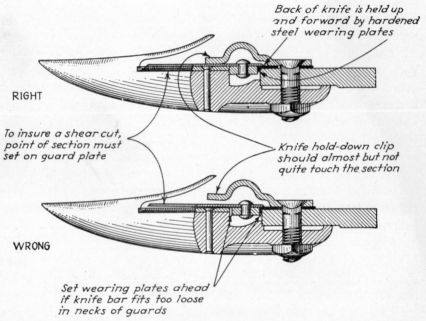

Fig. 505.—Right and wrong adjustments of hold-down clips and wearing plates.

Adjusting Other Cutter Bar Parts.   The cutter bar is the heavy
steel bar to which guards and other parts are attached.   It should
not be confused with the knife bar, which is the small bar or rib to
which sickle sections are riveted.

The parts of the cutter bar form a sort of groove or trough in
which the knife works back and forth.   Not only should the knife
be straight, but these parts on the cutter bar should be aligned and
form a straight place in which the knife can work.   Also, they should
be adjusted to fit the knife.   They should not fit too tightly and therefore bind.   Neither should they fit too loosely and allow the knife to
bounce or flop about.   The knife sections and the guard plates should

fit together snugly and form sharp shearing edges, just the same as the two blades of a pair of scissors should fit together reasonably tight (see Fig. 505).

The knife hold-down clips should almost but not quite touch the knife when it is resting on the guard plates. To adjust the clips, simply hammer them up or down, but be sure the knife is not in place under a clip when it is being hammered down (see Fig. 504). When a *thin* piece of tin can be just slipped under the clip, it may be considered in good adjustment.

The wearing plates, which support the back edge of the knife, are replaceable. When they become worn, replace them with new ones. Always adjust new wearing plates so their front edges just touch the back of the knife bar. The wearing plates are held in place with guard bolts, and the holes through which the bolts go are slotted. It is therefore a simple matter to loosen the bolts and adjust the plates forward or backward until they all line up.

Worn knife-head guides are a common cause of knife breakage, as well as poor cutting near the inner end of the knife. These parts should therefore be adjusted or replaced whenever looseness develops.

It is important that all guards and other cutter-bar parts be kept tight. Tight-fitting strong wrenches, such as socket wrenches, are best for tightening cutter-bar bolts.

Aligning a Cutter Bar. A cutter bar is in proper alignment if (1) the pitman is square with, or at a right angle to, the pitman drive shaft, and (2) if the pitman pushes and pulls straight on the sickle (see Fig. 506).

To offset the backward strain when cutting and to make the pitman and knife run straight, the cutter bar is given a certain amount of lead. That is, when the mower is standing still, the outer end of the cutter bar is slightly ahead of the inner end. The proper amount of lead is about $\frac{1}{4}$ in. per ft. of cut.

To check the pitman angle, place a square or other straight edge against the front face of the pitman wheel. If the pitman is parallel to the edge of the square, it is square with the pitman drive shaft. To change or adjust the pitman angle, adjust the tie rod in front of the pitman or the diagonal push bar behind it so as to move the inner shoe of the cutter bar forward or backward as may be needed. This adjustment will also affect the register of the knife. Therefore, check the register (see page 434) before making this adjustment. If some parts have been sprung and thus allow misalignment of the pitman with the pitman drive shaft, these parts may have to be straightened or replaced.

To check the lead of a cutter bar, place the mower on level ground, block the wheels, raise the end of the tongue 32 in. from the ground, and pull the outer end of the cutter bar back as far as it will go. Then tie a string to the pitman head, stretch it loosely over the center of the pitman, over the center of the knife head, and straight on out to the end of the cutter bar (see Fig. 506). Note how the back edge of

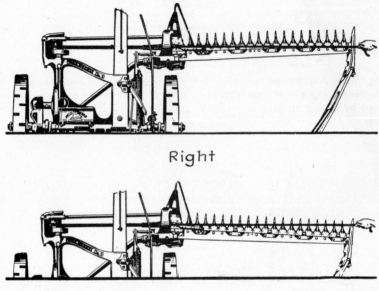

Right

Wrong

Fig. 506.—The outer end of the cutter bar should have "lead" to offset backward strain when cutting and to make the pitman and knife run straight. (Courtesy of Deere and Company.)

the knife (not the cutter bar) lines up with the string at the inner end and the outer end of the knife.

An eccentric bushing is provided on *some* mowers for adjusting cutter-bar alignment (see Fig. 507). On many mowers no adjustment is provided. On these mowers, and also frequently on mowers having an eccentric bushing, it will be necessary to determine just what causes the lag in the cutter bar, and then remove the cause. In many cases, it is wear on the hinge pins of the inner shoe (see Fig. 508). New hinge pins may need to be installed, or possibly the holes drilled oversize and oversize pins installed. Sometimes parts of the mower have been sprung, and these must be straightened or replaced.

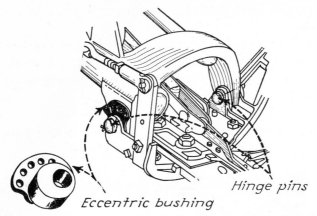

Eccentric bushing   Hinge pins

Fig. 507.—Adjustment for cutter bar lead. An eccentric bushing is provided on the rear hinge pin of some mowers. Lead on other mowers may generally be restored by taking up play at the hinge pins, or otherwise removing the cause of the lag.

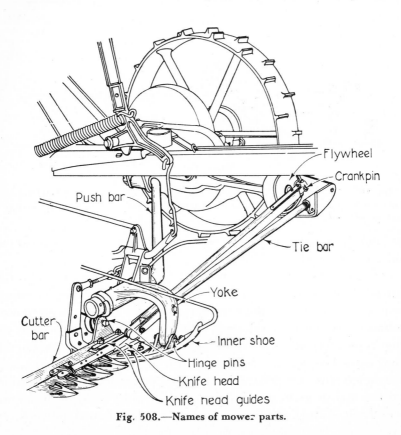

Fig. 508.—Names of mower parts.

Another cause of misalignment is worn bolt holes and bolts that fasten the cutter bar to the inner shoe. In such a case, the bar and inner shoe may be welded or brazed in proper position.

Lengthening the diagonal push bar behind the pitman or shortening the tie bar in front of the pitman is not satisfactory for restoring

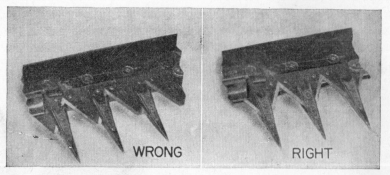

Fig. 509.—Knife sections should register or center in the guards at the ends of the pitman strokes. (Courtesy of Deere and Company.)

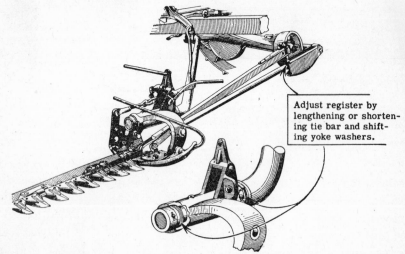

Adjust register by lengthening or shortening tie bar and shifting yoke washers.

Fig. 510.—A typical method of adjusting register. Similar methods are provided on other models of mowers. (Courtesy of Deere and Company.)

lead. Neither of these adjustments will improve the angle between the pitman and the sickle. Furthermore, they will change the angle between the pitman and the pitman drive shaft.

Adjusting Knife Register. By register of the knife is meant the centering of the knife sections in the guards at the ends of the knife strokes (see Fig. 509). If the knife does not register, the mower will do an uneven job of cutting, it will choke easily, and the draft will be heavy.   If a knife is out of register, first check to be sure the pitman straps are properly tightened and that there is not excessive play in the bearings at the ends of the pitman.   Then if the knife is still out of register, move the whole cutter bar in or out as may be necessary. To do this, shorten or lengthen the tie bar in front of the pitman, and also move the back of the inner shoe yoke in or out on the diagonal push bar the same amount.   Various methods are provided for adjust- ing the position of the yoke on the diagonal push bar.   On some mowers, washers may be shifted (see Fig. 510); on others, screw threads are provided; and on other mowers, still other methods are used.

3. Repairing and Adjusting Sprocket Chains

Chains and sprockets are used in many places on farm machines for transmitting power.   The more common type of chain is made

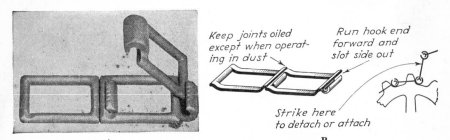

A   **B**
Fig. 511.—A, Malleable iron-link chain; B, steel-link chain.

of links that hook into each other.   These links are made of malleable iron or of pressed steel (Fig. 511).   In another type of chain, known as the *pintle* chain (Fig. 512), the links are fastened together by pins or rivets.   *Roller* chains (Fig. 513) are commonly used for heavy-duty applications.   Such chains have steel rollers working over steel pins that fasten the links together.

The sprocket wheels over which a chain runs should be kept in alignment.   Tighteners of some sort are commonly provided, and these should be kept in adjustment (see Fig. 514).   Chains should not be stretched too tight, yet they should not be so loose as to jump off or ride up on the sprocket teeth.

Chain links sometimes break and must be replaced.   Hook-link chains are taken apart and hooked together as indicated in Fig. 511.

When such chains are reinstalled on the sprockets, be sure to put them on *with the open part of the links* away from the sprockets and *with the hooked ends of the links* leading in the direction of travel.

When a chain becomes badly worn and must be replaced, it is almost certain that one or more of the sprockets will need replacing also. In fact, it is often a worn sprocket that causes the chain to wear

Fig. 512.—Pintle-link chain.

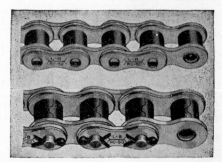

Fig. 513.—Two types of steel-roller chain.

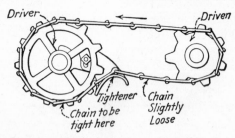

Fig. 514.—The proper method of running a hook-link chain on sprockets.

rapidly, and if only the chain is replaced, it will soon be ruined and have to be replaced again.

Keep sprocket chains oiled except when operating in dirt and dust.

4. Sharpening Plowshares

Three different kinds of plowshares are in common use, soft-center steel shares, crucible- or solid-steel shares, and chilled-iron shares. The methods of sharpening and hardening are different for the different kinds of material. The sharpening of steel shares in the forge is rather heavy work, although otherwise it is not so difficult.

Sharpening Steel Shares. Steel plowshares are sharpened by heating in the forge and drawing the edge by hammering. To heat a share, place it flat in the forge with the cutting edge over the center of the fire. Be careful to heat only a width of about two inches along the edge, and to heat at one time only as much as can be hammered out before it cools below a good working tem-

perature. Do not place the share in a vertical position with the edge down in the fire, as this will heat too much of the share and may cause warping.

Sharpen and shape the point of the share first. Heat it to a cherry red and upset it and bend it to give it the desired down and side suction as indicated in the following paragraph. After the point is sharpened, work back toward the heel, heating and hammering only a

Fig. 515.—To sharpen a steel plowshare, heat to a cherry red and hammer on top. Heat and hammer only a small section at a time.

small section at a time. Hammer on top of the share and be careful not to dent it with hammer marks more than necessary (see Fig. 515).

It is important for a plow to have suction, both downward and sideward (see Figs. 516 and 517). The proper amount of suction

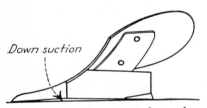

Fig. 516.—Plows must have down suction to secure penetration. The clearance at the point indicated should be about ¼ to ½ in.

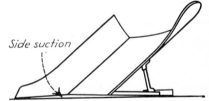

Fig. 517.—Side suction enables a plow to cut an even, full-width furrow. The clearance at the point indicated should be about ¼ in.

will depend upon the type of plow and upon the soil condition. Down suction should usually be about ¼ to ½ in. measured at the place shown in Fig. 516; and the side suction should usually be about ¼ in. measured at the place shown in Fig. 517. On most plows the suction is obtained by bending the point down and sidewise the desired

amount at the time of sharpening. On other plows the mounting of
the plow bottom on the plow beam is such that the suction will be
correct without bending the point.

In the case of a walking plow, shape the outer corner or wing of
the share so that a small flat surface bears on the ground and helps to
support the outer side of the plow. Sulky and tractor plows require
little or no flat bearing surface at the wing of the shares, as the plow
wheels furnish the stabilizing support needed.

Hardening Soft-center Steel Shares. A soft-center steel share
may be hardened by heating a strip about 2 in. wide along the cutting
edge to a uniform cherry red and then dipping it in clean, cold water,
cutting edge down. Solid- or crucible-steel shares should be hardened very little if at all. Hardening makes them brittle and therefore
subject to breakage.

Hard-surfacing Steel Shares. Different kinds of extremely hard
and long-wearing material, such as Stellite, can be applied to the
points and cutting edges of steel shares by means of the oxyacetylene
welding torch. The process of applying such material is practically
the same as bronze welding (see page 386), using a welding rod of the
particular material instead of one of bronze. These materials will
greatly increase the life of a share, and when carefully applied, their
use is generally satisfactory.

Sharpening Chilled-iron Shares. Chilled-iron shares cannot
be forged, owing to the nature of the iron. They must be sharpened
by grinding or chipping on the top side. Chilled-iron shares are
moderate in cost, and it is generally considered best to discard them
when they become badly worn and dull.

5. Sharpening Harrow Teeth

Spike-tooth harrow teeth that have sharp points and sharp edges
are much more effective than teeth that have become blunt and
rounded from long use. To sharpen harrow teeth, remove them
from the harrow and forge them at a red heat. Some harrow teeth
may be effectively hardened and tempered, and others not, depending
upon the kind of steel of which they are made. It is well to experiment on one or two teeth before hardening a whole set. It may be
possible to get them too hard, making them brittle and subject to
breakage in use. See pages 371 to 375 on hardening and tempering
tool steel.

6. Sharpening and Adjusting Ensilage-cutter Knives

To sharpen ensilage-cutter knives, grind them on a grinding wheel. Be careful to grind at the original bevel and with only moderate pressure to avoid drawing the temper.   If there are large nicks, they may be best removed by placing the knife on the work rest and moving it back and forth with the cutting edge square against the revolving wheel.   After removing the nicks, then grind the edge at the desired bevel.

If considerable grinding of ensilage-cutter knives is to be done, it may be worth while to build a small platform in front of the grinding wheel to serve as a work rest.   A small strip may be nailed to the platform to make it easy to place the knife against the wheel at just the right angle.

When the knives are put back on the cutter, be sure to adjust them carefully and bolt them on securely.   Adjust the knives to just clear the shear bar.   After considerable use, the shear bar itself will have become rounded and should be replaced or turned over to present a new square edge.

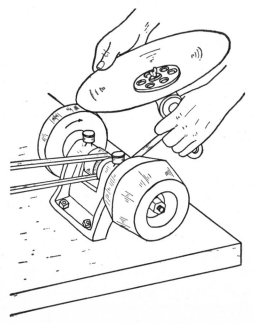

Fig. 518.—Disks and coulters may be ground on ordinary grinders.   Attachments to hold the disks in place are helpful.

7. Sharpening Disks and Coulters

There are various methods of sharpening disks and coulters.   In some sections of the country, blacksmiths have machines for rolling the edges of disks out thin and sharp.   Such machines sharpen disks by passing the cutting edges between large rollers under pressure. Disks and coulters may be sharpened also, by grinding or forging. Grinding is greatly facilitated by some sort of support that holds the disk in place against the grinding wheel (see Figs. 518 and 551). Such supports can be made in the farm shop.

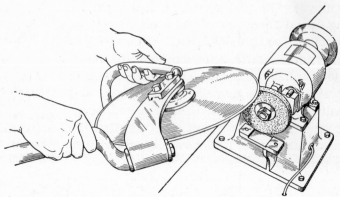

Fig. 519.—An excellent way to grind a coulter.   Adjust the coulter bearing to turn freely and hold the coulter with the bearing slightly to one side of the grinding wheel.

An excellent way to grind coulters is illustrated in Fig. 519. Adjust the bearing to turn freely, and hold the coulter with the bearing slightly to one side of the grinding wheel.   The coulter disk will then turn as it is being ground, and thus avoid overheating and drawing the temper.   The speed with which the coulter turns can be regulated by the distance the bearing is held to one side of the grinding wheel.

8. Sharpening and Setting Cultivator Shovels

When cultivator shovels become dull and blunt, they should be sharpened to as near the original shape as possible (see Fig. 520). This may be done by grinding or by forging.   After considerable use, cultivator shovels will become so worn that they should be repointed or replaced with new ones.

It is not a difficult job to draw out the points and edges of cultivator shovels by forging. Whether or not the shovels should be hardened and tempered after forging will depend upon the kind of steel used in the shovels. If the shovels are of soft-center steel, they may be hardened by heating about 2 in. along the cutting edge to a dull red and then dipping in water. Solid crucible-steel shovels should be hardened very little if at all. There is danger of breaking them during hardening. Also, it is easy to get them too hard and brittle, which may result in breakage in use.

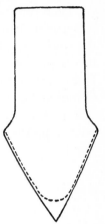

Fig. 520.—A cultivator shovel Dotted lines show shape when in need of resharpening. (Courtesy of Deere and Company.)

If cultivator shovels are to be sharpened by grinding, grind them on the back or under side.

Cultivators are provided with adjustments to change the angle or pitch of the shovels. If a shovel is set too straight (see Fig. 521), it will require considerable pressure to make it penetrate, and there will be a tendency for it to skip and jump. On the other hand, if a shovel is set too flat, the underneath part may be lower than the point, making penetration difficult.

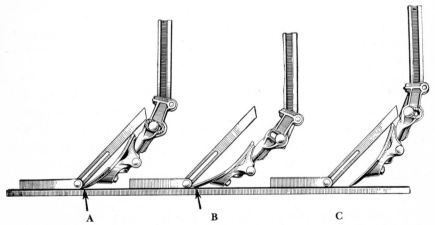

Fig. 521.—Correct and incorrect adjustments of pitch of cultivator shovels: A, correct adjustment; B, shovel set too flat, giving poor penetration; C, shovel set too straight, giving poor penetration and uneven operation. (Courtesy of Deere and Company.)

In order to throw soil into the row with the front shovels, they may be turned on their shanks. On some cultivators this will tend to make the gangs spread apart and work away from the row. This tendency may be overcome by turning some of the rear shovels on their shanks opposite to the way the front ones are turned.

9. Repairing Broken Tongues

Implement tongues often become broken and must be replaced or repaired. The better method is to replace them. A new tongue may be made by sawing, planing, or otherwise cutting suitable stock to the required size and shape. Then carefully mark all holes for bolts and other fastenings by putting the tongue in place on the implement and marking, or by marking from the old tongue. Painting of tongues and wooden parts greatly increases their useful life.

If a tongue is only split or cracked, it is often feasible to repair it by reinforcing with strap irons bolted in place. If only the end of a

Fig. 522.—Broken implement tongues may often be repaired by splicing on a new end by using an end-lap splice. Strap irons may also be bolted in place for reinforcement.

tongue is broken off, a new end may be made and attached with an end-lap splice or joint (see Fig. 522).

10. Protecting Machinery from Rust

Rusting of plows, disks, cultivators, and other machines should be avoided. A plow with a rusty moldboard, or a cultivator with rusty shovels, will require much more power than if in good condition. More fuel will be used (if tractor operated), more time will be lost, and a poorer quality of work will be done.

Used engine oil will prevent rusting for short periods. It may be applied with an old paint or whitewash brush, or with a hand sprayer. If an implement is not to be used for a few days, it is a good plan to oil the polished metal working parts. When implements are not to be used for long periods, heavy oils or greases should be applied. Special antirust greases or compounds are particularly good. Of course implements should be housed during seasons when they are not needed.

Metal parts of implements and machines, other than wearing surfaces, should be kept painted to avoid rusting. See page 137 on painting of metal surfaces.

11. Lubricating Machinery

Late models of farm machines are generally well protected against dust and dirt, and if they are operated and lubricated in accordance with instructions furnished with them, they may be expected to give long trouble-free service.

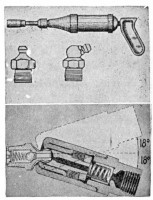

To be sure that the right kind of lubricant is used and that no points of lubrication have been overlooked, it is a good plan to consult the instruction book for that particular machine. Most instruction books contain lubrication charts.

Critical points on most new model farm machines are lubricated by grease forced through fittings by a pressure gun (see Fig. 523). Pressure guns provide positive lubrication. On most bearings if a little grease is kept working out around the spindle or shaft, dirt will not enter and there is assurance of adequate lubrication.

Fig. 523.—Above—pressure grease gun and fittings. Below—cross section of gun nozzle and fitting.

Installing Pressure-gun Fittings on Old Machines. Pressure-gun fittings may be installed on old machines where lubrication was originally provided by squirt can or hard oil cups. This is certainly advisable where such machines are to be used much. Where old hard oil cups were used, this is easily done. Simply unscrew the old cups, and screw in the new fittings. Bushings or reducers may be required in some cases.

To install pressure-gun fittings in plain oil holes used for squirt-can lubrication, either of two methods may be used. Special drive-type fittings may be obtained in some cases and installed by simply driving or wedging them into the old oil holes. In other cases, the old oil holes may be drilled and tapped with suitable threads (⅛- or ¼-in. pipe threads) and the new fittings screwed in. Grease-gun fittings and accessories are available through implement dealers and automobile supply houses.

Jobs and Projects

Do several repair jobs on farm machines and implements, working on equipment from your own farm if possible. The following list is suggestive.

1. Replace worn or broken sections in a mower knife.
2. Sharpen a mower knife.   Straighten it if it is bent.
3. Replace guard plates on a mower.
4. Align and adjust cutter-bar parts on a mower.
5. Align a cutter bar and adjust knife register on a mower.
6. Repair and adjust the sprocket chains on a binder.
7. Sharpen a plowshare.
8. Sharpen a set of harrow teeth.
9. Sharpen the disks and repair a disk harrow.
10. Repair a cultivator and sharpen and adjust the shovels.
11. Replace a broken tongue in an implement.
12. Repair a set of doubletrees and singletrees.
13. Install pressure-gun fittings on a machine or implement.
14. Completely overhaul and paint an implement, such as a mower, grain drill, corn planter, or manure spreader.

14. Maintaining Electrical Equipment

A<small>N</small> increasing amount of electrical equipment is used on farms, and while such equipment is generally more troublefree and foolproof than many other kinds of equipment, a certain amount of repair and maintenance work is necessary for best results. Although a complete treatment of the subject of working with electrical equipment is beyond the scope of this book, some of the more important and more common activities in this field are considered here.

MAJOR ACTIVITIES

1. Splicing Electric Wires

2. Attaching Wires to Terminals

3. Repairing Electric Cords

4. Replacing Fuses

5. Protecting Electric Motors against Overload

6. Cleaning and Lubricating Electric Motors

7. Rigging a Small Portable Electric Motor

8. Connecting Dry Cells

9. Charging a Storage Battery

10. Making Extensions to a Wiring System

11. Installing a Doorbell or Buzzer

12. Installing an Electric Fence

1. Splicing Electric Wires

Poor splices are a source of trouble and danger. If they are not mechanically strong, they may loosen or pull apart. If they do not make good electrical contact, there will be a drop of voltage, causing poor operation of appliances on the circuit; and the joints or splices will heat, possibly creating a fire hazard.

There are four main steps in making a splice: (1) removing the insulation and cleaning the wires; (2) twisting or fastening the wires together; (3) soldering the joint; and (4) insulating the splice by covering it first with rubber tape and then with friction tape.

Right   Wrong

Fig. 524.—In cutting insulation from a wire, hold the knife at an angle as in sharpening a pencil. Be careful not to nick the wire.

Removing Insulation. To remove insulation from a wire, cut it off with a knife or crush it with a pair of pliers and then strip it off. If it is to be cut off with a knife, be careful not to nick the wire. Cut at an angle as in sharpening a pencil (see Fig. 524).

If suitable pliers, such as electrician's pliers, are available, crushing and then stripping off the insulation will usually be quicker and easier than cutting with a knife. Crush the insulation with the heel of the

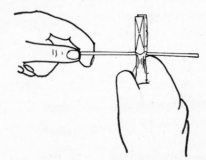

Fig. 525.—Crushing the insulation on a wire with pliers. Crushing and stripping off the insulation is generally easier than cutting it off with a knife.

pliers (see Fig. 525), and then strip it off. Next scrape the wires clean, using the back of a knife or some blunt-edged tool.

Making the Common Splice. To make the common splice, also called the *Western Union splice*, remove the insulation for about 4 in. on the end of each wire. Then place the wires together and hold them firmly as shown in Fig. 526A. Make five or six turns with one wire in one direction and five or six turns with the other wire in the other direction (see Fig. 526B and C). Then cut the ends off short and smooth them down with pliers to prevent damage to insulation which must be wrapped over the splice later.

Soldering the Splice. After the wires are cleaned and wrapped tightly together, apply a noncorrosive soldering flux, such as rosin or a noncorrosive paste, and then solder the splice. Under no conditions should acid be used as a flux. In soldering the splice, it is important that the wires themselves be heated hot enough to melt

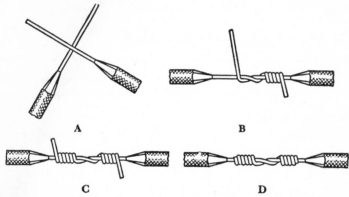

Fig. 526.—Steps in making the common, or Western Union, splice.

solder thoroughly. Be sure the solder penetrates into all the spaces between and around the wraps of the splice, and does not simply coat over the outside of the splice (see Fig. 527). (For further information on soldering, see pages 274 to 293.)

Insulating the Splice. After the splice is soldered, wrap it with rubber tape, starting on the tapered rubber insulation on one of the wires at one end of the splice. Wrap spirally toward the other end, keeping considerable tension on the tape and making the wraps overlap. When the end of the splice is reached, wrap back toward the other end in the

Fig. 527.—In soldering splices, be sure that the solder penetrates well into all spaces, as at A. A poor job is shown at B.

same manner. Wrap back and forth two or three times—at least until the thickness of rubber tape is equal to that of the original insulation. Be sure the wrapping of rubber tape covers all parts of the splice where the outer braid has been removed.

After the rubber tape is applied, then wrap at least two layers of friction tape over the rubber tape. Wrap it spirally in the same manner as the rubber tape was wrapped.

Splicing Duplex or Two-conductor Cord. To splice a two-conductor cord, splice each wire separately, using Western Union splices. Be careful to stagger the splices as shown in Fig. 528. Staggering minimizes the possibilities of short circuits and makes the

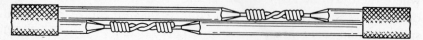

Fig. 528.—To splice a two-conductor cord, splice each wire separately, staggering the splices so they will not be side by side.

splice less bulky. Wrap each splice separately with rubber tape, and then wrap friction tape over the whole splice.

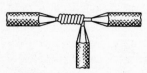

Fig. 529.—The tap splice is used for joining a branch wire to another wire.

Making a Tap Splice. A tap splice is used for joining a branch wire to another wire. To make it, remove about 4 in. of insulation from the end of the tap wire and about $1\frac{1}{4}$ in. from the main wire. Scrape the wires clean, and wrap the end of the tap wire around the main wire about five or six turns (see Fig. 529). Then solder the splice and insulate it by wrapping first with rubber tape and then with friction tape.

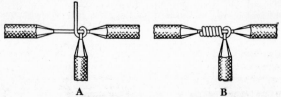

A   B

Fig. 530.—Steps in making a knotted tap splice. Use this kind of tap splice where there will be strain on the wires.

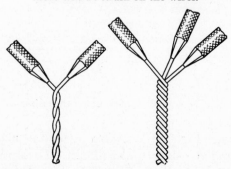

Fig. 531.—The pigtail splice is used for joining wires where there will be no strain on them.

Where there is likely to be pull or strain on the wires, then it is better to use the *knotted tap splice* (Fig. 530).

Making a Pigtail Splice. The pigtail splice, also called rattail splice (Fig. 531), is a quick and easy way to join the ends of wires where there will be no strain on them. It is generally used inside of outlet boxes in house wiring. Like all permanent splices, the pigtail splice should be soldered and then insulated by wrapping with rubber tape and friction tape.

2. Attaching Wires to Terminals

To attach a wire to a terminal like that shown in Fig. 532, first remove the insulation and then bend a hook as shown in Fig. 533.

Fig. 532.—A type of terminal commonly used for attaching wires.

Screw closes loop   Screw opens loop
RIGHT   WRONG

Fig. 533.—Right and wrong methods of hooking the end of a wire under a terminal screw. As the screw is tightened, it should close the loop, not open it.

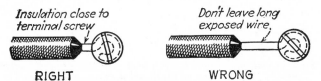

RIGHT   WRONG

Fig. 534.—Form the hook close to the end of the insulation to avoid too much bare or exposed wire.

Be sure to place the hook under the screw in such a manner that tightening the screw will tend to close the hook rather than open it. It is important that the hook be formed close to the end of the insulation on the wire (see Fig. 534). Small round-nose or long-nose pliers are convenient for forming such hooks.

Soldering lugs (see Fig. 535) are often used for attaching large wires, particularly stranded cables to terminals. To use such a lug, the end of the wire is soldered into the hollow end of the lug. This may be difficult unless the work is carefully done. Both the end of the wire and the inside of the lug, if not already tinned (coated with solder), must be cleaned, fluxed, and then tinned. The end of the wire may be tinned by dipping it into a small ladle of molden solder, after it has been cleaned and fluxed. To attach the lug to the wire, melt some solder into the lug, insert the end of the wire, and apply heat. Some additional solder may be required. It is important that the lug be completely filled. Otherwise, the joint may have high resistance which would cause heating and possibly melting of the

Fig. 535.—A type of soldering lug commonly used for attaching large wires, particularly stranded wires, to terminals.

Fig. 536.—A type of solderless connector.

solder and loosening of the joint. Be careful not to burn the insulation near the end of the wire.

Using Solderless Connectors. Special approved clamps and connectors that do not require soldering (see Fig. 536) are often available for making connections to electrical equipment. They are convenient and easy to use and are often better than splices or lugs that require soldering, particularly if the soldering is not well done.

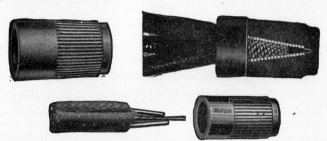

Fig. 537.—A type of solderless connector which is simply screwed onto the ends of the wires to be joined. (Courtesy of Ideal Commutator Dresser Company.)

Solderless connectors are now commonly used for connecting lighting-fixture wires to lighting circuits and for joining other wires where there will be little or no pull on the joint. Figure 537 shows such a connector. To join wires with one of these connectors, simply remove the insulation from the ends, clean them, and hold the ends side by side. Then place the connector over the wires and screw it on tight. It cuts its own threads and makes a tight, secure joint.

Such connectors are commonly made or covered with some insulating material and, therefore, do not require the use of tape for insulation.

3. Repairing Electric Cords

The attachment plugs on the ends of electric cords, like lamp cords, often become broken and need to be replaced.   Also, the end of a

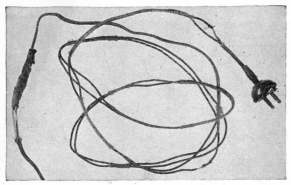

Fig. 538.—Frayed cords are dangerous particularly when used near good electrical grounds such as heat or water pipes or bathroom fixtures

cord which is attached to the lamp or other appliance sometimes becomes frayed, loose, or broken.   Repairs to such cords are easy to make, if a few simple rules are observed.   In time, electric cords

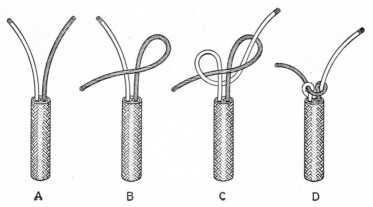

A   B   C   D

Fig. 539.—Steps in making the holding, or Underwriters', knot.

become so frayed and worn, however, that they are not safe to use and should be discarded.   Frayed and poorly insulated cords are dangerous, particularly when used near heat registers, radiators, water pipes, or bathroom fixtures.   Such cords have caused death from electric shock. Frayed or worn cords may also cause short circuits within themselves and possibly cause fires if the circuits upon which they are used are not properly protected by fuses.

Attaching a Plug to a Cord. Four points are important in attaching a plug to a cord: (1) tie a "holding knot" in the ends of the wires, (2) remove the insulation and clean the ends of the wires back just enough to hook around the terminal screws, (3) wrap the ends of the wires around behind the prongs better to withstand pulls on the cord, and (4) place the ends of the wires well under the screwheads and avoid fraying.

The holding knot, sometimes called the *Underwriters' knot*, is made as indicated in Fig. 539. If the knot would be too bulky to be drawn down well inside the attachment plug, it may be omitted. If it is not made, however, it is very important that the ends of the cord be wrapped behind the prongs better to resist pulls on the cord.

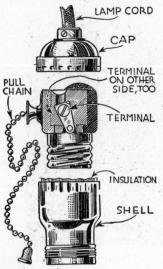

Fig. 540.—A lamp socket taken apart to show the various parts.

Attaching a Lamp Socket to a Cord. To attach a lamp socket to a cord, first remove the cap from the socket, usually by pressing at a point indicated on the socket shell and then prying. Then insert the end of the cord through the cap, tie the holding knot, remove the insulation from the ends of the wires and attach them to the terminal screws in the same manner as attaching a plug (see Fig. 540). Be sure to leave no loose frayed ends of the wire. It is important that the insulating bushing in the top of the cap, and the insulation inside the shell, be in place and in good condition.

4. Replacing Fuses

A fuse is a protective device used to limit the current that may flow in a circuit. It consists essentially of a short length of metal wire or ribbon of low melting temperature, enclosed in a suitable plug or cartridge. The size of fuse wire or link in a fuse is such that it will carry a given current continuously, but will melt or "blow" in case a larger current flows. When a fuse blows, it opens the circuit and stops the flow of current.   A fuse is therefore a safety device, and the importance of using fuses of the proper size cannot be overemphasized.

In case an oversize fuse is installed, currents larger than the safe capacity of the wires may flow, overheating the wires, and possibly causing fires.   In new wiring installations, nontamperable fuses are required according to the National Electrical Code.[1]   A nontamperable fuse prevents the use of a larger size fuse once a fuse of the proper size

Fig. 541.—Before replacing a blown fuse, stand on dry footing and open the main switch ahead of the fuse box if it is accessible.   (Courtesy of General Electric Company.)

has been installed.   In older wiring systems, nothing prevents one from using any size of fuse up to 30 amp., since all plug fuses up to this size are interchangeable.

A blown or burned-out plug fuse can usually be identified by a smoked or smudged window in the cover.   Before replacing a blown fuse, open the main switch if it is near and is easily accessible.   This eliminates the possibility of shock.   This precaution is important in case of old-style fuse boxes with exposed current-carrying parts.   On some of the newer style fuse boxes, only the tops of the fuse plugs

[1] The National Electrical Code is a set of regulations governing the installation of electric wiring and electrical equipment.   It is issued by the National Board of Fire Underwriters.

themselves are exposed. In such boxes, the fuses may be replaced without danger of shock if one is at all careful. In opening or closing power-line switches, stand on dry footing in order to avoid the possibility of shock.

After the main switch is opened, remove the blown fuse and replace it with one of the proper size, usually 15 amp. for ordinary house-wiring circuits. Then reclose the main switch. As mentioned

Fig. 542.—Replacing a blown fuse after opening the main switch. Be sure to use a new fuse of the proper size. (Courtesy of General Electric Company.)

in an earlier paragraph, it is exceedingly important that a fuse of the proper size be used.

5. Protecting Electric Motors against Overload

Electric motors cannot be protected by ordinary fuses. A motor draws a much higher current while starting than it does after it is started and is operating under normal load. A fuse that would allow the starting current to flow would also allow a much higher than normal full load current to flow. Thus a motor might become overloaded and burn out without burning out the fuse.

A motor should be protected by a delayed-action fuse or by a thermal-over load protective device. A delayed-action fuse (see Fig. 543) allows considerably higher than normal current to flow momentarily while the motor is starting, but would open the circuit in case of an overload lasting more than a few minutes.   Such a fuse should be placed at or near the motor, so that only the current supplying the motor flows through it.   It would not do to place the delayed-action fuse in the fuse box supplying two motors or one motor and other appliances.

The delayed-action fuse shown in Fig. 543 has a small pot of solder or other metal that melts at low temperature and through which the current flows.   A comparatively large current can flow momentarily without melting this small pot of metal, but if a current even slightly more than normal flows continuously the solder soon melts, and thus opens the circuit.

Fig. 543.—A delayed-action type of fuse which will carry temporary overloads without blowing.

A thermal overload protective device is commonly attached to the motor or to the starting switch for the motor.   It gives the same sort of protection to a motor as a delayed-action fuse. It will allow higher than normal currents to flow temporarily, but will trip and open the circuit in case of overloads that cause the motor to heat above safe operating temperatures.   After a thermal-overload device opens the circuit and the motor cools, then the device may be reset and the motor started again.

Fig. 544.—Testing a motor for overheating.   If the bare hand can be held on the motor for at least 10 sec., it is not too hot.   (Courtesy of General Electric Company.)

To install a delayed action fuse in a motor circuit, simply wire a fuse receptacle in the circuit.   It is best to have the fuse receptacle inside a metal box.   Be sure to use a fuse of a size that is not more than 15 or 20 per cent larger than the normal full-load current of the motor, as indicated on the motor name plate.

A thermal-overload protective device to be installed on a motor should likewise be suited in size to the current rating of the motor.

A motor running under full load will soon become quite warm. If one can hold his hand on it for at least 10 sec., however (see Fig. 544), the motor is not too hot.

6. Cleaning and Lubricating Electric Motors

Motors operating in dusty places often become very dirty and should be cleaned occasionally. Dust and dirt may obstruct the flow

Fig. 545.—A badly neglected motor. Dust and dirt may obstruct air passages and cause overheating or may get into moving parts and cause undue wear. (Courtesy of General Electric Company.)

Fig. 546.—An end plate removed from a dirty motor so that it may be easily and thoroughly cleaned. (Courtesy of General Electric Company.)

of air around and through a motor and cause it to overheat. Also dirt may get in the bearings or on the brushes and cause undue wear. Air from a tire pump, a blower attachment on a vacuum cleaner, or from a compressed air line is very effective in removing dirt from the inside parts of a motor.   If the motor is extremely dirty, remove one or both end plates to give better access to the insides (see Figs. 546 and 547).   To remove the end plates, take the nuts off the bolts that hold the plates in place and pry gently and carefully, or jar the end plates with a small hammer and block of wood.

Fig. 547.—A motor being cleaned by blowing with air under pressure.   Both end plates and the rotor have been removed to make thorough cleaning easy.   (Courtesy of General Electric Company.)

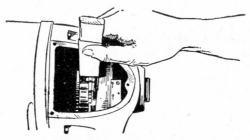

Fig. 548.—Cleaning a motor commutator. A piece of fine sandpaper (No. 00) is held lightly against the commutator by means of a small piece of wood while the motor is running.   (From "Protect Your Electric Motors," Central Hudson Gas and Electric Corporation.)

Cleaning the Commutator.   On the end of the armature or rotating part of some motors, there is a commutator.   Stationary carbon brushes, held in brush holders, ride on the commutator, and after long service it becomes dirty.   To clean a commutator, fold a piece of fine sandpaper (No. 00) around the end of a thin narrow piece of wood, and hold it against the commutator while the motor is running (see Fig. 548).   *Do not use emery cloth.*   Emery is a conductor of electricity and particles may become imbedded between the commutator bars and cause trouble. If the commutator is rough or pitted, or ridged from long use, remove the armature from the motor and take it to a shop equipped to turn the commutator on a lathe.

Oiling a Motor. The bearings of an electric motor require very little oil. Either overlubricating or underlubricating may cause trouble. Excess oil may work into the windings, causing early deterioration of the insulation and failure of the motor.

Some electric motors are equipped with plain bearings (see Fig. 549), and a small oil reservoir is provided at each bearing. A few

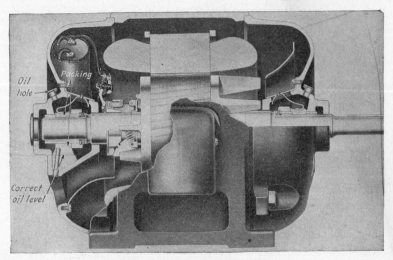

Fig. 549.—A plain-bearing motor equipped with oil reservoirs packed with waste or yarn. A few drops of oil (not squirts) every few weeks is usually adequate. (Courtesy of General Electric Company.)

drops (not squirts) of oil every few weeks in each bearing is usually adequate.

Other motors are equipped with ball bearings that are intended to be cleaned and repacked with a light cup grease or petroleum jelly after long periods of service. Some motors are equipped with ball bearings that were greased and sealed when they were made and that require no further attention. Still other motors are equipped with bearings that are to be lubricated occasionally with grease from a pressure gun.

7. Rigging a Small Portable Electric Motor

A small motor may easily be made portable so that it may be quickly and easily moved from one job to another (see Fig. 550). To rig up such a motor, twist together some short pieces of insulated wire to serve as a handle and fasten the ends under nuts on the end of the motor. Then drill and bolt short lengths of pipe to the motor base. The motor is then ready to be taken from job to job. To hold the

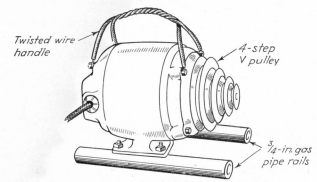

Fig. 550.—A small portable electric motor.

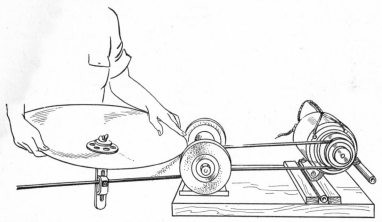

Fig. 551.—Sharpening a disk with a portable electric motor. Note the cleats on the grinder base board which hold the motor in place. Part of the weight of the motor helps keep the belt tight.

motor in proper position at its various stations, nail or screw narrow cleats to the motor stands or supports (see Fig. 551).

8. Connecting Dry Cells

Dry cells are often used to operate such appliances as telephones and electric fences. Two to four cells are commonly used, and they are practically always connected in series. To connect cells in series, connect the positive or center terminal of one cell to the negative or outside terminal of the next (see Fig. 552). This leaves a positive connection at one end of the set of cells and a negative connection at the other. Connect these two terminals of the battery to the terminals of the appliance or into the circuit in which it is to be used.

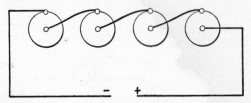

Fig. 552.—Dry cells connected in series. The positive or center terminal of one cell is connected to the negative or outside terminal of the next.

If there are instructions on the appliance indicating which terminal is to be connected to the positive side of the battery and which to the negative, be sure to observe them. Make all connections tight.

9. Charging a Storage Battery

The state of charge of a storage battery is best tested with a hydrometer (see Fig. 553). The specific gravity of the electrolyte (acid) in a battery varies with its state of charge, and the hydrometer is used to measure specific gravity. When a battery is fully charged, the specific gravity will be 1.250 to 1.300; and when discharged, about 1.100 to 1.150.

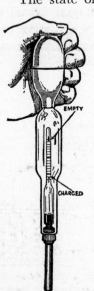

Fig. 553.—A hydrometer for testing the state of charge of storage batteries.

Storage batteries must be charged by direct current. Alternating current cannot be used without first rectifying it. There are on the market different types of battery chargers, however, which operate on alternating current and deliver direct current to a battery. To use such a charger, simply connect it to the storage battery according to directions on the charger and plug it into an a.c. outlet.

Another way often used to charge storage batteries is to drive an automobile type of generator with a small gas engine or a $\frac{1}{4}$ or $\frac{1}{3}$ hp. electric motor. The generator is then connected to the battery in the same manner as in an automobile or on a tractor. The positive terminal of the generator must be connected to the positive terminal of the battery and the negative terminal of the generator to the negative terminal of the battery. The generator should be equipped with a reverse-current cutout, just the same as an automobile generator, to prevent current from flowing backward from the battery to the generator in case the generator stops running. An automobile

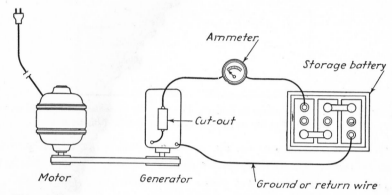

Fig. 554.—A battery-charging outfit made by driving an automobile-type generator with an electric motor.

type of ammeter should also be connected in the charging circuit to indicate the rate of charge. The charging rate of the generator can be regulated by some method, depending upon the type of generator, but most commonly by shifting the third brush on the commutator. Usually a small screw or bolt on the end of the generator may be loosened or turned to allow the third brush to be shifted on the commutator. Shifting the brush in the direction of rotation of the armature increases the charging rate. A generator speed of 1,000 to 1,500 r.p.m. is usually satisfactory.

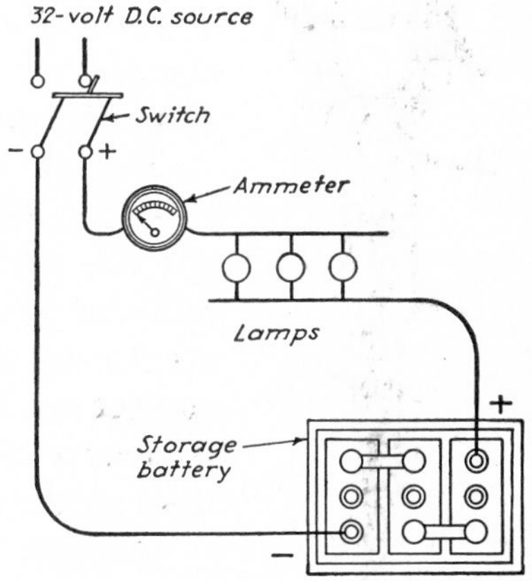

If a d.c. farm light plant is available, it may be used for charging an automobile or tractor storage battery by connecting a suitable

Fig. 555.—An arrangement for charging a storage battery from a 32-volt d.c. farm light plant.

resistance in the charging circuit. Suitable resistance units for charging batteries from 32-volt light plants are usually available through dealers in such plants. Lamp bulbs may also be used for resistance (see Fig. 555). The more lamps used, the greater the charging current. A 50-watt 32-volt lamp will allow about 1½ amp.

to flow; two such lamps, about 3 amp.; etc.   It is always best to have
an ammeter in the charging circuit to indicate the charging rate.   A
suitable ammeter can usually be obtained from some junked automobile, or from a dealer in auto parts.

A battery should not be charged at a rate so high that it becomes
hot or gases (bubbles) excessively.   Batteries may be charged at
higher rates when first put on charge than later after they become
nearly charged.   The maximum charging rate varies with the size
of the battery, but for most 6-volt automobile and tractor batteries,
10 to 15 amp. is usually high enough.   A longer charging period at
a moderate rate is usually better than a higher rate for a shorter
period.

Adding Water to Storage Batteries.   Add water to each cell
of a storage battery occasionally to replace that lost by evaporation
and that lost due to the action of the charging current.   The charging
current decomposes some of the water, forming gases which escape
through the vents in the caps on top of the cells.

Only clean water approved by the dealer selling the battery, or
distilled water, should be added to a storage battery.   Keep battery
water in a glass bottle or jug.   If approved battery water or distilled
water is difficult to get, melted artificial ice or filtered rain water that
has not been in contact with metal may be used.   A good way to
collect rain water is to place an earthenware jar out in the rain after
it has been raining about ten minutes.

Do not allow the level of the electrolyte in a storage battery to
fall so low as to expose the plates, and do not add too much water.
Usually a level about $\frac{3}{8}$ in. above the plates is high enough.   Adding
too much water will cause the cells to overflow when the battery
becomes warm and the electrolyte expands.   This not only causes a
loss of electrolyte, but may cause serious corrosion of battery terminals.

Cleaning Corroded Terminals.   If electrolyte is spilled and left
on battery terminals, a greenish deposit will form.   To stop the action
of the electrolyte and to clean the terminals, wash the parts thoroughly
with a solution of common baking soda or ammonia.   Brushing with
a stiff brush is helpful.   Then wash with water and wipe dry, and
finally apply a light coat of vaseline or cup grease.

10. Making Extensions to a Wiring System

Wiring a building for electric service, or making major extensions
or changes to a wiring system, should be done only by an experienced

electrician or someone familiar with the regulations of the National
Electrical Code.   By using extreme care, however, one with less experience may safely make minor extensions, such as installing a convenience outlet, or a lamp to be controlled by a pull chain, or even a lamp
to be controlled by a wall switch.   If in doubt as to procedures in
doing such work, consult a book or manual on wiring or someone
experienced in wiring.   *Make sure that the electric current is turned off
before making any extensions or repairs.*

Cutting and Attaching Armored Cable to Outlet Boxes.   Flexible armored cable (see Fig. 556) containing two or more insulated

Fig. 556.—Armored cable is widely used in wiring buildings.   (Courtesy of National
Electric Products Corporation.)

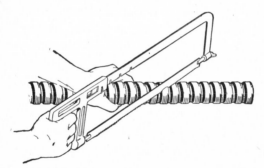

Fig. 557.—To cut armored cable, saw at an angle of about 45 deg.   (Courtesy of
National Electric Products Corporation.)

wires is widely used in wiring of buildings.   Splices in the wiring are
made inside of outlet boxes.   To cut armored cable, saw at an angle
of about 45 deg. with a hack saw as shown in Fig. 557.   Then twist
and pull to separate the spiral steel armor (see Fig. 558).

To attach the cable to an outlet box, cut and remove the armor
back about 6 in. from the ends of the wires.   Then unwrap the heavy
paper found between the wires and the armor, and jerk it so as to
tear it off back a short distance within the cable (see Figs. 559 and
560).   Next, insert a fiber bushing inside the end of the armor to
prevent the metal from cutting the insulation on the wires (see Fig.
561).   After inserting the fiber bushing, attach a connector to the

end of the armor (see Fig. 562) and then fasten the connector to the outlet box (see Fig. 563).

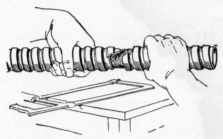

Fig. 558.—After sawing, twist and pull to separate the steel armor. (Courtesy of National Electric Products Corporation.)

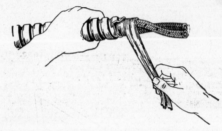

Fig. 559.—Unwrapping the paper between the wires and the armor. (Courtesy of National Electric Products Corporation.)

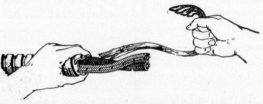

Fig. 560.—Jerking the paper wrapper to tear it off back a short distance inside the armor. (Courtesy of National Electric Products Corporation.)

Fig. 561.—Inserting a fiber bushing to prevent the rough edges of the metal armor from damaging the insulation on the wires. (Courtesy of National Electric Products Corporation.)

Using Nonmetallic Sheath Cable. Nonmetallic sheath cable (see Fig. 564) is widely used in wiring farm buildings. It consists of two or more insulated wires inside a nonmetallic outer cover which has been treated to make it moisture- and fire-resistant. It is connected to outlet boxes with suitable connectors in much the same way as armored cable.

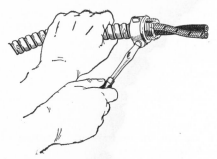

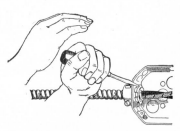

Fig. 562.—Attaching a cable connector to the cable. (Courtesy of National Electric Products Corporation.)

Fig. 563.—Fastening the connector to an outlet box. (Courtesy of National Electric Products Corporation.)

Fig. 564.—Nonmetallic sheath cable is widely used for wiring farm buildings. (Courtesy of National Electric Products Corporation.)

When electric wires and cables are exposed as in barns, basements, and attics, they must be properly supported and given protection against mechanical injury. For details of such support and protection when making minor extensions to a wiring system, examine the methods used in other parts of the system. In case of doubt, consult an experienced electrician or the National Electrical Code.

Installing an Outlet and Pull-chain Fixture. This type of fixture is commonly used in such places as closets, basements, and attics. To install a pull-chain fixture, it will be necessary to run two wires or a two-wire cable from a source of current to a metal outlet box (see Fig. 565) mounted

Fig. 565.—One of many types of metal outlet boxes used in wiring buildings. (Courtesy of National Electric Products Corporation.)

at the location of the new fixture. This cable must be supported or mounted on the walls or other parts of the building in an approved manner. All connections and splices must be made inside metal outlet boxes. Current may be supplied from a branch circuit already in place, if the addition of this extra outlet does not overload the circuit.

Mount the new outlet box on or in the ceiling or wall at the place selected for the new fixture. Securely fasten the box in place. Then

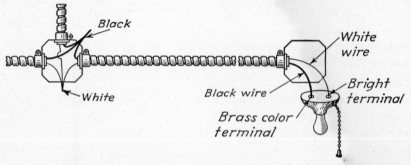

Fig. 566.—The wiring connections for a pull-chain fixture using armored cable.

open the main switch or the switch at the fuse and distribution panel; and open a metal outlet box that is near the new outlet and has two live wires running into it, that is, two wires that come direct from the fuse panel without the interruption of a switch. Then run a two-wire cable from this box to the new box just installed. Fasten the ends of the cable into openings in the boxes with suitable cable connectors, and fasten the cable to walls or other supports in an approved manner. Splice the new wires onto the old ones in the old outlet box, connecting black wire to black and white wire to white (see Fig. 566). Then connect the wires at the new box to the terminal screws of the pull-chain fixture, connecting the white wire to the bright screw and the black wire to the brass colored screw. In case of a fixture having pigtail wires in-

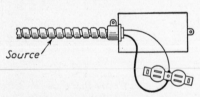

Fig. 567.—Installing a convenience outlet.

stead of terminal screws, splice the pigtail wires to the supply wires. Finally, attach the fixture to the outlet box, the fixture serving as a cover for the box.

Installing a Convenience or Appliance Outlet. To install a convenience or appliance outlet, first mount a metal outlet box in the baseboard or on the wall at the location for the new outlet, and securely fasten it in place.   Then run a two-wire cable from this new box to a suitable source of current, probably another outlet box in the wiring system, and connect the wires in the manner outlined in the preceding paragraph (see Fig. 567).

Installing a Fixture Controlled by a Wall Switch.   To install a fixture and a wall switch to control it, mount a metal outlet box for the fixture and also one for the wall switch.   Then run a two-wire cable between these two boxes, and also between the fixture box and a source of current, in the manner outlined for installing a pull-chain fixture. Make the connections as shown in Fig. 568.

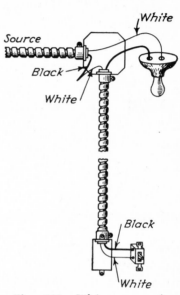

Fig. 568.—Wiring connections for a fixture controlled by a wall switch.

11. Installing a Doorbell or Buzzer

Current supplied from dry cells or from a low-voltage transformer is commonly used for operating door-bells and buzzers.   Annunciator wire or bell wire is commonly used to wire bell or buzzer systems.   It is fastened to walls or other supports with insulated staples.   Since such wires carry only low-voltage current,

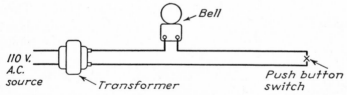

Fig. 569.—Wiring connections for a door-bell system using a transformer as a source of current.

they do not require as heavy insulation as wires supplying current for lights and 110-volt appliances.   A small low-voltage transformer mounted on a metal outlet box of the house-lighting system, is a much more dependable source of current for a bell or buzzer than a dry-cell battery.

The simplest doorbell system consists of a push button at the door, a bell, and a source of current wired as shown in Fig. 569. It is easy, however, to wire the system so that a button at one door rings a bell and a button at a second door rings a buzzer (see Fig. 570).

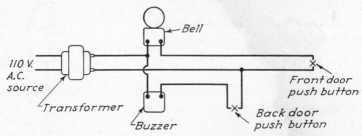

Fig. 570.—Connections for operating both a bell and a buzzer from the same transformer.

12. Installing an Electric Fence

Electric fences have proved both economical and practical, particularly for temporary fencing. Such a fence is quickly and easily installed and often makes it feasible to utilize certain temporary pastures that might otherwise go to waste. An electric fence consists essentially of one or two fence wires fastened to posts by means of porcelain insulators and connected to a fence charger or controller. The controller is a special unit, operated from a dry-cell or a storage battery or from a 110-volt lighting circuit. It may be mounted in the barn or other building and connected to the fence, or it may be mounted somewhere away from buildings if it is battery operated.

Although it is possible to build a homemade battery-operated fence charger, it is usually more satisfactory and cheaper in the end to buy a unit from a reputable dealer or manufacturer. Under no conditions is it advisable to build a homemade unit to be operated from 110-volt lighting circuits. Unless properly made and approved by the Underwriters' Laboratories, or other testing and approving agencies, such units may be dangerous. Fatal accidents, both to human beings and to animals, have been caused by the use of unapproved fence chargers operated from 110-volt power and light circuits.

Setting the Posts and Attaching the Wire. In building an electric fence, brace the corner posts well, and space the line posts two to three rods apart. Posts may have to be placed closer together in going over uneven ground. Fasten the wire on the posts at a height about three-fourths the height of the animals to be confined within the fence.   One wire 30 to 40 in. above the ground is usually satisfactory for horses and cattle.   In case of hogs of different sizes,

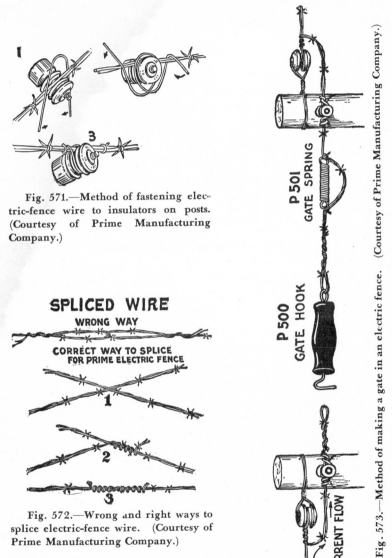

Fig. 571.—Method of fastening elec-tric-fence wire to insulators on posts. (Courtesy of Prime Manufacturing Company.)

SPLICED WIRE

WRONG WAY

CORRECT WAY TO SPLICE
FOR PRIME ELECTRIC FENCE

1

2

3

Fig. 572.—Wrong and right ways to splice electric-fence wire. (Courtesy of Prime Manufacturing Company.)

P 501 GATE SPRING

P 500 GATE HOOK

CURRENT FLOW

Fig. 573.—Method of making a gate in an electric fence.   (Courtesy of Prime Manufacturing Company.)

two wires are usually better, one about 16 to 18 in. from the ground for the large hogs, and one 6 to 8 in. for the smaller ones.

Stretch the wire tight enough to prevent sagging between the posts, and attach it to insulators on the posts (see Fig. 571).   On corner and end posts use heavy-duty or strain insulators. Use the Western Union type of splice in splicing the fence wire, so as to ensure a good electrical contact at the joint (see Fig. 572).

Gates are easily made as indicated in Fig. 573.

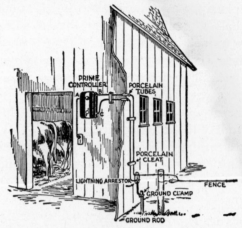

Fig. 574.—An electric-fence controller located in a barn and connected to the fence. (Courtesy of Prime Manufacturing Company.)

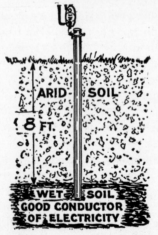

Fig. 575.—Be sure the ground rod or pipe extends down to permanently moist soil. (Courtesy of Prime Manufacturing Company.)

Connecting the Fence Unit or Controller. Mount the fence controller in a place where it will be kept dry and clean, and connect one terminal to the insulated fence, and the other to a ground connection (see Fig. 574). It is important that the ground rod or pipe extend down to permanently damp soil (see Fig. 575). Also, connect **a lighting arrestor between the insulated fence line and the ground rod.**

Training Livestock to Respect an Electric Fence. Always train animals before attempting to confine them with an electric fence. Otherwise, they may not respect it and go right through it. To train the animals, fence off a corner of a feed lot or corral with electric fence, place some attractive feed just beyond the fence, and allow the animals to come in contact with it a few times. Most animals can be trained in 30 min. to an hour. Sheep, being better protected by their wool than most animals, are harder to control and may require a training period of several hours. Ears of corn hung to the training fence with small wires are sometimes used in training animals. If the ground is dry, it may be advisable to moisten it under and in front of the training wire, in order to give better ground connection and a stronger shock or sting during training.

It is important not to let weeds and brush grow up and touch the insulated fence wire, as this would ground the wire and make the fence ineffective.

Points on Working with Electrical Equipment

1. Do not handle connected or "live" electrical equipment with wet hands or when standing on wet footing.
2. When doing wiring, be sure to do it in a safe and approved manner, as prescribed by the National Electrical Code. When in doubt consult an electrician or an authoritative book or manual on wiring.
3. Make all splices and connections in electric wires in an approved and workmanlike manner. Loose, dirty, or poorly insulated splices are not only a source of trouble, but of danger.
4. Pull on the plug, not the cord, when disconnecting an appliance. (On some of the newer style outlets, it is necessary to twist the plug to the left before pulling it out.)
5. Do not use cords with broken or defective insulation.
6. Keep motors and other electrical equipment free from dirt and moisture.
7. Always turn the current off at the main switch when working on a circuit.
8. Always replace a blown fuse with one of the proper size.
9. Do not investigate or work around a power transformer or a high-voltage line. In case of trouble, call the power company or utility office.
10. See that motors are properly lubricated, but not overlubricated.
11. See that motors are properly protected by delayed-action fuses or by suitable thermal-overload protective devices. Ordinary fuses do not protect motors.

Jobs and Projects

1. Inspect all desk lamps and floor lamps about your home, and repair or replace any cords that are not in good condition.

2. Make an extension cord or trouble lamp. Use a bakelite or weather-proof socket to avoid the possibility of shocks when using it on damp floors.

3. Inspect the fuse cabinets about your home or farm, and make sure the proper sizes of fuses are used. Get a few extra fuses and make a place to keep them near the fuse box, so that replacements may be made quickly and easily and with the proper sizes of fuses.

4. Oil the electric motors about your home or farm. Wipe and clean them as much as possible without taking them apart. Make a calendar or schedule for oiling them in the future, and tack it up near one of the motors.

5. Disassemble and clean an electric motor. Blow out any accumulated dust and dirt with air under pressure if it is available. Clean the commutator, if it has one, and smooth it with fine (No. 00) sandpaper if it is only slightly rough. If the commutator is badly pitted or grooved, take it to a shop and have it turned true. Be careful in assembling the motor not to damage any parts.

6. Rig a small electric motor to make it portable, and arrange places where it is to be used on various jobs, so that it can be easily and quickly set in place and belted to the load.

7. Make a battery-charging outfit from an old automobile generator, arranging it to be driven with a portable electric motor.

8. Inspect all the electric motors about your home or farm to make sure they are properly protected against overload. If any motor is not properly protected, install a delayed-action fuse of the proper size (not more than 15 to 20 per cent larger than the current rating on the motor name plate) or a suitable overload protective device.

9. Inspect the wiring system about your home or farm, and make a list of short desirable extensions that you could install. Install one or more of these extensions, first being sure you know how to do the work safely and in an approved manner. In case of doubt, consult your instructor or an experienced electrician. Have the work approved by an electrical inspector.

10. Install a doorbell or buzzer in your home, or a set of bells and buzzers.

11. Install and operate an electric fence. Carefully read and observe all instructions furnished with the fence unit or controller.

12. Make an electric chick brooder, or an electric pig brooder, according to plans obtained from a bulletin or book in your school library.

DEFINITIONS OF SHOP TERMS

acetylene, a fuel gas commonly used in welding.

aggregate, material such as sand and gravel or rock to be mixed with Portland cement and water to form concrete.

annealing, softening of metal.

arc, a part of a curved line.

A sustained spark caused by flow of electricity across a gap.

arc welding, welding of metals by means of heat supplied for an electric arc.

arris, a sharp edge formed by the meeting of two surfaces.

awl, a pointed instrument or tool for marking or piercing materials.

backfire, extinguishing of the flame of a welding torch with a loud snap.

bank-run gravel, the natural mixture of sand and gravel as taken from a gravel bed.

bead, a ridge of metal deposited on the work in welding.

bit, a cutting edge or part of a cutting tool, or the blade of a cutting tool.

block, a case or shell containing a grooved pulley, or set of them, over which rope is run.

blowpipe, a gas welding torch.

boiled oil, linseed oil that has been treated with drying compounds to make it dry faster.

boxing, mixing paint by pouring it from one container to another several times.

branch circuit, that part of the circuit or wiring system beyond the last fuses.

bronze welding, a kind of oxyacetylene welding in which parts are joined by coating with bronze instead of by melting or fusion.

buffing wheel, a wheel covered with cloth, leather, or similar material and used for polishing metal.

burnish, to make smooth by rubbing usually with a hard smooth piece of steel called a burnisher.

burr, a rough thin edge left on the work by a grinding or cutting tool.

A smooth thin edge formed by burnishing a scraper.

A small washer put on the end of a rivet before bradding.

casein, a product obtained from milk. It is used as a principal ingredient in some glues.

chalk line, a cord coated with chalk dust and used to mark a straight line between two points on a surface.

A straight line marked on a surface by a chalk line or cord.

chamfer, a straight flat surface formed by cutting away the sharp edge at the meeting of two surfaces.

473

chuck, a mechanism for holding and turning a tool or for holding and turning work in a machine.

circuit, the path of flow of electric current, consisting usually of two wires, one to bring the current from the source of supply and the other to take it back.

clearance, the space or distance between parts.

clinch, to fasten securely, especially with nails by bending the points after they have been driven through.

counterboring, making a hole larger in diameter for a short distance near its mouth.

crater, a hole or depression made by an electric arc in welding.

crosshatch, to mark with parallel lines more clearly to define parts of an object on a drawing or sketch.

current, movement or flow of electric charges in a circuit.

cut acid, zinc chloride or the liquid produced by the action of muriatic acid on zinc. Used as a soldering flux.

dado, a groove across a board to receive the end or edge of another board.

degree, a unit of measure of angles; $\frac{1}{90}$th of a right angle, or $\frac{1}{360}$th of a circle.

delayed-action fuse, a fuse that will allow momentary overloads on a circuit, but that will blow and open a circuit in case of a continuous overload even slightly greater than normal.

die, a tool for cutting screw threads on a rod or bolt.

dowel, a round pin fitting into holes in adjoining pieces to prevent slipping.

drier, a substance mixed with paints to make them dry faster.

drawfile, to file by pushing a file sidewise rather than lengthwise.

drawing metal, making a piece thinner, or smaller in cross section, and longer.

end grain, the ends of the fibers of wood as on the ends of boards.

fall rope, the free end of a rope (in a set of blocks and tackle) to which the pull is applied.

ferrule, a ring or sleeve, usually of metal, put on the end of a tool handle to prevent splitting.

filler rod, a rod of steel or other metal used in building up a weld in oxy-acetylene welding (also in carbon-arc type of electric-arc welding).

fillet, a concave junction between two surfaces that meet at an angle to each other.

fillet weld, a weld made in the angle or corner between two surfaces.

finishing concrete, troweling or otherwise working the surface of concrete before it hardens to give it the desired texture or smoothness.

flange, a rim or rib or disklike projection on a part for strength, for guiding, or for attachment to another part.

flashback, a burning of gasses back inside a welding torch.

flush, on a level with adjacent surfaces.

fuse, a short length of metal that melts and opens an electric circuit in case of overload.

fusion welding, welding of metal parts by melting and flowing together.

flux, a material applied to metals to aid in soldering or welding.

gain, a notch cut in a board or piece to receive another member.

gasket, material, usually in sheet form, used between surfaces to ensure a tight or leakproof joint.

glaziers' points, small flat, thin triangular or diamond-shaped pieces of metal used to hold panes of glass in place in a frame.

grain, the fiber of wood or other fibrous material, or the direction or arrangement of fibers in wood.

green coal, fresh coal added to a forge fire before it has changed to coke.

gullet, the notch or space between two adjacent teeth in a saw blade.

gumming, enlarging the gullets or spaces between teeth of some kinds of saws.

hack saw, a metal-cutting saw.

hardy, a cutting tool with a shank for insertion in a hole in an anvil.

inhibitor, a material used to restrain or slow up a process.

jamb, an upright piece forming the side of an opening, as a door or window.

jointing, straightening and fitting the edges of boards or parts to be joined together.

Filing or dressing down the ends of saw teeth so the points of all teeth will be in line with each other.

kerf, a notch or groove made by a cutting tool, such as a saw.

lacquer, a kind of varnish with or without color.

lagging a pulley, covering the face of the pulley with leather or other belting material to reduce belt slippage.

laying out, measuring and marking material or parts to be cut or otherwise worked upon.

mechanical advantage, the increase in leverage, pull, or force gained by use of a tackle, device, or machine.

mortise, a hole cut into or through one piece to receive another piece.

oilstone, a sharpening stone upon which oil is used.

oxidizing flame, a flame with an excess of oxygen and, therefore, usually with a tendency to burn or oxidize the metal being heated.

oxyacetylene welding, a kind of welding wherein the heat is supplied by the burning of acetylene in the presence of oxygen.

parting strip, a thin piece of wood in a window frame to form the partition between the grooves in which the window sashes slide up and down.

peen, the end of the head of a blacksmith's or machinist's hammer opposite the flat face.

peening, working or beating metal with the peen of a hammer.

pigment, the finely divided particles of solid material in paint, such as white lead, zinc oxide, or titanium.

plumb, vertical.

A weight attached to a line and used to indicate a vertical direction.

priming coat, the first coat of paint which serves as a foundation for succeeding coats.

pumice stone, a light volcanic stone used, especially in powdered form, for smoothing and polishing.

rabbet, a groove cut in the edge or face of one piece usually to receive another member, as a panel.

raker, a kind of tooth on timber or log saws.

raw oil, linseed oil in its natural state.

ream, to enlarge or dress out a hole with a tool called a reamer.

reducing flame, a flame with insufficient oxygen to burn the fuel completely.

reeving blocks, passing or threading a rope through blocks to form a tackle.

reversed polarity, an arrangement of direct-current welding leads wherein the electrode is positive and the work is negative.

r.p.m., revolutions per minute.

sash, a frame in which panes of glass are set.

serving (a rope splice), wrapping tightly with strong cord to improve appearance and durability.

shielded-arc welding, a process of electric-arc welding wherein the arc and molten metal are protected from the atmosphere by a gaseous shield.

silt, fine particles of soil suspended in water or deposited on material.

slag, nonmetallic material trapped in a weld.

soft paste lead, a paste containing white lead and small amount of raw linseed oil and turpentine and used as a principal ingredient of some kinds of paint.

sprocket, a toothed wheel to engage a chain.

stock, material such as wood or iron to be used in making parts.

A handle of a tool, such as for holding and turning a tap.

straight polarity, an arrangement of direct-current welding leads wherein the work is positive and the electrode is negative.

striking an arc, starting an arc in electric-arc welding.

strop, to sharpen by rubbing on a leather strap (or strop).

stuffing box, a device to prevent leakage around a piston rod.

suction, the set or tendency of a tool to engage or penetrate.

tackle (also called blocks and tackle), a set of blocks and rope for raising, lowering, or moving heavy objects.

tang, a projecting shank on a tool for connecting to a handle, as on a file or a pitch fork.

tap, a tool for cutting screw threads inside a hole.

temper, to bring to the desired degree of hardness, usually by heat-treatment and cooling.

tenon, an end of a piece specially shaped, usually with a shoulder, to fit into a mortise or hole.

thinner, a liquid, usually highly volatile, such as turpentine, added to paint to thin it.

thong, a small strap or strip of leather for fastening parts together, as in belt lacing.

tinning, (in soldering) coating surfaces of a soldering iron, or metals to be soldered, with a smooth clean coat of solder.
(In brazing or bronze welding), coating surfaces to be brazed or bronze welded with brass or bronze.

true surface, a surface that is a true plane or that is straight and without depressions, humps, warp, twist, or wind.

upsetting metal, making a piece thicker and shorter.

vehicle, the liquid portion of a paint, usually oil, that serves as a carrier of the pigment or solid particles.

voltage, electrical force or pressure that sends current through a circuit.

welding electrode, a welding rod used in electric-arc welding.

whet, to sharpen by rubbing on or with a sharpening stone.

whipping (the end of a rope), binding by wrapping with strong cord.

window stop, a thin strip of wood along a side of a window frame to hold the window in place and to form one edge of the groove in which the window slides up and down.

wire edge, a burr or rough edge produced on a tool by grinding or heavy whetting.

work, material or parts being cut, shaped, or otherwise worked upon.

work rest, the part of a grinder or other machine upon which the work is held or supported while being ground, cut, or shaped.

working edge, the edge of a piece from which measurements are made or gaged.

working surface, the surface from which measurements are made or gaged.

INDEX